Drug Design and Discovery

Surya K. De

Drug Design and Discovery

Inhibitors of Mitogen-Activated Protein Kinases

Springer

Surya K. De
Conju-Probe
San Diego, CA, USA

ISBN 978-3-031-75681-8 ISBN 978-3-031-75682-5 (eBook)
https://doi.org/10.1007/978-3-031-75682-5

This Springer imprint is published by the registered company Springer Nature Switzerland AG
The registered company address is: Gewerbestrasse 11, 6330 Cham, Switzerland

Preface

After proteases, the protein kinase family is the second largest enzyme family and the fifth largest gene family in humans. Protein kinases are the second largest group of drug targets after G-protein coupled receptors (GPCR) and they account for almost 20–30% drug development pipelines. The human genome has almost 518 kinase genes and 106 pseudogenes. Protein kinases comprise 2% of the mammalian genome and catalyze the transfer of the γ-phosphoryl group of the ATP to specific protein substrates. This covalent post-translational modification is a crucial component of normal cellular communication and maintenance of homeostasis. The signals can initiate inside or outside the cells and pass through a series of steps with multiple enzymes and messengers. The MAPK family comprises three major groups such as ERKs, JNKs, and p-38.

Each chapter embedded in this book covers important information regarding different approaches in designing small molecules for MAP kinase inhibitors. There are four strategies utilized by different groups in developing MAP kinase inhibitors such as ATP site inhibitor, non-ATP site inhibitor, bi-dentate inhibitor, and covalent inhibitor. The verification of the ability of small molecules or small peptides to modulate MAPK activity as a diabetes, cancer, and Parkinson disease or as a diagnostic and /or prognostic tool is now possible and probable. The medicinal chemists could use these factors in laying out new strategies for developing novel MAPK inhibitors. Nevertheless, multiple approaches to identify MAPK inhibitors with higher selectivity and safety windows may lead to more efficient and faster therapies in the years to come. We anticipate that more studies describing MAPK functions in human health make this field more attractive and will continue to be of substantial intrinsic merit. This book offers the principles of drug design and successful implementation of MAP kinase inhibitors from hit, and latest development in this field.

San Diego, CA, USA

Surya K. De

Contents

About the Author

Surya K. De is a Director at Conju-Probe, San Diego, California, USA. Surya K. De obtained his Ph.D. in Organic Chemistry from Jadavpur University, Kolkata (Calcutta), India. He worked at the University of Washington (Seattle), Purdue University (West Lafayette, Indiana), the Scripps Research Institute, and Sanford-Burnham Medical Discovery Institute prior to joining a biotech company. He has published over 120 papers in reputed international journals, covering a broad array of specialized topics in science, and he holds 15 US and international patents. He has published four solo-authored books. He has been a fellow of the Royal Society of Chemistry (London, UK), the Royal Society of Medicine (London, UK), the American Institute of Chemists (USA), and the Indian Chemical Society. He was a recipient of P. R. Vijayalaxmi Award for Innovation in Cancer Research (2023, USA). He has been an elected councilor at the American Chemical Society since 2022.

Common Abbreviations and Acronyms

AA	Anaplastic anemia
ABC	Advanced breast cancer
ACE	Angiotensin converting enzyme
ABMT	Autologous bone marrow transplant
AD	Alzheimer's disease
ADR	Adverse drug reaction
ADP	Adenosine diphosphate
AFP	Alpha-fetoprotein
AKT	AKR mouse thymoma kinase
ALBL	Acute lymphoblastic B-cell leukemia
ALAT	Alanine aminotransferase
ALCL	Anaplastic large-cell lymphoma
ALK	Anaplastic lymphoma kinase
AML	Acute myeloid leukemia
ALL	Acute lymphoblastic leukemia
ALT	Alanine aminotransferase
AMKL	Acute megakaryocytic leukemia
AMP	Adenosine monophosphate
APDS	Activated PI3K delta syndrome
AP-1	Activator protein 1
ATP	Adenosine triphosphate
BBB	Blood-brain barrier
BCC	Basal cell carcinoma
BPDCN	Blastic plasmacytoid dendritic cell neoplasm
Bcl2	B-cell lymphoma 2
BMI	Body mass index
BMX	BMX nonreceptor tyrosine kinase
BTK	Bruton's tyrosine kinase
BRCA1	Breast cancer gene 1
BRCA2	Breast cancer gene-2
BRAF	v-raf murine sarcoma viral oncogene homolog B1

CAMK	Calcium/calmodulin-dependent protein kinase
CAT	Computerized axial tomography
CAR-T	Chimeric antigen receptor T cells
CBC	Complete blood count
CDC	Center for Disease Control
CDK	Cyclin-dependent kinase
CT	Computed tomography
CLL	Chronic lymphocytic leukemia
CTCL	Cutaneous T cell lymphoma
CTLA-4	Cytotoxic T lymphocyte-associated antigen-4
CK1	Casein kinase 1
CML	Chronic myelogenous leukemia
CAT	Computerized axial tomography (scan)
CMML	Chronic myelomonocytic leukemia
CNS	Central nervous system
CHF	Congestive heart failure
cAMP	cyclic AMP
COPD	Chronic obstructive pulmonary disease
CPK	Creatine phosphokinase
CPR	Cardiopulmonary resuscitation
CRC	Colorectal cancer
CRF	Chronic renal failure
CR	Complete remission
CXR	Chest X-ray
DM	Diabetes mellitus
DNA	Deoxyribonucleic acid
DLBCL	Diffuse large B-cell lymphoma
DLCL	Diffuse large-cell lymphoma
DNAPK	DNA-dependent protein kinase
EBV	Epstein-Barr virus
EGFR	Epidermal growth factor receptor
ERK	Extracellular signal-regulated kinase
ECG/EKG	Electrocardiogram
FDA	U.S. Food and Drug Administration
FGFR	Fibroblast growth factor receptor
FL	Follicular lymphoma
FSH	Follicle-stimulating hormone
FLT3	fms-related tyrosine kinase 3
GERD	Gastroesophageal reflux disease
GIST	Gastrointestinal stromal tumor
GBM	Glioblastoma multiforme
GI	Gastrointestinal
GFR	Glomerular filtration rate
GPCR	G protein-coupled receptor
GSK3	Glycogen synthase kinase 3

HAV	Hepatitis A virus
HBV	Hepatitis B virus
HER1	Human epidermal growth factor receptor 1
HPC	Hepatitis C virus
HIV	Human immunodeficiency virus
HCC	Hepato-cellular carcinoma
HCL	Hairy cell leukemia
HL	Hodgkin lymphoma
HPV	Human papillomavirus
HDL	High-density lipoprotein
HGB	Hemoglobin
HCK	Hematopoietic cell kinase
HCL	Hairy cell leukemia
HL	Hodgkin's lymphoma
HRT	Hormone replacement therapy
HSP	Heat shock protein
HTS	High-throughput screening
IC_{50}	Concentration required to obtain 50% inhibition
IBD	Inflammatory bowel disease
IBS	Irritable bowel syndrome
ICD	Implantable cardioverter defibrillator
ICU	Intensive care unit
IDDM	Insulin-dependent diabetes mellitus
IM	Intramuscular
IP	Intraperitoneal
IT	Intrathecal
IV	Intravenous
IL	Interleukin
IUD	Intrauterine device
IVP	Intravenous pyelogram
IgH	Immunoglobulin heavy chain
iMiD	Immunomodulatory drugs
IKK	IκB kinase
ILK	Integrin-linked kinase
IRS	Insulin receptor signaling protein
JAK	Janus kinase
JNK	c-Jun N-terminal kinase
JIP	JNK-interacting protein
JBD	JNK-binding domain
kDa	kilodalton
KD	Kinase domain
LDL	Low-density lipoprotein
LBD	Ligand-binding domain
LE	Ligand efficiency
LFT	Liver function tests

LCH	Langerhans cell histiocytosis
LDH	Lactic dehydrogenase
LMM	Lentigo Maligna Melanoma
LN	Lymph node
LP	Lumbar puncture
LRRK1/2	Leucine-rich repeat kinase 1/2
MAPK	Mitogen-activated protein kinase
MAP2K	Mitogen-activated protein kinase kinase
MAP3K	Mitogen-activated protein kinase kinase kinase
MAP4K	Mitogen-activated protein kinase kinase kinase kinase
MAPKAPK	Mitogen-activated protein kinase-activated protein kinase
MKP1	Mitogen-activated protein kinase phosphatase 1
MET	Mesenchymal epithelial transition factor
MEK	MAPK/ERK kinase
MRI	Magnetic resonance imaging
MDS	Myelodysplastic syndromes
MCL	Mantle cell lymphoma
MM	Multiple myeloma
MDR	Multi-drug resistant
MI	Myocardial infarction (heart attack)
MMR	Measles, mumps, and rubella
MRI	Magnetic resonance imaging (scan)
MRT	Malignant rhabdoid tumor
MRSA	Methicillin-resistant *Staphylococcus aureus*
MS	Multiple sclerosis
MTD	Maximum tolerated dose
mTOR	mammalian target of rapamycin
mTORC1	mammalian target of rapamycin complex 1
mTORC2	mammalian target of rapamycin complex 2
NIDDM	Non-insulin-dependent diabetes mellitus
NKDA	No known drug allergies
NSAID	Non-steroidal anti-inflammatory drug
NSCLC	Non-small cell lung cancer
NHL	Non-Hodgkin lymphoma
NPC	Nasopharyngeal carcinoma
NRSTS	Non-rhabdomyosarcoma soft tissue sarcoma
NF	Neurofibromatosis
NFκB	Nuclear factor κB
NLK	Nemo-like kinase
NMR	Nuclear magnetic resonance
NTRK	Neurotrophic tyrosine receptor kinase
OCD	Obsessive-compulsive disorder
ORR	Overall response rate
OS	Overall survival
PAD	Peripheral arterial disease

PAT	Paroxysmal atrial tachycardia
PAP	Papanicolaou (a test for cervical cancer)
PD	Parkinson's disease
PET	Positron emission tomography
PFT	Pulmonary function test
PID	Pelvic inflammatory disease
PI-3K	Phosphoinositide 3-kinase
PMS	Premenstrual syndrome
PPD	Purified protein derivative
PT	Prothrombin time
PTH	Parathyroid hormone
PTSD	Post-traumatic stress syndrome
PTT	Partial thromboplastin time
PTEN	Phosphatase and tensin homolog
PSA	Prostate-specific antigen
PARP	Poly ADP ribose polymerase
PD1	Programmed cell death protein-1
PD-L1	Programmed death-ligand 1
PO	Per os (by mouth)
PUD	Peptic ulcer disease
PVC	Premature ventricular contraction
PFS	Progression-free survival
PD-1	Programmed cell death protein-1
PR	Partial remission
PDGFR	Platelet-derived growth factor receptor
PDK1	Phosphoinositide-3-dependent kinase 1
PET	Positron emission tomography
PH	Pleckstrin homology (PIP3 binding) domain
Ph+	Philadelphia chromosome positive
PIP2	Phosphatidylinositol (4,5)-bisphosphate
PIP3	Phosphatidylinositol (3,4,5)-trisphosphate
PK	Pharmacokinetics
PKA/B/C	Protein kinase A/B/C
PTB	Phospho-tyrosine binding domain
RA	Rheumatoid arthritis
RCT	Randomized control trial
RNA	Ribonucleic acid
RBC	Red blood cell/red blood count
RSV	Respiratory syncytial virus
RCC	Renal cell carcinoma
RANK	Receptor activator of nuclear factor-kappa B
RET	Rearranged during transfection
RTK	Receptor tyrosine kinase
RGC	Receptor guanylyl cyclase
SAD	Seasonal affective disorder

SIDS	Sudden infant death syndrome
SLE	Systemic lupus erythematosus
SC	Subcutaneous
SCC	Squamous cell carcinoma
SCC	Small cell carcinoma
SCLC	Small cell lung cancer
SLL	Small lymphocytic leukemia
STD	Sexually transmitted disease
SCNSL	Secondary central nervous system lymphoma
SH1	src homology 1 domain (kinase domain)
SH2	src homology 2 domain
SH3	src homology 3 domain
STAT	Signal transducer and activator of transcription
SYK	Spleen tyrosine kinase
TB	Tuberculosis
TAH	Total abdominal hysterectomy
TIA	Transient ischemic attack (stroke)
TIBC	Total iron binding capacity
TCC	Transitional cell carcinoma
TFG-β	Transforming growth factor beta
TMJ	Temporomandibular joint
TNF-α	Tumor necrosis factor alpha
TNBC	Triple-negative breast cancer
TBI	Total body irradiation
TCP	Thrombocytopenia
TKL	Tyrosine kinase-like kinase
TK	Tyrosine kinase
TYK2	tyrosine kinase 2
TSH	Thyroid-stimulating hormone
TURP	Transurethral resection of prostate gland
US	Ultrasound (scan)
URI	Upper respiratory infection
UTI	Urinary tract infection
UVR	Ultraviolet radiation
VEF	Ventricular ejection fraction
VEGFR	Vascular endothelial growth factor receptor
VHL	Von Hippel Lindau
WBC	White blood cell count
WCC	White cell count
WM	Waldenstrom's macroglobulinemia
WT	Wild type
XRT	Radiotherapy (external)
YST	Yolk sac tumor

Chapter 1
An Overview of Protein Kinase Inhibitors

Abstract Protein kinases become one of the most successful drug targets for the treatment of various diseases. Protein kinase catalyzes the transfer of the γ-phosphoryl group of the ATP to specific protein substrates which is an important cellular signal communication and maintenance of homeostasis. Most kinases share a common protein kinase domain. Some kinase inhibitors bind to the ATP site and some inhibitors bind to the allosteric site. There are 82 kinase inhibitors in the market. Here chemical structures of approved kinase inhibitors, their binding modes, medical uses, and names of drug developers are described. Humans have four typical MAP kinases such as extracellular signal-regulated kinases (ERKs), c-Jun N-terminal kinases (JNKs), p38 MAP kinases, and nemo-like kinase (NLK), and their structure and activation, cellular signaling, and pathological implications are highlighted.

Keywords Roles of kinase activation · FDA-approved kinase inhibitors · Mitogen-activated kinases · Structure · Function

1.1 Introduction

The protein kinase family is the second largest enzyme family (after proteases) and the fifth largest gene family in humans. Protein kinases are the second largest group of drug targets after G-protein-coupled receptors (GPCR) and they account for an estimated 20–30% drug development pipeline. The human genome has almost 518 kinase genes and 106 pseudogenes. Among 518 known kinases, 478 are typical and 40 are atypical kinases [1–16]. Protein kinases comprise 2% of the mammalian genome and catalyze the transfer of the γ-phosphoryl group of the ATP to specific protein substrates (Fig. 1.1). This covalent post-translational modification is a crucial component of normal cellular communication and maintenance of homeostasis. The signals can initiate inside or outside the cells and pass through a series of steps with multiple enzymes and messengers. Reversible protein phosphorylation and dephosphorylation are the most prominent post-transcriptional modifications to

S. K. De, *Drug Design and Discovery*,
https://doi.org/10.1007/978-3-031-75682-5_1

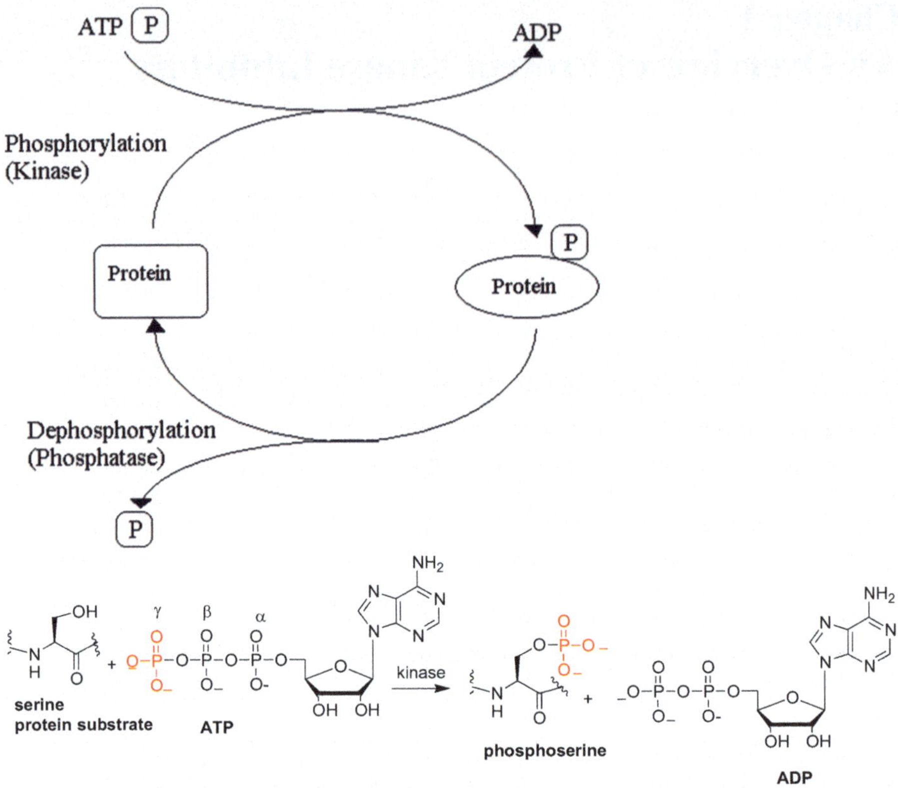

Fig. 1.1 Protein phosphorylation and dephosphorylation and protein kinase-catalyzed transfer of a γ-phosphoryl group from ATP onto serine hydroxyl group of a protein substrate (showing with red color)

proteins discovered by Edmond Fisher and Edwin Krebs (Fig. 1.1), who shared the Nobel Prize in 1992.

Protein kinases are enzymes that add inorganic phosphate to proteins and protein phosphatases are enzymes that remove inorganic phosphate from protein. The amino acids in proteins mostly phosphorylated are serine, threonine, and tyrosine and less common amino acid phosphorylated is histidine. There are 82 small molecule protein kinase inhibitors and 4 antibody kinase inhibitors FDA-approved for treating cancers and other diseases (see Tables 1.1 and 1.2). There are significant numbers of small molecule kinase inhibitors in various phases of clinical trials.

Table 1.1 FDA-approved kinase inhibitors and their applications, binding mode, and PDB code

Drug name	Target kinase	Protein substrate Phosphorylation site	Binding mode	Medical use	PDB Code and Refs.
Abemaciclib	CDK	Serine/threonine	Type I	Breast cancer	5L2S [17]
Abrocitinib	JAK	Tyrosine	Type 1	Atopic dermatitis	6BBU [18]
Acalabrutinib	BTK	Tyrosine	Covalent	CLL/MCL	8FD9 [19]
Afatinib	EGFR HER2	Tyrosine	Covalent	Non-small cell lung carcinoma (NSCLC)	4G5J [20]
Alectinib	ALK	Tyrosine	Type I	NSCLC	3AOX [21]
Alpelisib	PI3K	Inositol/serine/threonine	Type I	Breast cancer	4JPS [22]
Apatinib	VEGFR2	Tyrosine	Type I	Gastric carcinoma	ND
Axitinib	PDGFR	Tyrosine	Type II	Renal cell carcinoma (RCC)	4AG8 [23]
Avapritinib	PDGFR	Tyrosine	NR	Gastrointestinal stromal tumors	NR
Baricitinib	JAK	Tyrosine	NR	Rheumatoid arthritis, alopecia areata	6VN8 [24]
Binimetinib	B-RAF/MEK	Serine/threonine	Type III	Melanoma	7M0U [25]
Brigatinib	ALK/EGFR	Tyrosine	Type I	NSCLC	6MX8 [26]
Bosutinib	Bcr-Abl/Src	Tyrosine	Type II	CML	3UE4 [27]
Cabozantinib	FLT3/VEGFR/c-MET	Tyrosine	Type II	RCC/medullary thyroid cancer	4MXC [28]
Capmatinib	c-MET	Tyrosine	NR	NSCLC	NR
Ceritinib	ALK	Tyrosine	Type I	NSCLC	4MKC [29]
Cobimetinib	MEK	Serine/Threonine/ Tyrosine	Type III	Melanoma	4AN2 [30]
Copanlisib	PI-3 K	Inositol/Serine/ Threonine	Type I	Follicular lymphoma (FL)	5G2N [31]
Crizotinib	ALK	Tyrosine	Type I	NSCLC	3ZBF [32]
Dasatinib	ABL/KIT	Tyrosine	Type I	CML/ALL	3G6D [33]

(continued)

Table 1.1 (continued)

Drug name	Target kinase	Protein substrate Phosphorylation site	Binding mode	Medical use	PDB Code and Refs.
Dabrafenib	BRAF	Serine/Threonine	Type I 1/2	Melanoma	5CSW [34]
Dacomitinib	EGFR	Tyrosine	Covalent	NSCLC	4I23 [35]
Deucravacitinib	TYK2	Tyrosine	Type III	Psoriasis	NR
Duvelisib	PI-3 K	Inositol/Serine/ Threonine	NR	CLL/SLL/FL	NR
Encorafenib	MAPK/BRAF	Serine/Threonine	Type I	Melanoma	NR
Entrectinib	TRK	Tyrosine	Type I	NSCLC	5FTO [36]
Erdafitinib	FGFR	Tyrosine	Type I	Urothelial carcinoma	5EW8 [37]
Erlotinib	EGFR	Tyrosine	Type I	NSCLC/pancreaticcancer	4HJO [38]
Everolimus	mTOR	Serine/Threonine		Renal cell carcinoma	6L3A [39]
Fedratinib	JAK2	Tyrosine	Type I	Myelofibrosis	6VNE [24]
Fostamatinib	SYK	Tyrosine	Type I	Autoimmune/thrombocyto-penia	4O75 [40]
Futibatinib	FGFR	Tyrosine	Covalent	Cholangiocarcinoma	6MZW [41]
Gefitinib	EGFR	Tyrosine	Type I	NSCLC	5Y7Z [42]
Gilteritinib	FLT3	Tyrosine	Type I	AML	6JQR [43]
Ibrutinib	BTK	Tyrosine	Covalent	MCL/ CLL/SLL	5P9J [44]
Idelalisib	PI-3 K	Inositol/Serine/ Threonine	Type I	CLL/SLL	4XE0 [45]
Imatinib	ABL/KIT	Tyrosine	Type II	CLL/ALL	2HYY [46]
Infigratinib	FGFR	Tyrosine	Type I	Cholangiocarcinoma	3TT0 [47]
Lapatinib	EGFR/HER2	Tyrosine	Type I 1/2	Breast cancer	3BBT [48]
Larotrectinib	TRK	Tyrosine	Type I	Solid tumors with NTRK fusion	NR
Leniolisib	PI3Kδ	Inositol/Serine/ Threonine	Type I	APDS	5O83 [49]
Lenvatinib	FGFR/KIT	Tyrosine	Type I	Thyroid cancer	3WZD [50]

Drug name	Target kinase	Protein substrate Phosphorylation site	Binding mode	Medical use	PDB Code and Refs.
Lorlatinib	ALK	Tyrosine	Type I	NSCLC	4UXL [51]
Midostaurin	FLT3/KIT	Tyrosine	Type I	NSCLC	4NCT [52]
Mobocertinib	EGFR	Tyrosine	Covalent	NSCLC	7A6K [53]
Neratinib	EGFR	Tyrosine		Covalent	2JIV [54]
Netarsudil	ROCK	Serine/Threonine	Type I	Glaucoma	NR
Nilotinib	ABL/KIT	Tyrosine	Type II	CML	3CS9 [55]
Nintedanib	Src/Lck/PDGFR/VEGFR	Tyrosine	Type I	Idiopathic pulmonary fibrosis/NSCLC	3C7Q [56]
Osimertinib	EGFR	Tyrosine	Covalent	NSCLC	6JXT [57]
Pacritinib	JAK2/FLT3	Tyrosine	Type I	Myelofibrosis	5LBZ [58]
Palbociclib	CDK	Serine/Threonine	Type I	Breast cancer	5L2I [17]
Pazopanib	VEGFR/KIT	Tyrosine	Type I	Renal cancer	NR
Pemigatinib	FGFR	Tyrosine	Type I	Cholangiocarcinoma	7WCL [59]
Pexidartinib	KIT/FLT3	Tyrosine	Type II	Tenosynovial giant cell tumor	4R7H [60]
Pirtobrutinib	BTK	Tyrosine	Type I	MCL	8FLL [61]
Ponatinib	ABL/FLT3	Tyrosine	Type II	CML/ALL	3OXZ [62]
Pralsetinib	RET	Tyrosine	Type I	NSCLC	7JU5 [63]
Radotinib	ABL	Tyrosine	Type II	CML	NR
Rapamycin	mTOR	Serine/Threonine	NR	RCC/kidney transplantation	7SQ9 [64]
Regorafenib	VEGFR/KIT	Tyrosine	Type II	Colorectal cancer	NR
Ribociclib	CDK4/6	Tyrosine/Serine	Type I	Breast cancer	5L2T [17]
Ripretinib	PDGFR/KIT	Tyrosine	Type II	Gastrointestinal stromal tumor	6MOB [65]
Ritlecitinib	JAK3	Tyrosine	Covalent	Severe alopecia	NR
Ruxolitinib	JAK	Tyrosine	Type I	Myelofibrosis	6VGL [24]

(continued)

Table 1.1 (continued)

Drug name	Target kinase	Protein substrate Phosphorylation site	Binding mode	Medical use	PDB Code and Refs.
Selpercatinib	RET	Tyrosine	Type I	NSCLC/thyroid cancer	7JU6 [63]
Selumetinib	MAPK/MEK	Serine/Threonine	Type III	Neurofibromatosis type 1	4U7Z [66]
Sorafenib	VEGFR/FLT3/KIT	Tyrosine	Type II	RCC, HCC, thyroid cancer	4ASD [23]
Sunitinib	PDGFR/FLT3/KIT	Tyrosine	Type I	Renal cancer	3G0E [67]
Tepotinib	c-MET	Tyrosine	Type I	NSCLC	4R1V [68]
Tofacitinib	JAK	Tyrosine	Type I	Rheumatoid arthritis	3LXK [69]
Trametinib	MEK	Serine/Threonine	Type III	Melanoma	7JUR [70]
Trilaciclib	CDK4/6	Serine/Threonine	NR	Bone marrow suppression	NR
Tucatinib	HER2	Tyrosine	NR	Breast cancer	NR
Umbralisib	PI3K	Inositol/Serine/ Threonine	Type I	Marginal zone lymphoma (MZL), follicular lymphoma	NR
Upadacitinib	JAK	Tyrosine	NR	Rheumatoid arthritis	NR
Vandetanib	VEGFR/EGFR/RET	Tyrosine	Type I	Thyroid cancer	2IVU [71]
Vemurafenib	BRAF	Serine/Threonine	Type I	Melanoma	3OG7 [72]
Volasertib	PLK1	Serine/Threonine	Type I	AML	3FC2 [73]
Zanubrutinib	BTK	Tyrosine	Covalent	MCL	6J6M [74]

NR not reported

Table 1.2 FDA-approved kinase inhibitors' name, structure, MW, and company information

Year	Drug structure and generic name (trade name)	MW	Company
1999	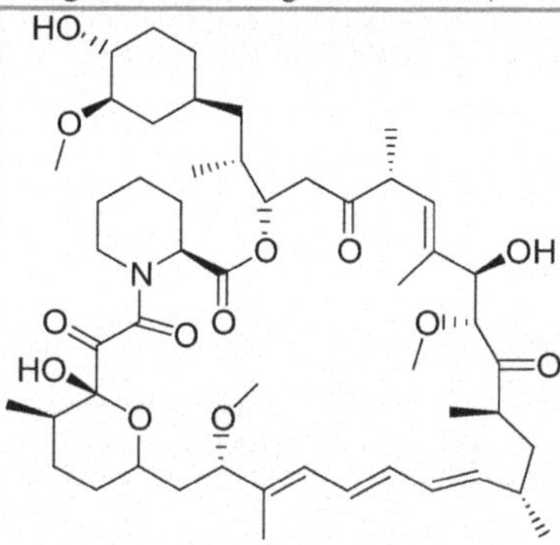 Rapamycin (Rapamune, Sirolimus)	914	Wyeth
2001	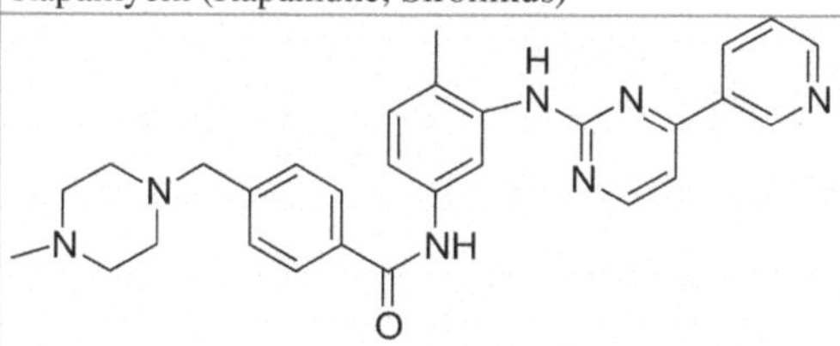 Imatinib (Gleevec)	494	Novartis
2003	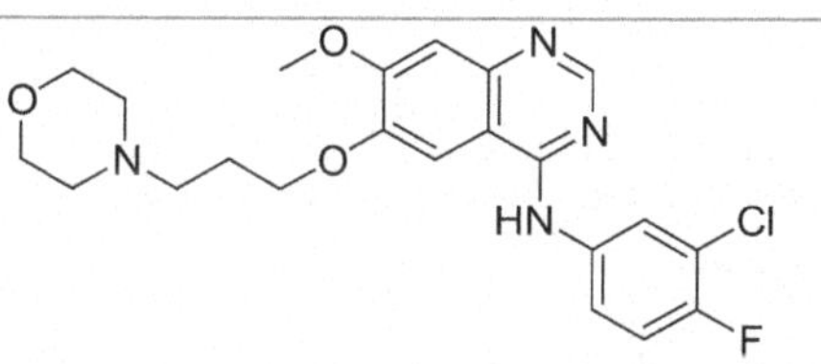 Gefitinib (Iressa)	447	Astra Zeneca
2004	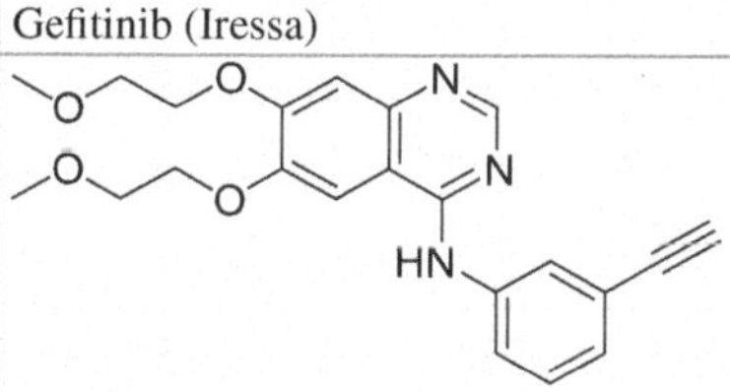 Erlotinib (Tarceva)	393	Genentech
2005	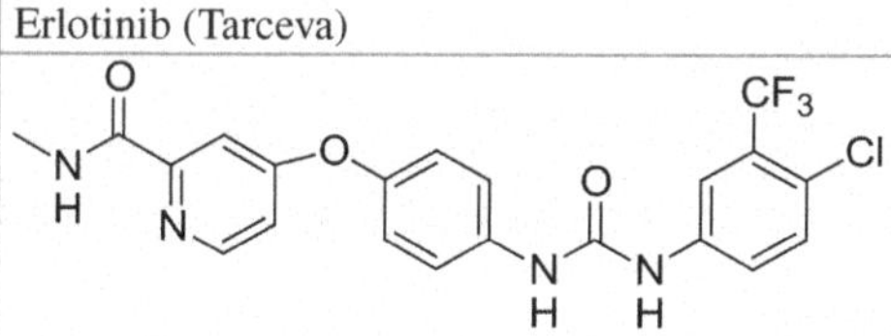 Sorafenib (Nexavar)	465	Onyx

(continued)

Table 1.2 (continued)

Year	Drug structure and generic name (trade name)	MW	Company
2006	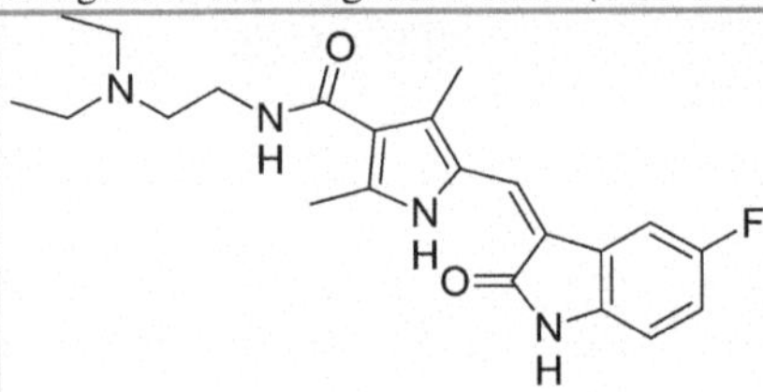Sunitinib (Sutent)	398	Pfizer
2006	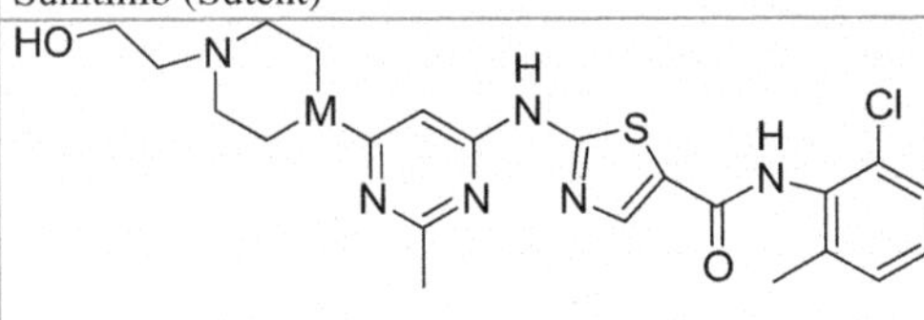Dasatinib (Sprycel)	488	Bristol Myers Squibb (BMS)
2007	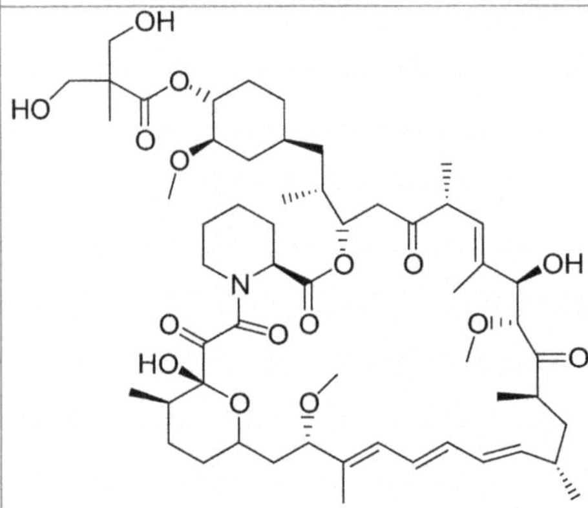Temsirolimus (Torisel)	1024	Wyeth
2007	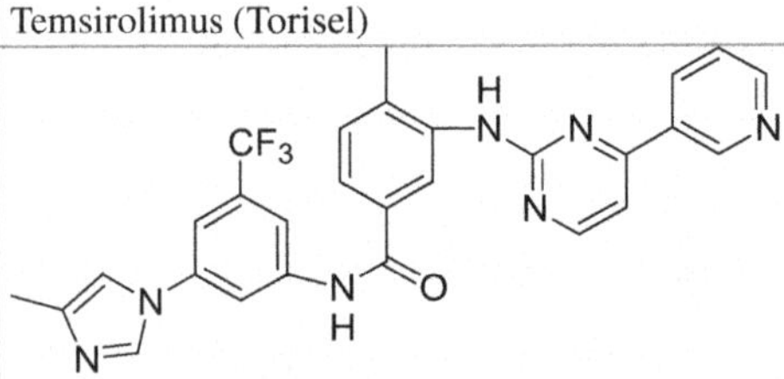Nilotinib (Tasigna)	530	Novartis

(continued)

Table 1.2 (continued)

Year	Drug structure and generic name (trade name)	MW	Company
2007	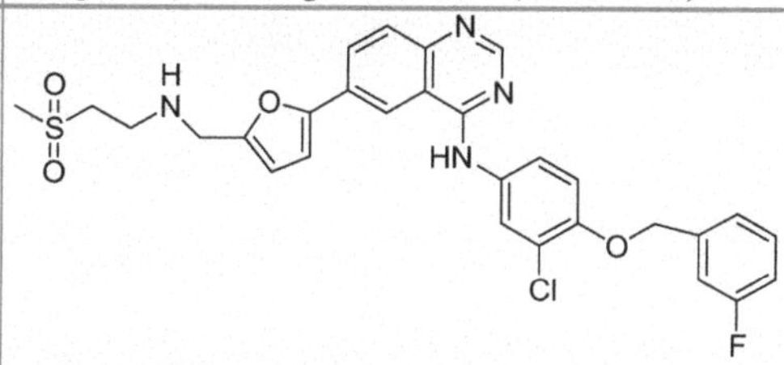 Lapatinib (Tykerb)	581	GSK
2009	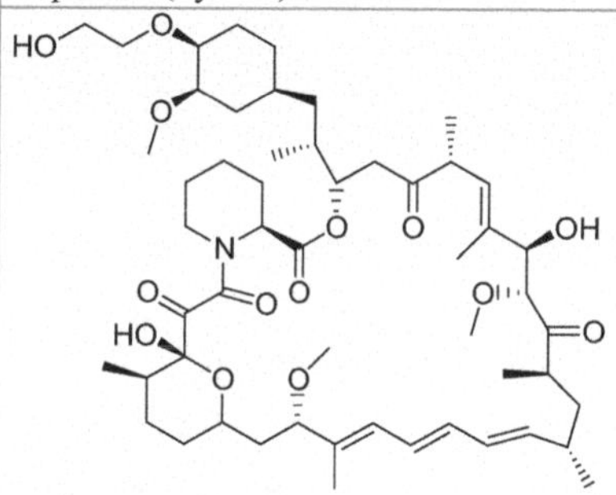 Everolimus (Afinitor)	958	Novartis
2009	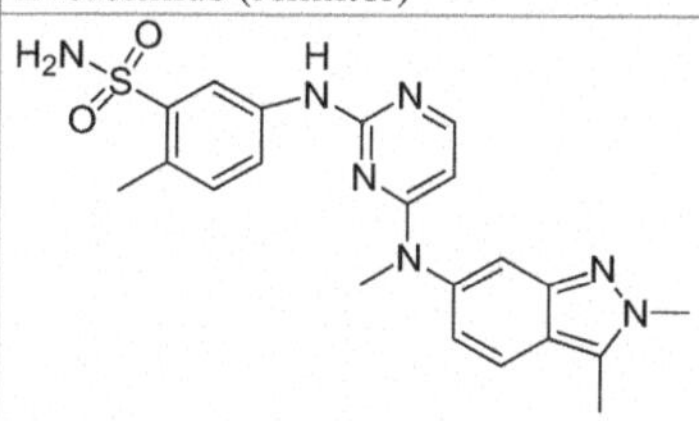 Pazopanib (Votrient)	437	GSK
2011	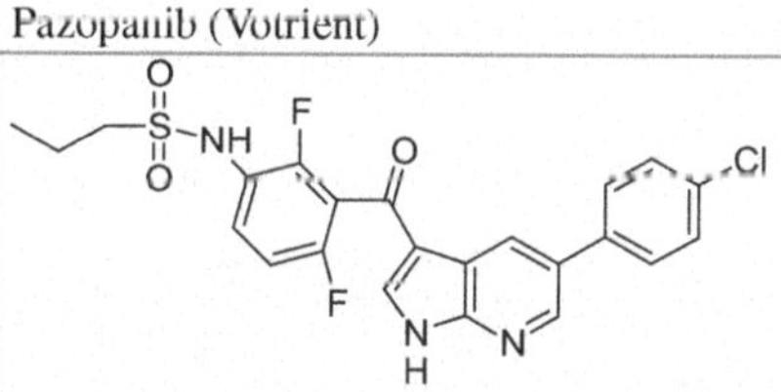 Vemurafenib (Zelboraf)	489	Genentech
2011	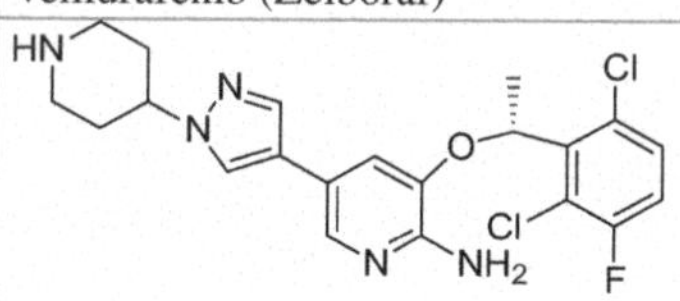 Crizotinib (Xalkori)	450	Pfizer

(continued)

Table 1.2 (continued)

Year	Drug structure and generic name (trade name)	MW	Company
2011	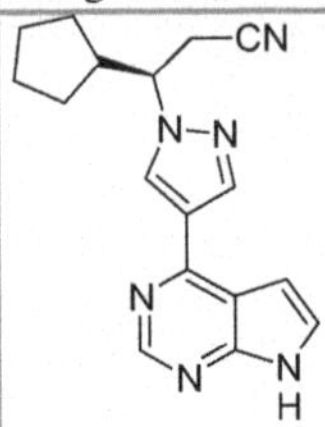 Ruxolitinib (Jakafi)	306	Incyte
2011	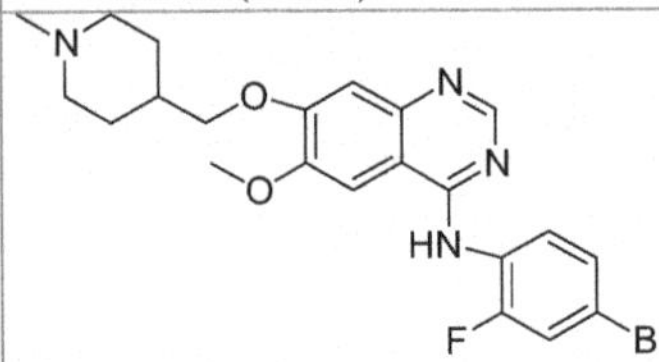 Vandetanib (Caprelsa)	475	Astra Zeneca
2012	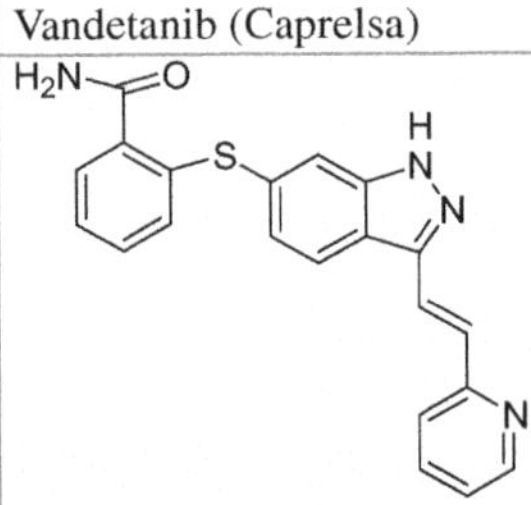 Axitinib (Inlyta)	386	Pfizer
2012	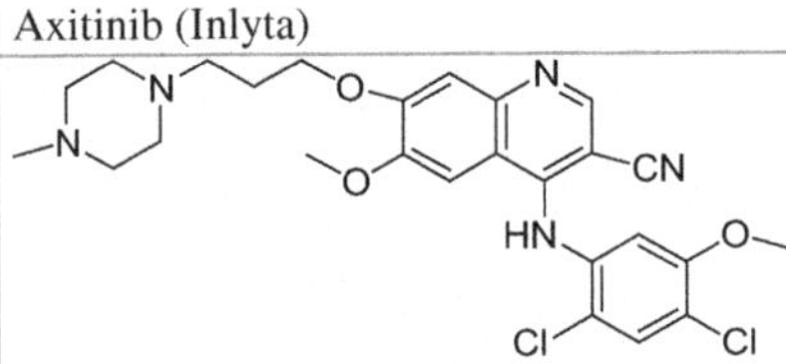 Bosutinib (Bosufil)	530	Wyeth
2012	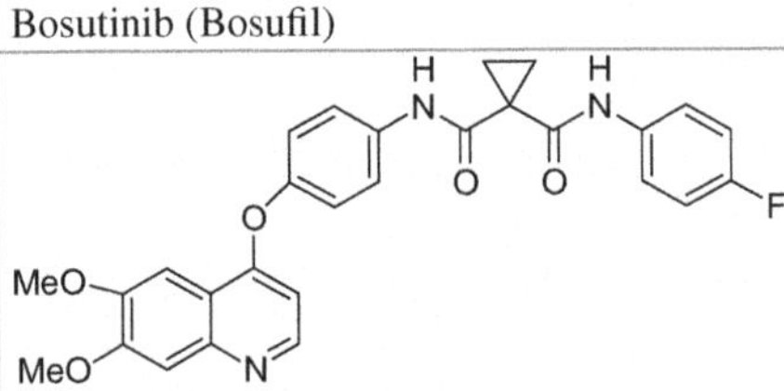 Cabozantinib (Cometriq)	501	Exelixis
2012	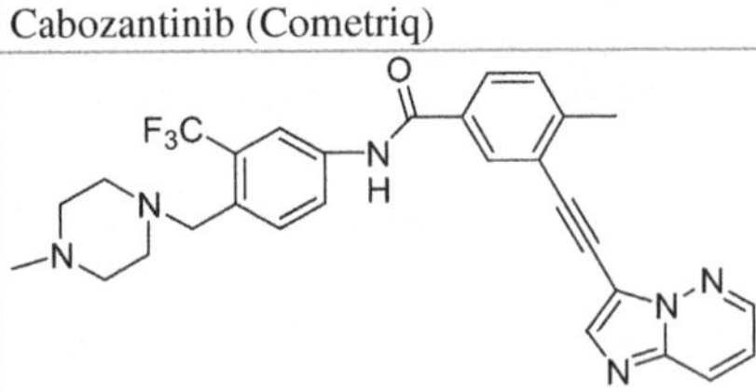 Ponatinib (Iclusig)	533	Ariad

(continued)

Table 1.2 (continued)

Year	Drug structure and generic name (trade name)	MW	Company
2012	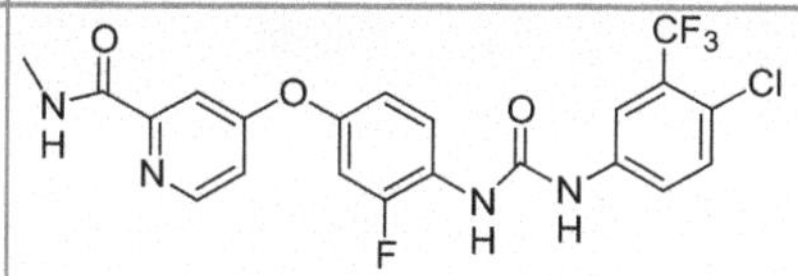 Rigorafenib (Stivarga)	483	Bayer
2012	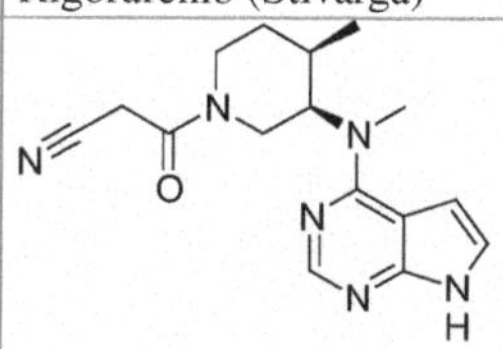 Tofacitinib (Xeljanz)	312	Pfizer
2013	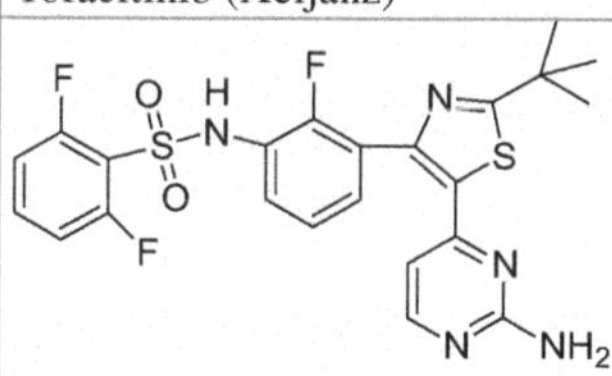 Dabrafenib (Tafinlar)	520	GlaxoSmithKline (GSK)
2013	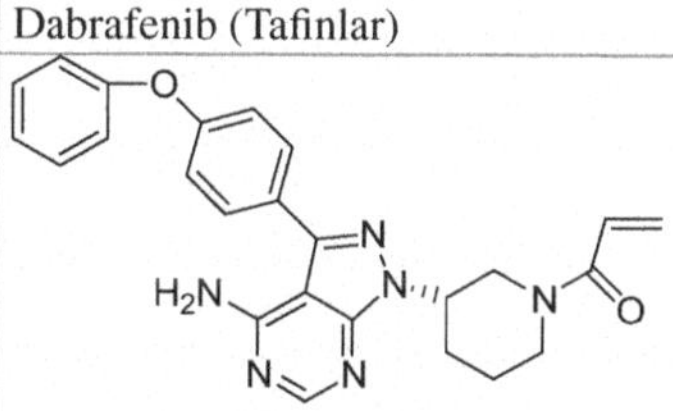 Ibrutinib (Imbruvica)	441	Pharmacyclics/Johnson & Johnson
2013	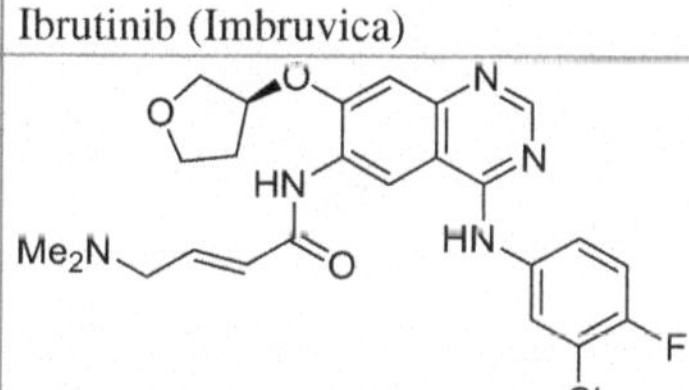 Afatinib (Gilotrif)	486	Boeh-ringer Ingelheim (BI)
2013	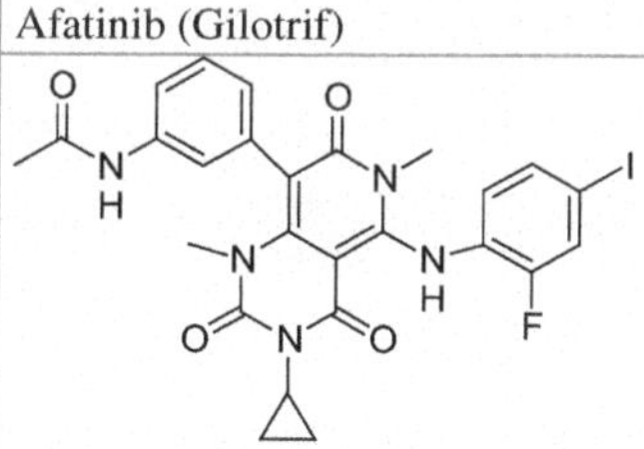 Trametinib (Mekinist)	615	GSK

(continued)

Table 1.2 (continued)

Year	Drug structure and generic name (trade name)	MW	Company
2014	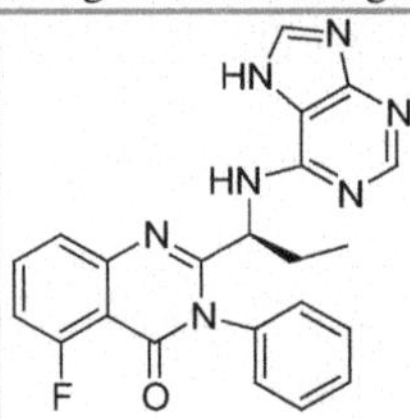 Idelalisib (Zydelig)	415	Gilead
2014	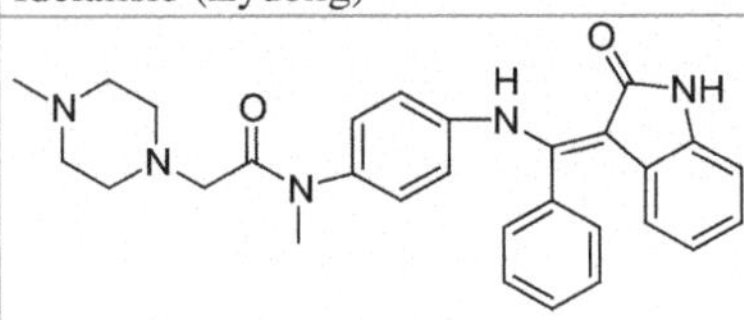 Nintedanib (Ofev/Vargatef)	540	B. I.
2015	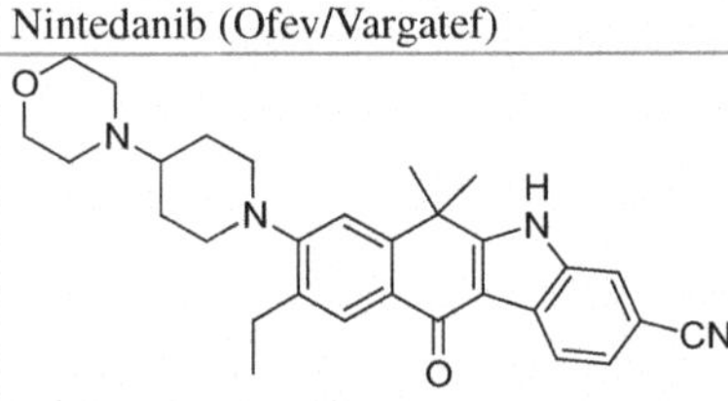 Alectinib (Alecensa)	483	Hoffman LaRoche
2015	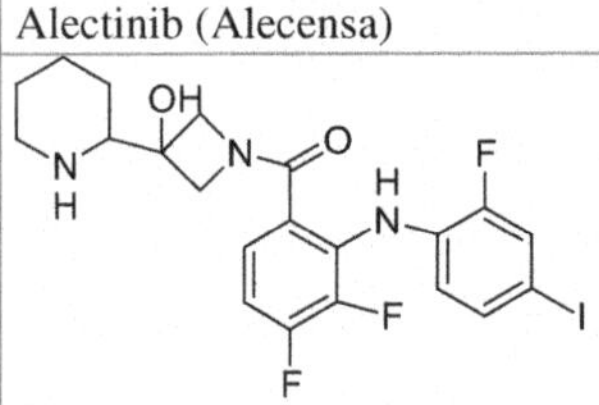 Cobimetinib (Cotellic)	531	Genentech
2015	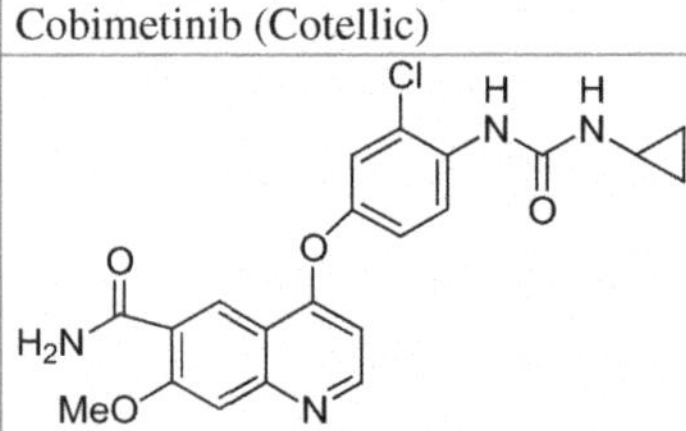 Lenvatinib (Lenvima)	427	Eisai
2015	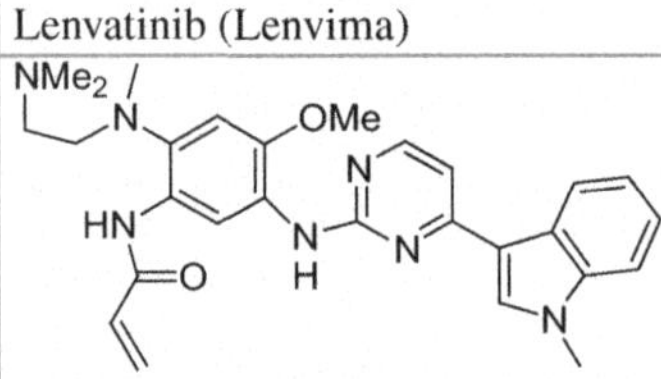 Osimertinib (Tagrisso)	500	Astra Zeneca

(continued)

Table 1.2 (continued)

Year	Drug structure and generic name (trade name)	MW	Company
2015	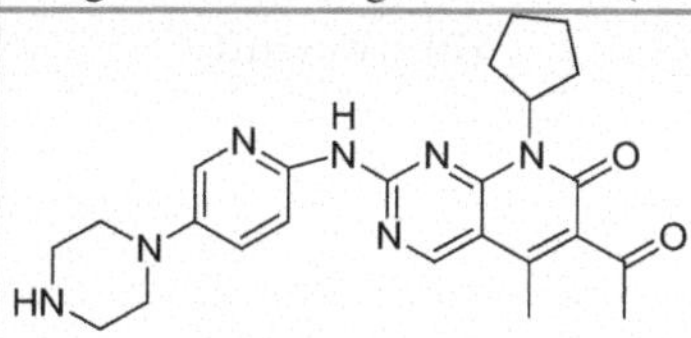 Palbociclib (Ibrance)	447	Park Davis
2017	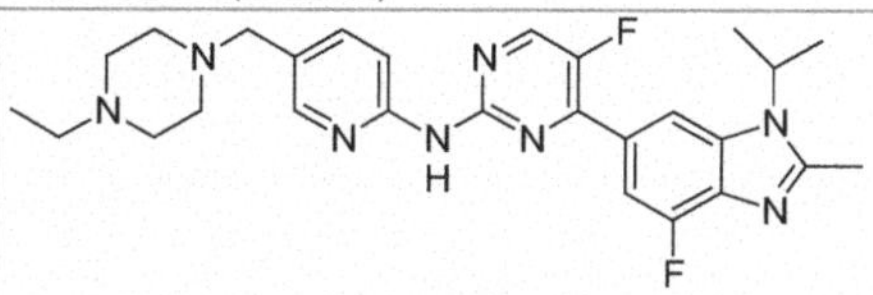 Abemaciclib (Verzenio)	507	Eli Lilly
2017	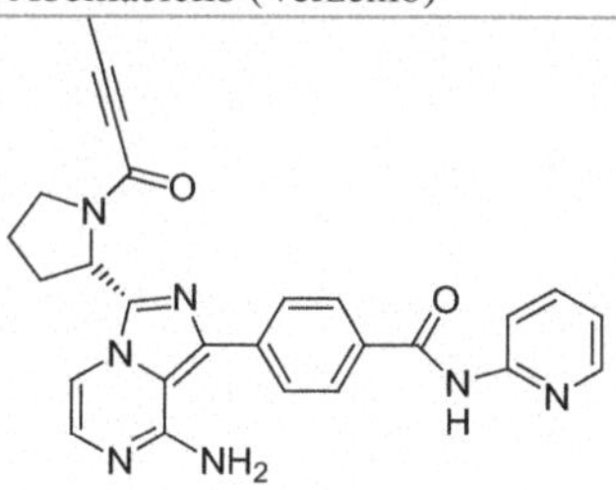 Acalabrutinib (Calquence)	466	Acerta Pharm
2017	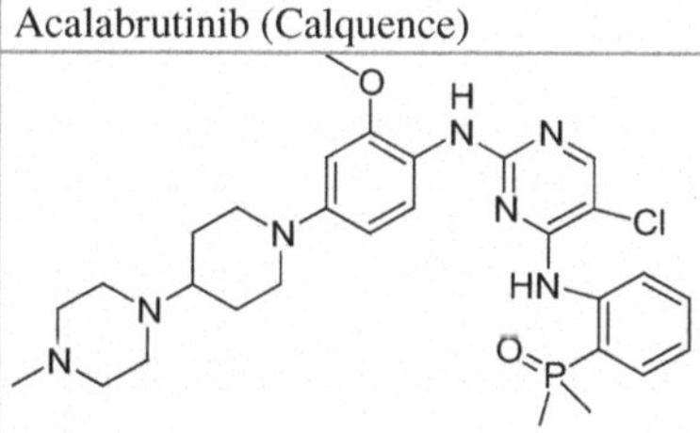 Brigatinib (Alunbrig)	584	Ariad
2017	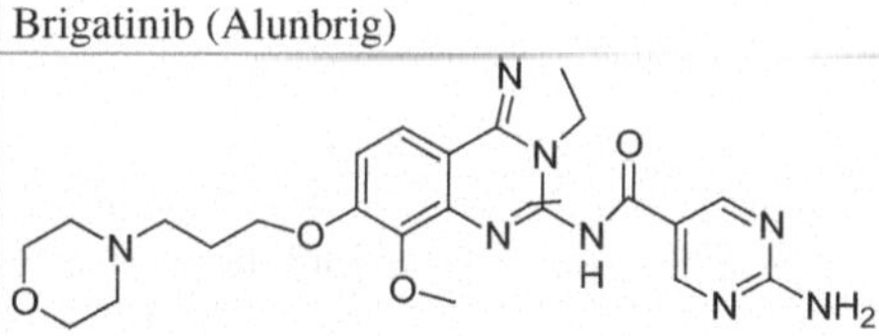 Copanlisib (Aliqopa)	480	Bayer

(continued)

Table 1.2 (continued)

Year	Drug structure and generic name (trade name)	MW	Company
2017	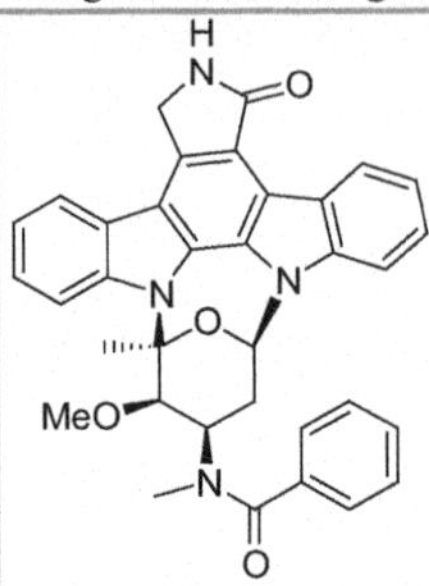 Midostaurin (Rydapt)	570	Novartis
2017	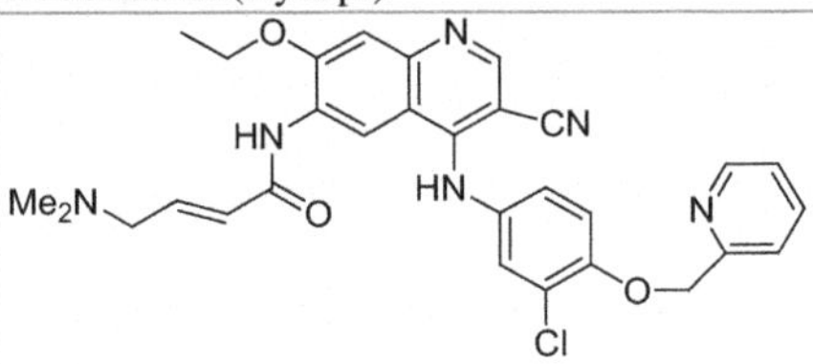 Neratinib (Nerlynix)	557	Puma
2017	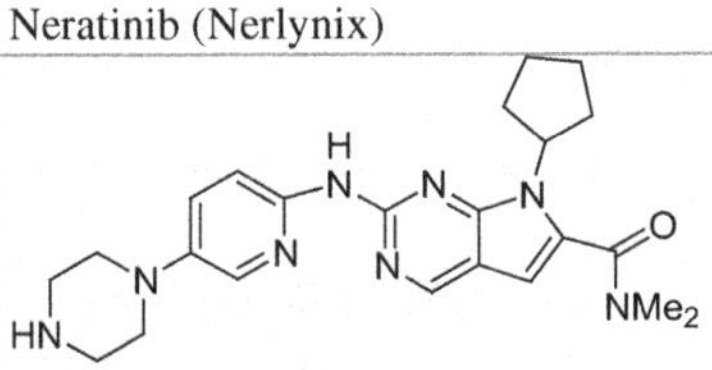 Ribociclib (Kisqali)	435	Novartis
2018	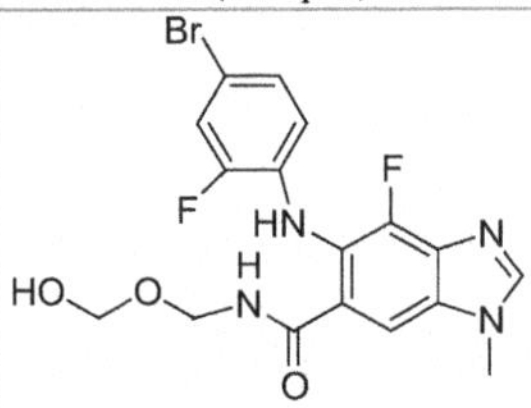 Binimetinib (Mektovi)	441	Array
2018	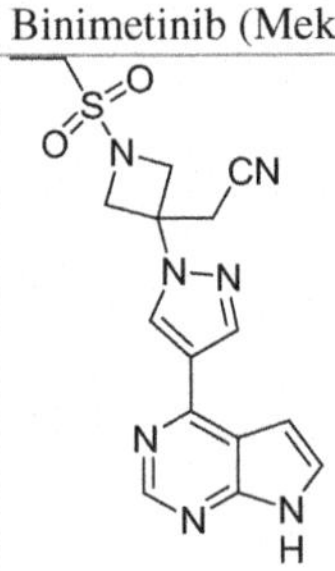 Baricitinib (Olumiant)	371	Eli Lilly

(continued)

Table 1.2 (continued)

Year	Drug structure and generic name (trade name)	MW	Company
2018	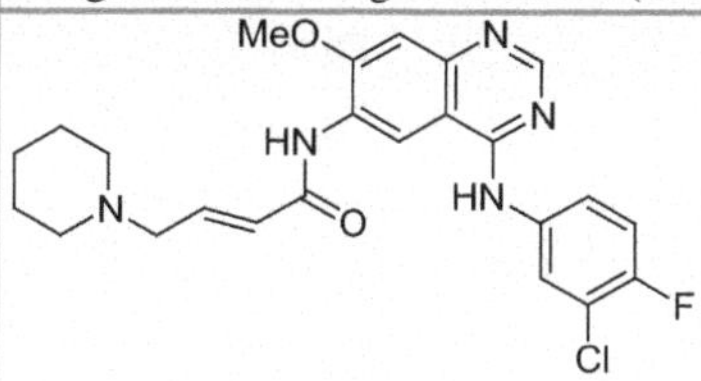 Dacomitinib (Visimpro)	470	Pfizer
2018	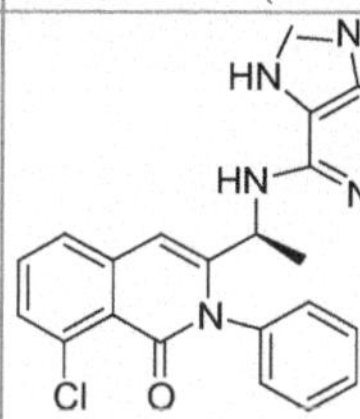 Duvelisib (Copiktra)	416	Verastem
2018	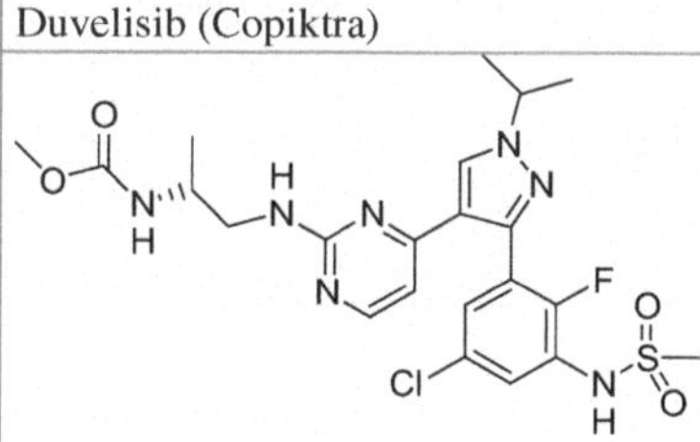 Encorafenib (Braftovi)	540	Array
2018	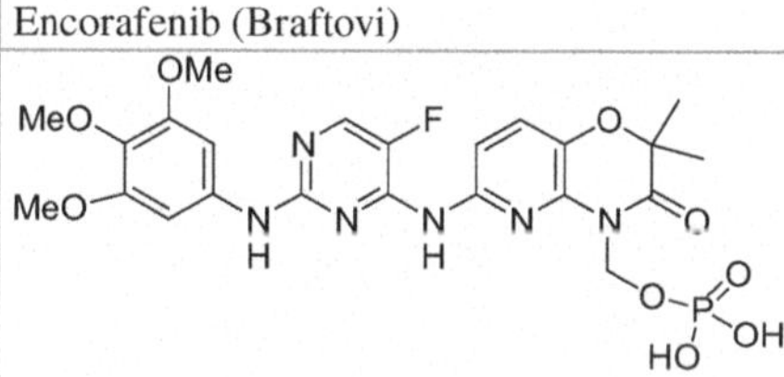 Fostamatinib (Tavalisse)	580	Rigel
2018	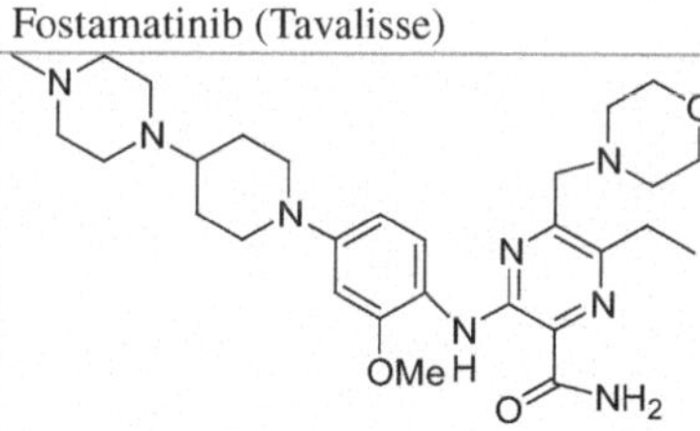 Giltertinib (Xospata)	553	Astellas

(continued)

Table 1.2 (continued)

Year	Drug structure and generic name (trade name)	MW	Company
2018	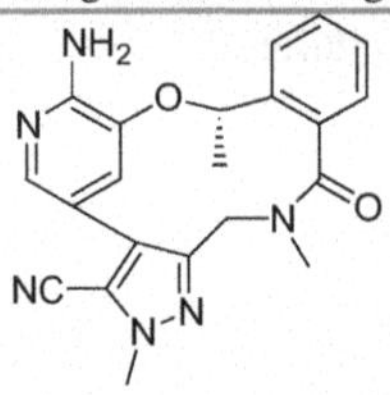 Lorlatinib (Lorbrena)	406	Pfizer
2018	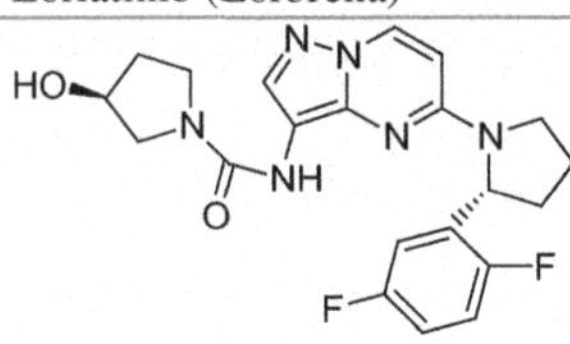 Larotrectinib (Vitrakvi)	428	Bayer
2018	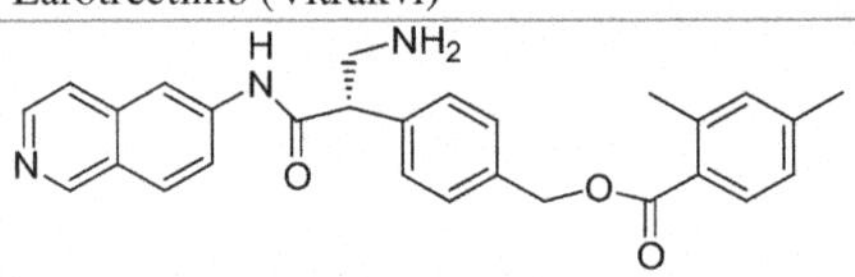 Netarsudil (Rhopressa)	453	Aerie Pharm
2019	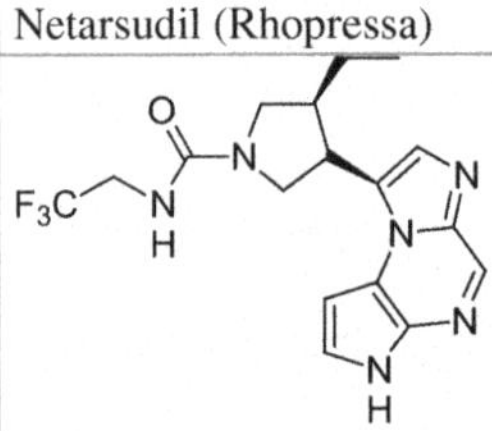 Upadacitinib (Rinvoq)	380	AbbVie
2019	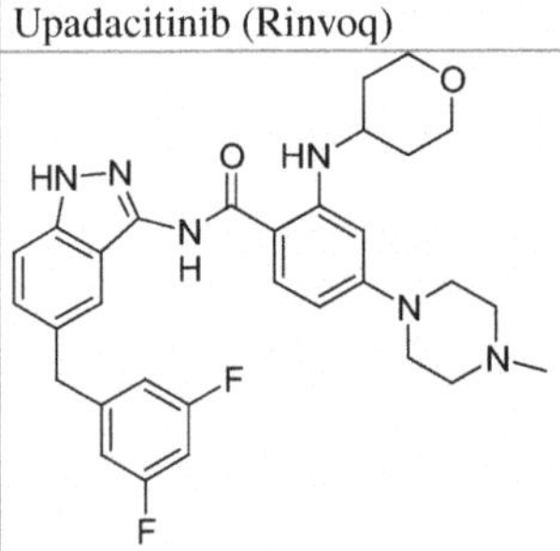 Entrectinib (Rozlytrek)	561	Genentech
2019	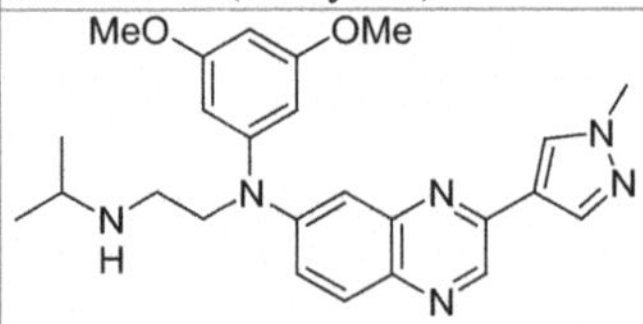 Erdafitinib (Balversa)	446	Janssen

(continued)

Table 1.2 (continued)

Year	Drug structure and generic name (trade name)	MW	Company
2019	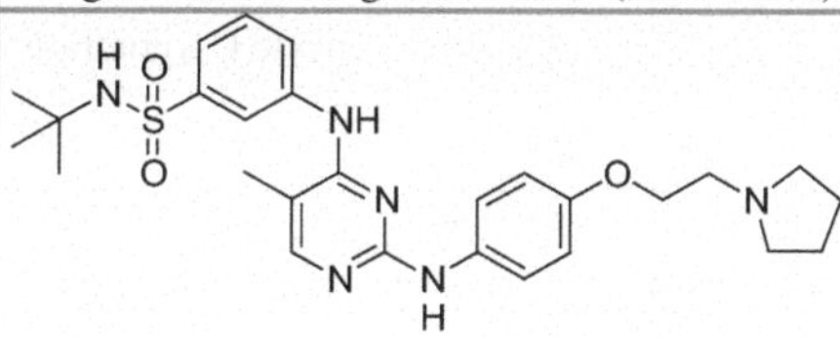 Fedratinib (Inrebic)	515	Impact Biomedicines
2019	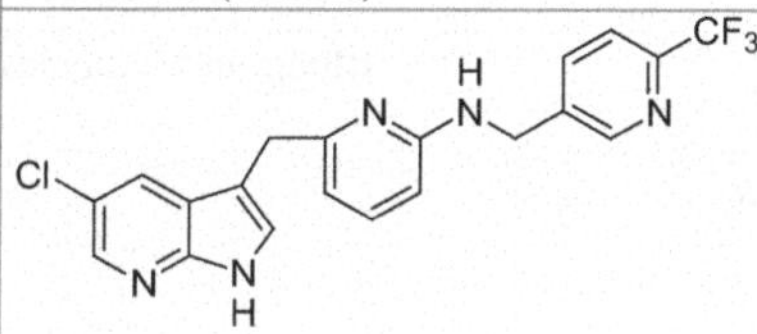 Pexidartinib (Turalio)	417	Daiichi Sankyo
2019	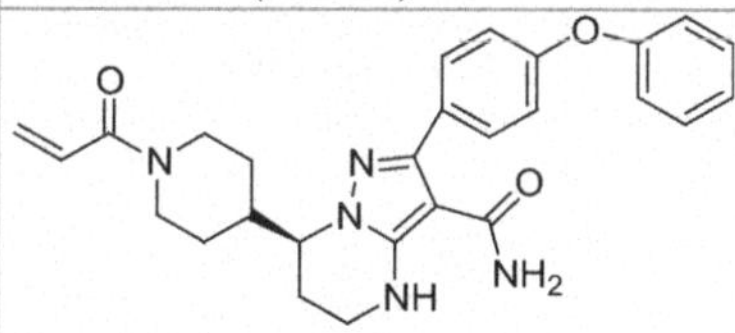 Zanubrutinib (Brukinsa)	472	BeiGene USA Inc
2020	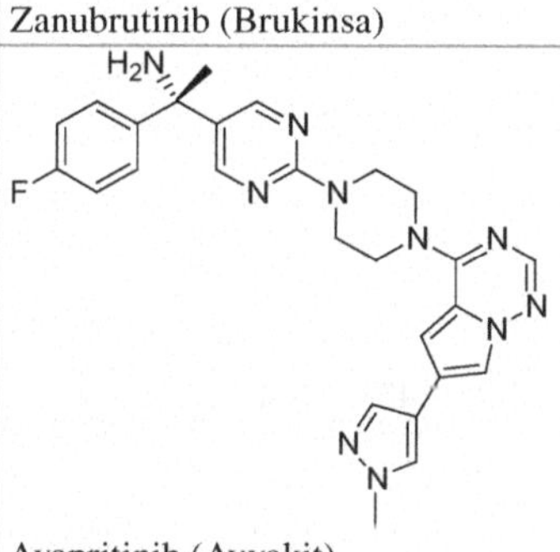 Avapritinib (Ayvakit)	499	Blueprint Medicines Corporation
2020	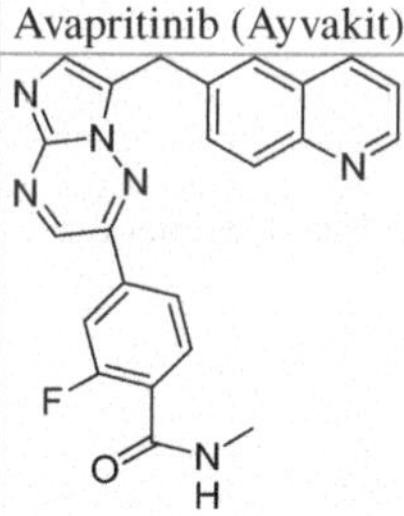 Capmatinib (Tabrecta)	412	Novartis

(continued)

Table 1.2 (continued)

Year	Drug structure and generic name (trade name)	MW	Company
2020	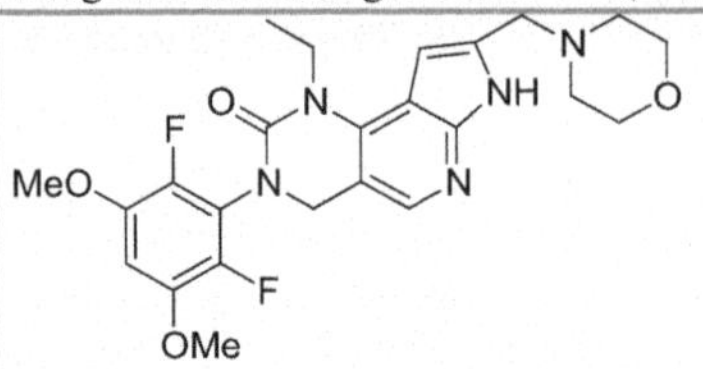 Pemigatinib (Pemazyre)	487	Incyte Corporation
2020	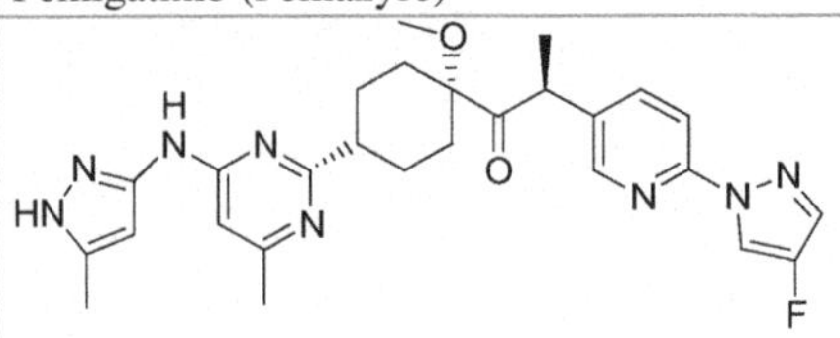 Pralsetinib (Gavreto)	534	Blueprint Medicines
2020	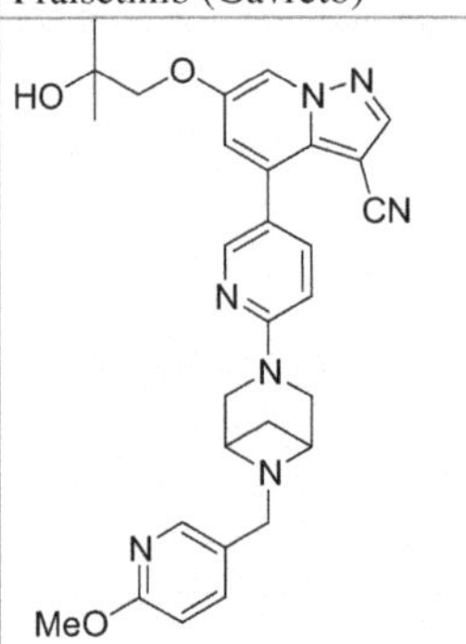 Selpercatinib (Retevmo)	526	Eli Lilly
2020	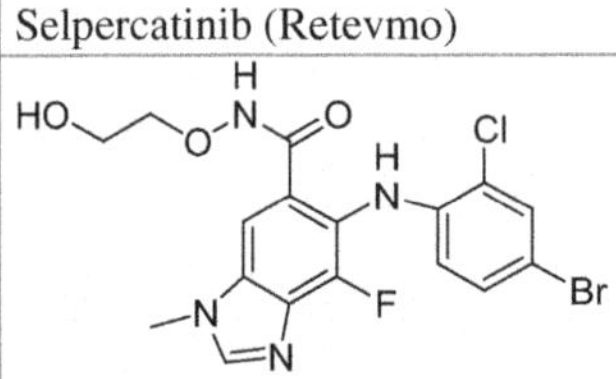 Selumetinib (Koselugo)	458	AstraZeneca
2020	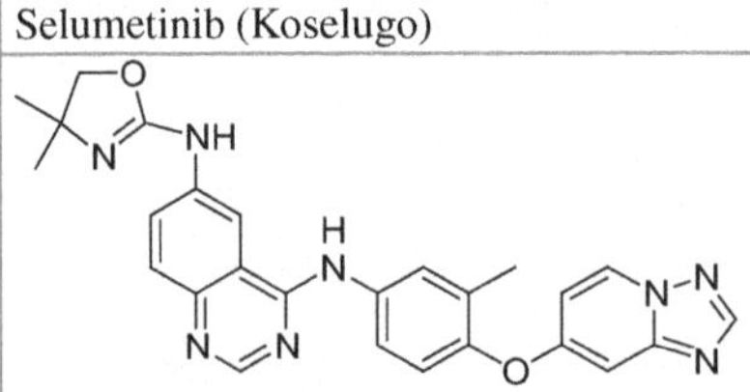 Tucatinib (Tukysa)	481	Seattle Genetics

(continued)

Table 1.2 (continued)

Year	Drug structure and generic name (trade name)	MW	Company
2020	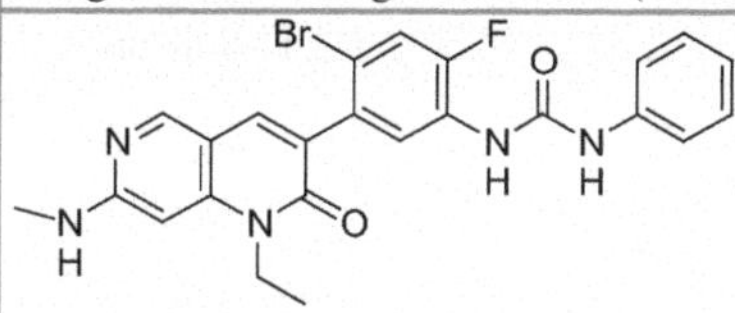 Ripretinib (Qinlock)	510	Deciphera Pharmaceuticals
2021	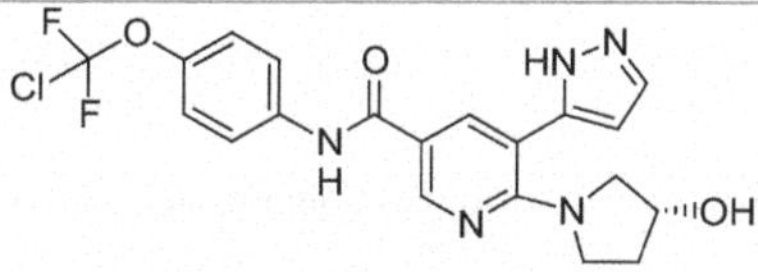 Asciminib (Scemblix)	450	Novartis
2021	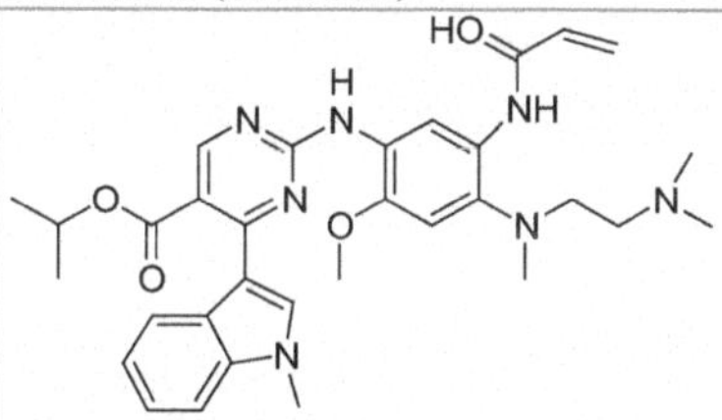 Mobocertinib (Exkvity)	585	Ariad
2021	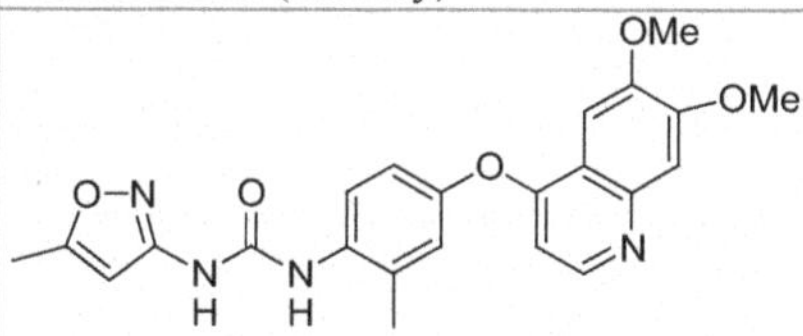 Tivozanib (Fotivda)	454	AVEO Oncology
2021	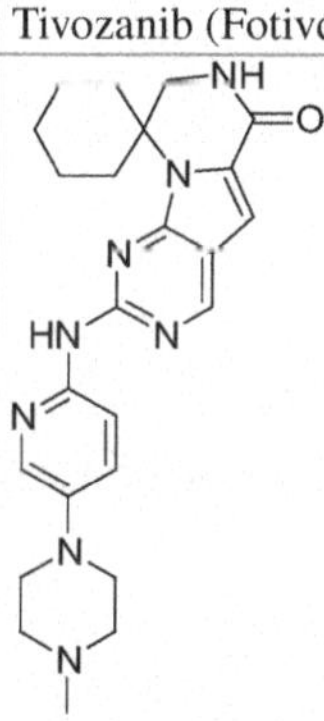 Trilaciclib (Cosela)	446	G1 Therapeutics
2021	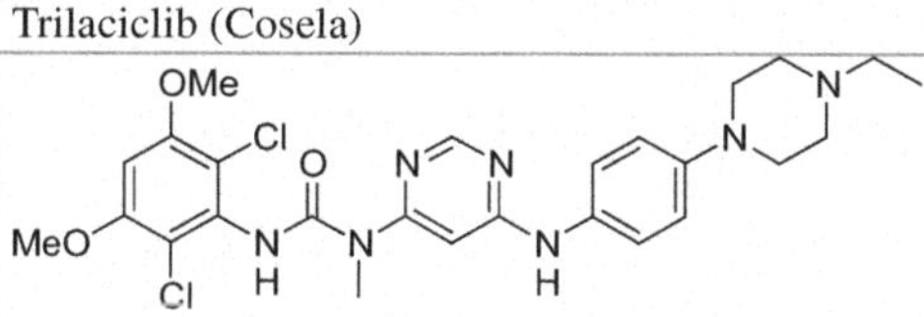 Infigratinib (Truseltiq)	560	Novartis

(continued)

Table 1.2 (continued)

Year	Drug structure and generic name (trade name)	MW	Company
2021	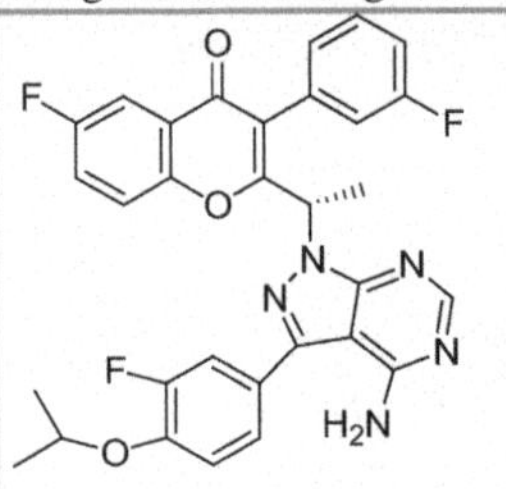 Umbralisib (Ukoniq)	571	TG Therapeutics
2021	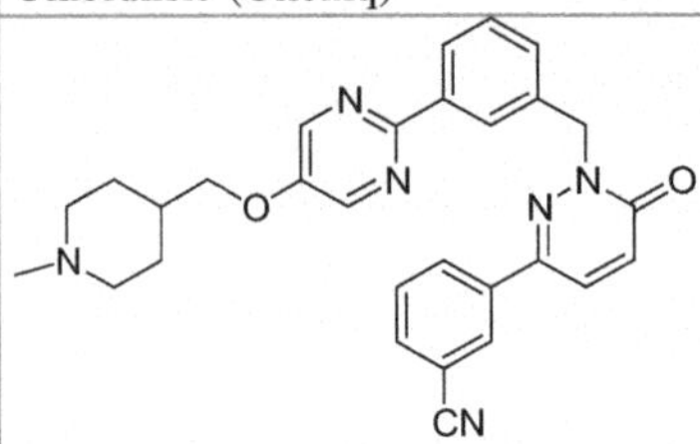 Tepotinib (Tepmetko)	492	EMD Serono
2021	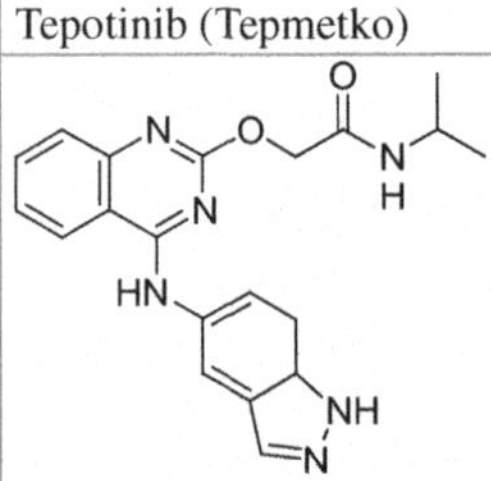 Belumosudil (Rezurock)	452	Kadmon pharma
2022	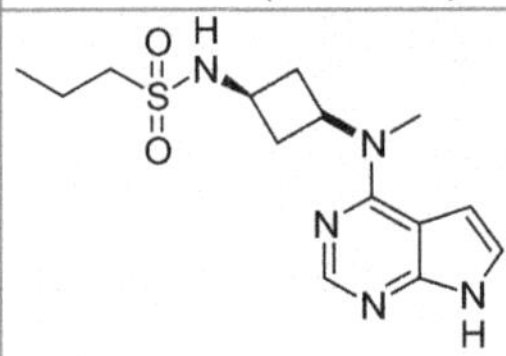 Abrocitinib (Cibinqo)	323	Pfizer
2022	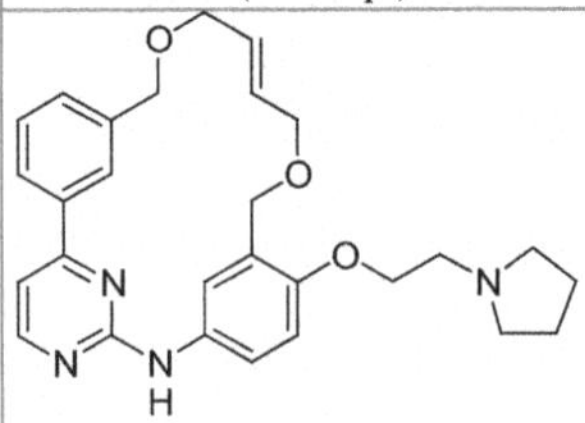 Pacritinib (Vonjo)	472	CTO Biopharma

(continued)

Table 1.2 (continued)

Year	Drug structure and generic name (trade name)	MW	Company
2022	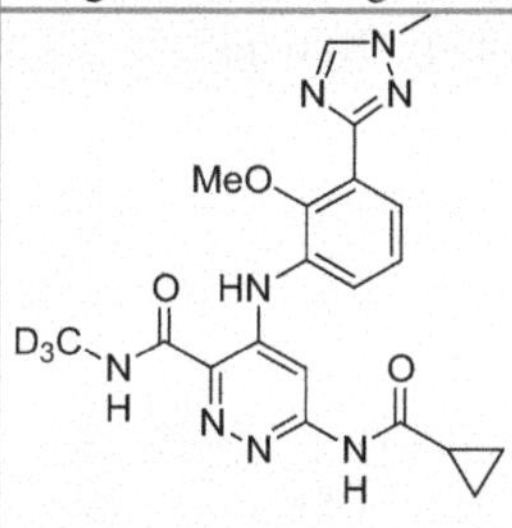 Deucravacitinib (Sotyktu)	425	BMS
2022	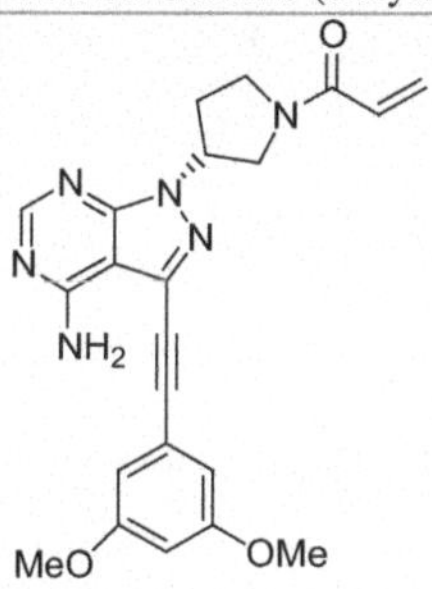 Futibatinib (Lytgobi)	418	Taiho
2023	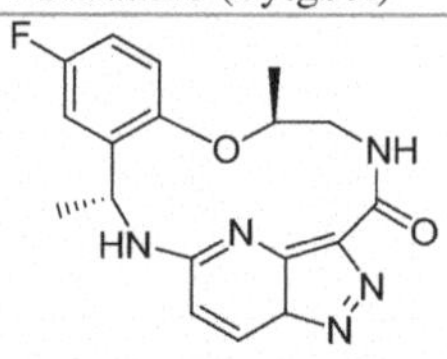 Repotrectinib (Augtyro)	355	TP Therapeutics
2023	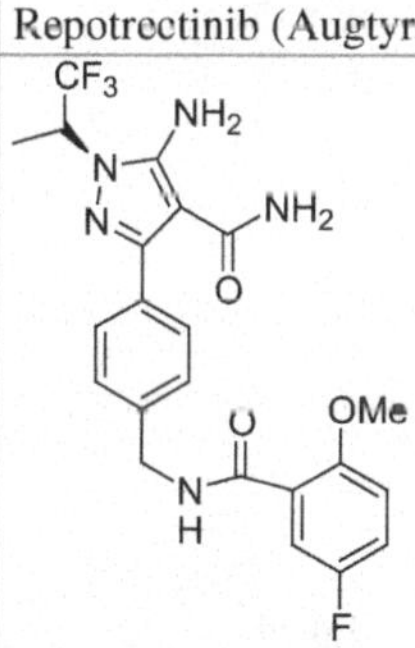 Pirtobrutinib (Jaypirca)	479	Eli Lily
2023	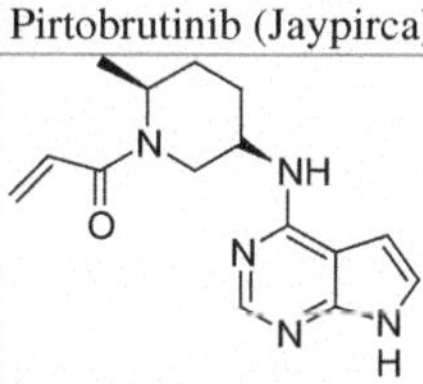 Ritlectinib (Litfulo)	285	Pfizer

(continued)

Table 1.2 (continued)

Year	Drug structure and generic name (trade name)	MW	Company
2023	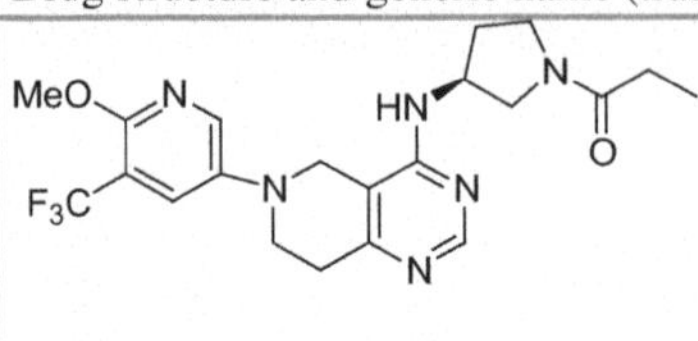 **Leniolisib** (Joenja)	450	Novartis
2023	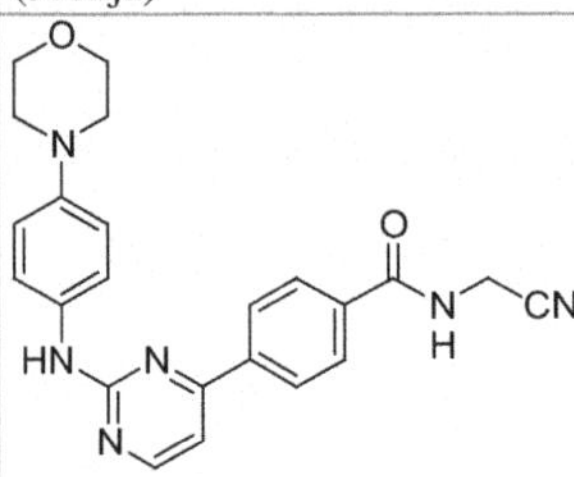 Momelotinib (Ojjaara)	414	GSK
2023	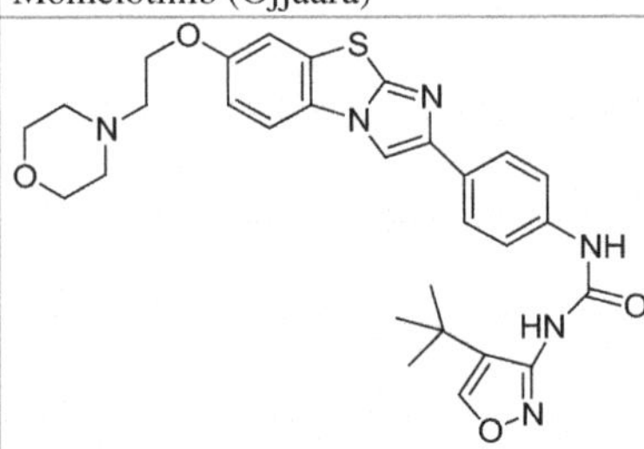 Quizartinib (Vanflyta)	560	Daiichi Sankyo
2023	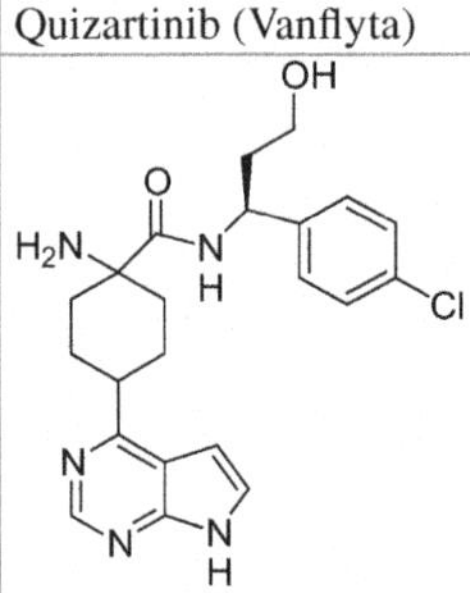 Capivasertib (Truqap)	428	AstraZeneca
2023	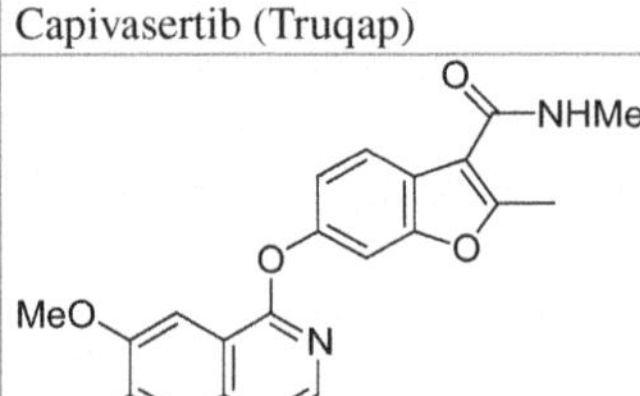 Fruquintinib (Fruzaqla)	393	Takeda

1.1.1 *Classification of Kinases*

According to the phosphorylation of amino acids, protein kinases are classified into the following groups:

1. Ser/Thr protein kinases.
    (a) Cyclic nucleotide-dependent.
    (b) Calcium-calmodulin-dependent.
    (c) Protein kinase C (PKC).
    (d) Mitogen-activated protein kinases (MAPK).
    (e) G-protein-coupled/Rhodospin/BARK.
2. Ser/Thr/Tyr protein kinases.
    (a) MAP kinase kinase (MAPKK).
3. Tyrosine protein kinases.
    (a) Cytosolic tyrosine kinases or non-receptor tyrosine kinases (32 members).
    (b) Receptor tyrosine kinases (58 members).

1.1.2 *Protein Kinase Domains*

478 typical kinases share a eukaryotic protein kinase domain. 40 atypical kinases do not have this protein kinase domain, but still show kinase activity.

Based on the sequence similarity of kinase domains, 478 typical kinases are divided into eight major groups:

(i) TK (tyrosine kinase, 90 members).
(ii) TKL (tyrosine kinase-like kinase, 43 members).
(iii) STE (STE7-, STE11-, and STE20-related kinase, 47 members).
(iv) CK1 (casein kinase 1, 12 members).
(v) AGC (protein kinase A/G/C-related kinase, 63 members).
(vi) CAMK (calcium/calmodulin-dependent kinase, 74 members).
(vii) CMGC (CDK/MAPK/GSK/CDK like-related, 61 members).
(viii) RGC (receptor guanylyl cyclase, 5 members).

40 atypical protein kinases are pyruvate dehydrogenase kinases, bromodomain kinases, B cell receptors (BCRs), and PI3 kinase-related kinases.

Knighton et al. published the first crystal structure of protein kinase A (PKA) in 1991 (Fig. 1.2a, b). The catalytic domain of a kinase comprises two lobes, a smaller *N*-terminal lobe (N-lobe: almost 80 amino acids) and a larger C-terminal lobe (*C*-lobe: almost 180 amino acids). As shown in Fig. 1.2, N-lobe has 5-stranded β-sheets (β1–β5). The cleft between the two lobes lies the ATP-binding site. The phosphates of ATP hold under the Gly-rich loop. A conserved Lys residue on β3 is connected to the C-lobe via a divalent cation, generally with Mg^{2+} (as shown in

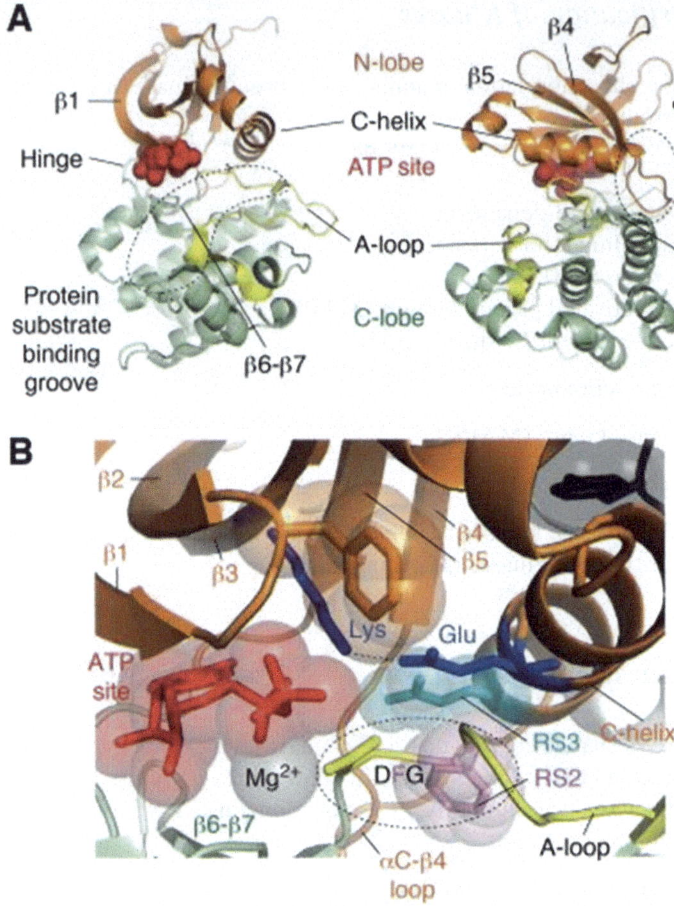

Fig. 1.2 (**a**) and (**b**): Key structural features of a kinase domain N-lobe (orange), C-lobe (green), ATP (red): Images are based on the structure of Aurora-A kinase domain (Protein Data Bank entry: 1OL5). (Adapted from Arter et al. [14])

Fig. 1.2b). The lobes are connected by a short linker sequence generally called the hinge region. The hinge region forms two important hydrogen bonding interactions with the adenine base of the ATP molecule. The central residue that mostly conserved the 3-residue sequence is the DFG motif, located at the *N*-terminus of the A-loop, recognizing the protein substrate and controlling kinase activity.

1.1.3 Types of Kinase Inhibitors

Based on binding mode, kinase inhibitors are divided into the following groups (see Table 1.1):

(a) Type I kinase inhibitor: An inhibitor that binds to the active site of ATP pocket of a kinase, DFG-in and αC-in.
(b) Type II: An inhibitor that binds to the inactive conformation of a kinase DFG-out and αC-out.
(c) Type I ½ inhibitor: It belongs to between type I and type II, DFG-in and αC-out inactive conformation.
(d) Type III: An inhibitor binds to an allosteric pocket next to the ATP-binding site, with no formation of hydrogen bonding interaction with the hinge residues of a kinase.
(e) Type IV: An inhibitor that binds to any allosteric sites distant from the ATP pocket in the inactive conformation of a kinase.
(f) Type V or covalent inhibitor: An inhibitor that targets an active site cysteine residue in or around the ATP binding site of a kinase.

1.1.4 FDA-Approved Kinase Inhibitors

Table 1.1 summarizes approved kinase inhibitors, their name, target kinase, phosphorylation site, binding mode, medical uses, and protein data bank code with reference.

1.1.5 Chemical Structure of FDA-Approved Kinase Inhibitors

Table 1.2 describes chemical structure, year of approval, molecular weight, and drug developer name.

1.1.6 MAP Kinases

Mitogen-activated protein kinases (MAP kinases or MAPKs) are a family of protein kinases that play crucial roles in various cellular processes, including cell growth, differentiation, proliferation, gene expression, mitosis, apoptosis, and response to extracellular signals. MAP kinases are part of a larger signaling cascade that transmits signals from the cell surface to the nucleus, where they regulate gene expression and other cellular responses [75–82]. Dysregulation of MAP kinase signaling has been implicated in various diseases, including cancer, inflammatory disorders, neurodegenerative diseases, and cardiovascular diseases. Therefore, understanding the mechanisms of MAP kinase signaling and developing targeted therapies to modulate these pathways represent important areas of research in biomedicine and drug discovery [75–82].

The core MAP kinase signaling pathway typically involves a cascade of three kinases:

MAP Kinase Kinase Kinase (MAP3K) Also known as MAPKKK or MEKK, these proteins are activated in response to extracellular signals such as growth factors, cytokines, or stress stimuli. MAP3Ks phosphorylate and activate MAP kinase kinases (MAP2Ks) in the next step of the cascade.

MAP Kinase Kinase (MAP2K) Also known as MAPKK or MEK, these proteins are activated by phosphorylation by MAP3Ks. MAP2Ks, in turn, phosphorylate and activate MAP kinases (MAPKs) in the final step of the cascade.

MAP Kinase (MAPK) MAPKs are serine/threonine protein kinases that are activated by phosphorylation on specific threonine and tyrosine residues by MAP2Ks. Activated MAPKs phosphorylate downstream targets, including transcription factors, other kinases, and cytoplasmic proteins, to propagate the signal and elicit cellular responses.

Humans have four typical MAP kinases, including.

A. Extracellular signal-regulated kinases (ERKs)
B. c-Jun N-terminal kinases (JNKs)
C. p38 MAP kinases
D. Nemo-like kinase (NLK)

1.1.6.1 Extracellular Signal-Regulated Kinases (ERKs)

Extracellular signal-regulated kinases (ERKs) are a subgroup of the mitogen-activated protein kinase (MAPK) family that play essential roles in various cellular processes, including cell proliferation, differentiation, survival, and migration [83–87]. They are key components of signaling pathways that transduce extracellular stimuli into intracellular responses, regulating gene expression and cellular behaviors.

Here are some important features and functions of extracellular signal-regulated kinases (ERKs):

Structure and Activation ERKs are serine/threonine protein kinases characterized by a conserved structure, including a kinase domain responsible for phosphorylation of target proteins. Activation of ERKs typically occurs through a cascade of phosphorylation events involving a series of upstream kinases, including Raf, MEK (MAPK/ERK kinase), and ultimately ERK itself. Human ERK1 has 379 amino acid residues and ERK2 has 360 amino acid residues. ERK1 and ERK2 share almost 84% identical sequences. The N-terminal lobe has five β-strands ($\beta1 - \beta5$), a conserved αC-helix, and a glycine-rich loop. The C-terminal lobe has an α-helix with six conserved segments ($\alpha D - \alpha I$) and four conserved β-strands ($\beta6 - \beta9$).

Cellular Signaling ERKs are activated by a wide range of extracellular signals, including growth factors (e.g., EGF, PDGF), cytokines, hormones, and environmental stressors. Upon activation, ERKs translocate to the nucleus where they phosphorylate and activate transcription factors, leading to changes in gene expression and cellular responses.

Role in Cell Proliferation and Differentiation ERK signaling is essential for regulating cell proliferation and differentiation in various cell types and tissues. Activation of ERKs promotes cell cycle progression, cell growth, and differentiation by regulating the expression of genes involved in these processes.

Cell Survival and Apoptosis ERK signaling can also promote cell survival by inhibiting apoptosis (programmed cell death) and promoting cell survival pathways. ERK activation can modulate the expression and activity of anti-apoptotic proteins and pro-survival factors, helping cells adapt to extracellular cues and stressors.

Regulation of Developmental Processes ERK signaling is critically involved in embryonic development, tissue morphogenesis, and organogenesis. It regulates key developmental processes such as embryonic patterning, neural tube formation, limb development, and organ development.

Pathological Implications Dysregulation of ERK signaling has been implicated in various human diseases, including cancer, neurodevelopmental disorders, cardiovascular diseases, and inflammatory conditions. Aberrant activation of ERKs can promote tumor growth, metastasis, and drug resistance in cancer cells, making them attractive targets for cancer therapy.

Therapeutic Targeting Due to the central role in disease pathogenesis, ERKs have emerged as promising therapeutic targets for drug development. Small molecule inhibitors targeting components of the ERK signaling pathway are being investigated for their potential in treating cancer and other diseases characterized by dysregulated ERK signaling.

In summary, extracellular signal-regulated kinases (ERKs) are critical mediators of cellular signaling involved in a wide range of physiological and pathological processes. Understanding the complex regulation and functions of ERKs provides insights into normal cellular physiology and offers opportunities for therapeutic intervention in disease states.

1.1.6.2 c-Jun N-terminal Kinases (JNKs)

c-Jun N-terminal kinases (JNKs) are a group of protein kinases that belong to the mitogen-activated protein kinase (MAPK) family. They play crucial roles in various cellular processes such as cell proliferation, differentiation, apoptosis, and inflammation. JNKs are involved in responding to stress signals, including UV radiation, heat shock, cytokines, and environmental toxins [80, 88–96].

Here are some key points about c-Jun N-terminal kinases:

Structure JNKs are serine/threonine kinases and are composed of a conserved catalytic domain and regulatory domains. They are activated by phosphorylation on threonine and tyrosine residues within a conserved TXY motif in the activation loop.

Activation JNKs are activated by a variety of upstream signaling pathways, including the MAPK cascades. The most well-known pathway involves the activation of JNKs by MAP kinase kinases (MKKs), specifically MKK4 and MKK7, through phosphorylation. MKK4 and MKK7 are activated by various cellular stresses and cytokines.

Substrates JNKs phosphorylate a wide range of substrates, including transcription factors such as c-Jun, ATF2, and Elk-1. Phosphorylation of these transcription factors regulates gene expression, leading to the activation or repression of specific target genes involved in stress responses, apoptosis, and proliferation.

Functions JNKs play important roles in regulating cell survival, apoptosis, inflammation, and cellular responses to stress. They are involved in diverse physiological processes such as immune response, neuronal plasticity, and development.

Pathological Implications Dysregulation of JNK signaling has been implicated in various diseases, including cancer, neurodegenerative disorders, diabetes, and inflammatory diseases. In cancer, aberrant JNK signaling can promote tumor progression by regulating cell proliferation, survival, and metastasis.

Therapeutic Potential Due to their involvement in various disease processes, JNKs have been explored as potential therapeutic targets. Small molecule inhibitors targeting JNKs have been developed and studied for their efficacy in preclinical models of cancer, inflammation, and other diseases [80, 88–96].

Overall, JNKs are key regulators of cellular responses to stress and play important roles in diverse physiological and pathological processes.

1.1.6.3 p-38 MAK Kinases

The p38 MAP kinases, also known as p38 mitogen-activated protein kinases, are a subgroup of the larger MAP kinase family. It was identified first as a 38 KDa (kilodaltons) protein hence it is called p38. It has four subtypes namely p38α (MAPK14), p38β (MAPK11), p38γ (MAPK12), and p38δ (MAPK13). The p38 MAPK plays crucial roles in cellular responses to various stress stimuli, inflammatory signals, and environmental cues [79, 82, 87, 97–100]. Here are some important features and functions of p38 MAP kinases:

Structure and Activation p38 MAP kinases are serine/threonine protein kinases characterized by a conserved structure, including a kinase domain responsible for phosphorylating target proteins. The p-38α is comprised of 360 amino acid residues and it has aTGY motif. The phosphorylation of Thr180 and Tyr182 residues results in the activation of p38 MAPK. The p38β contains 364 amino acid residues and its 75% homology is identical with p38α. The p38γ and p38δ are composed of 367 and 365 amino acid residues respectively. All isoforms have protein kinase domain and the activation motif TXY.

Activation of p38 MAP kinases typically occurs through phosphorylation by upstream kinases, such as MKK3 and MKK6, in response to extracellular signals. The p38 MAPK is activated through Thr-Gly-Tyr bis-phosphorylation in the activation loop, resulting in increased production of pro-inflammatory cytokines such as TNF-α, IL6, and IL-1β.

Substrates of p38 MAPK p38 MAP kinase controls transcription factors, transcription regulators, and other downstream protein kinases. It has several substrates. It activates MAPK-activated protein kinase (MK2/3/5), MAPKs-interacting protein kinases, mitogen- and stress-activated kinase (MSK1/2), p21-activated kinase, glycogen synthase kinase, protein kinase C epsilon (PKCε), heat shock protein (HSP), p53, and others.

Cellular Signaling p38 MAP kinases are activated in response to diverse extracellular stimuli, including pro-inflammatory cytokines (e.g., TNF-alpha, IL-1), environmental stressors (e.g., UV radiation, oxidative stress), osmotic shock, and microbial pathogens. Once activated, p38 MAP kinases transduce these signals into intracellular responses by phosphorylating downstream substrates, including transcription factors, other kinases, and cytoskeletal proteins.

Regulation of Inflammatory Responses p38 MAP kinases play central roles in regulating inflammatory responses by controlling the expression of pro-inflammatory cytokines, chemokines, and adhesion molecules. They are involved in the activation of immune cells, such as macrophages, neutrophils, and dendritic cells, and contribute to the initiation and propagation of inflammatory cascades.

Stress Response and Cell Survival p38 MAP kinases are critical components of cellular stress response pathways that protect cells from various stressors, including oxidative stress, DNA damage, and metabolic stress. They regulate cell cycle arrest, DNA repair mechanisms, and apoptosis (programmed cell death) in response to cellular damage or environmental insults.

Tissue Homeostasis and Regeneration p38 MAP kinases are involved in maintaining tissue homeostasis and regulating cellular processes essential for tissue repair and regeneration. They contribute to wound healing, tissue remodeling, and stem cell function in various organs and tissues.

Role in Development and Differentiation p38 MAP kinases play roles in embryonic development, organogenesis, and tissue differentiation. They regulate cell fate decisions, lineage commitment, and morphogenetic processes during embryonic development and tissue morphogenesis.

Pathological Implications Dysregulation of p38 MAP kinase signaling has been implicated in various human diseases, including inflammatory disorders, autoimmune diseases, neurodegenerative diseases, cardiovascular diseases, and cancer. Aberrant activation of p38 MAP kinases can contribute to disease pathogenesis by promoting chronic inflammation, tissue damage, and aberrant cell proliferation and survival.

Therapeutic Targeting Given their central role in disease pathogenesis, p38 MAP kinases have emerged as attractive therapeutic targets for drug development. Small molecule inhibitors targeting p38 MAP kinase activity are being investigated for their potential in treating inflammatory diseases, neurodegenerative disorders, and cancer.

In summary, p38 MAP kinases are versatile signaling molecules that regulate diverse cellular processes involved in normal physiology and disease pathogenesis. Understanding the complex regulation and functions of p38 MAP kinases provides insights into their roles in health and disease and offers opportunities for therapeutic intervention in various pathological conditions.

1.1.6.4 Nemo-Like Kinase (NLK)

Nemo-like kinase (NLK) is a serine/threonine protein kinase that belongs to the MAP kinase family. NLK was first identified in Drosophila melanogaster as a homolog of the mammalian MAP kinase, ERK (extracellular signal-regulated kinase). It plays diverse roles in various cellular processes, including development, cell fate determination, proliferation, and differentiation [97–121].

Here are some important features and functions of nemo-like kinase (NLK):

Conservation and Structure NLK is highly conserved across species, from flies to humans. Structurally, NLK contains characteristic domains found in MAP kinases, including a kinase domain responsible for phosphorylation and activation of target proteins.

Regulation of Gene Expression NLK regulates gene expression by phosphorylating transcription factors and transcriptional co-regulators. It can either activate or repress gene transcription, depending on the context and cellular signaling pathways involved.

Cellular Signaling Pathways NLK participates in various signaling pathways, including the Wnt signaling pathway, TGF-β signaling pathway, and Notch signaling pathway. It acts downstream of these pathways to modulate gene expression and cellular responses.

Developmental Processes NLK plays essential roles in embryonic development, tissue morphogenesis, and organogenesis. In drosophila, NLK is involved in patterning of the embryonic epidermis, segmentation, and wing development. In mammals, NLK is implicated in neural development, skeletal development, and organogenesis.

Cell Cycle Regulation NLK regulates cell cycle progression by phosphorylating cell cycle regulators and checkpoint proteins. It can influence cell proliferation, cell cycle arrest, and apoptosis in response to cellular stress or developmental cues.

Cancer and Disease Dysregulation of NLK signaling has been implicated in various human diseases, including cancer, neurodevelopmental disorders, and cardiovascular diseases. Aberrant NLK activity can promote tumor growth, metastasis, and drug resistance in cancer cells.

Therapeutic Target Due to its involvement in multiple cellular processes and its dysregulation in disease states, NLK has appeared as a potential therapeutic target for drug development. Small molecule inhibitors targeting NLK activity are being explored for their therapeutic potential in cancer and other diseases [122–137].

In summary, nemo-like kinase (NLK) is a multifunctional protein kinase involved in diverse cellular processes and signaling pathways. Its roles in development, gene expression, and disease pathogenesis make NLK an important focus of research in cell biology, developmental biology, and therapeutic intervention. Overall, MAPK became an attractive drug target in various diseases [138, 139].

1.1.6.5 MAP Kinase Family

MAPK1 (ERK2), MAPK3 (ERK1), MAPK7 (ERK5),

MAPK8 (JNK1), MAPK9 (JNK2), MAPK10 (JNK3).

MAPK11 (p38-beta), MAPK12 (p38-gamma), MAPK13 (p38-delta), MAPK14 (p38-alpha).

MAPK4 (ERK4: atypical MAPK).

MAPK6 (ERK3: atypical MAPK).

MAPK15 (ERK7/ERK8: atypical MAPK).

NLK (nemo-like kinase: atypical MAPK) (Table 1.3).

A common docking domain in MAP kinases composed a negatively charged amino acids such as aspartic acid and glutamic acid that may interact with the positively charged residues such as histidine, lysine, and arginine in kinase interaction motif (KIM) through electrostatic interactions.

Table 1.3 Kinase domain/phosphorylation site motif of MAP kinases

MAP kinase	Kinase domain/phosphorylation site motif
JNK1, 2, 3	TPY
ERK1, 2, 5, 7, 8	TEY
ERK3α, β	SEG
ERK4	SEG
p38 α, β, γ, δ	TGY
NLK	TQE

References

1. Roskoski R Jr. Classification of small molecule protein kinase inhibitors based upon the structures of the drug-enzyme complexes. Pharmacol Res. 2016;103:26–48.
2. Roskoski R Jr. Properties of FDA-approved small molecule protein kinase inhibitors: a 2023 update. Pharmacol Res. 2023;187:106552. https://doi.org/10.1016/j.phrs.2022.106552. Epub 2022 Nov 17
3. Morphy R. Selectively nonselective kinase inhibition: striking the right balance. J Med Chem. 2010;53:1413–37.
4. Ferguson FM, Gray NS. Kinase inhibitors: the road ahead. Nat Rev Drug Discov. 2018;17:353–77.
5. Roskoski R Jr. Properties of FDA-approved small molecule protein kinase inhibitors. Pharmacol Res. 2019;144:19–50.
6. Kung JE, Jura N. Prospects for pharmacological targeting of pseudokinases. Nat Rev Drug Discov. 2019;18:501–26.
7. Kannaiyan R, Mahadevan D. A comprehensive review of protein kinase inhibitors for cancer therapy. Expert Rev Anticancer Ther. 2018;18:1249–70.
8. Bhullar KS, Lagarón NO, McGowan EM, Parmar I, Jha A, Hubbard BP, Rupasinghe HPV. Kinase-targeted cancer therapies: progress, challenges and future directions. Mol Cancer. 2018;17:48.
9. Barf T, Kaptein A. Irreversible protein kinase inhibitors: balancing the benefits and risks. J Med Chem. 2012;55:6243–62.
10. Dar AC, Shokat KM. The evolution of protein kinase inhibitors from antagonists to agonists of cellular signaling. Annu Rev Biochem. 2011;80:769–95.
11. Wang B, Wu H, Hu C, Wang H, Liu J, Wang W, Liu Q. An overview of kinase down regulators and recent advances in discovery approaches. Signal Transduct Target Ther. 2021;6(1):423. https://doi.org/10.1038/s41392-021-00826-7. PMID: 34924565; PMCID: PMC8685278
12. Huang K, Wang YH, Brown A, Sun G. Identification of N-terminal lobe motifs that determine the kinase activity of the catalytic domains and regulatory strategies of Src and Csk protein tyrosine kinases. J Mol Biol. 2009;386(4):1066–77. https://doi.org/10.1016/j.jmb.2009.01.012. PMID: 19244618; PMCID: PMC2768531
13. Hubbard SR, Till JH. Protein tyrosine kinase structure and function. Annu Rev Biochem. 2000;69:373–98. https://doi.org/10.1146/annurev.biochem.69.1.373.
14. Arter C, Trask L, Ward S, Yeoh S, Bayliss R. Structural features of the protein kinase domain and targeted binding by small-molecule inhibitors. J Biol Chem. 2022;298(8):102247. https://doi.org/10.1016/j.jbc.2022.102247. Epub 2022 Jul 10. PMID: 35830914; PMCID: PMC9382423
15. Knighton DR, Zheng JH, Ten Eyck LF, Ashford VA, Xuong NH, Taylor SS, Sowadski JM. Crystal structure of the catalytic subunit of cyclic adenosine monophosphate-dependent protein kinase. Science. 1991;253(5018):407–14. https://doi.org/10.1126/science.1862342.

16. Attwood MM, Fabbro D, Sokolov AV, Knapp S, Schiöth HB. Trends in kinase drug discovery: targets, indications and inhibitor design. Nat Rev Drug Discov. 2021;20(11):839–61. https://doi.org/10.1038/s41573-021-00252-y. Epub 2021 Aug 5. Erratum in: Nat Rev Drug Discov. 2021 20(10):798
17. Chen P, Lee NV, Hu W, Xu M, Ferre RA, Lam H, Bergqvist S, Solowiej J, Diehl W, He YA, Yu X, Nagata A, VanArsdale T, Murray BW. Spectrum and degree of CDK drug interactions predicts clinical performance. Mol Cancer Ther. 2016;15(10):2273–81. https://doi.org/10.1158/1535-7163.MCT-16-0300. Epub 2016 Aug 5
18. Vazquez ML, Kaila N, Strohbach JW, Trzupek JD, Brown MF, Flanagan ME, Mitton-Fry MJ, Johnson TA, TenBrink RE, Arnold EP, Basak A, Heasley SE, Kwon S, Langille J, Parikh MD, Griffin SH, Casavant JM, Duclos BA, Fenwick AE, Harris TM, Han S, Caspers N, Dowty ME, Yang X, Banker ME, Hegen M, Symanowicz PT, Li L, Wang L, Lin TH, Jussif J, Clark JD, Telliez JB, Robinson RP, Unwalla R. Identification of N-{cis-3-[methyl(7H-pyrrolo[2,3-d]pyrimidin-4-yl)amino]cyclobutyl}propane-1-sulfonamide (PF-04965842): a selective JAK1 clinical candidate for the treatment of autoimmune diseases. J Med Chem. 2018;61(3):1130–52. https://doi.org/10.1021/acs.jmedchem.7b01598. Epub 2018 Jan 23
19. Lin DY, Andreotti AH. Structure of BTK kinase domain with the second-generation inhibitors acalabrutinib and tirabrutinib. PLoS One. 2023;18(8):e0290872. https://doi.org/10.1371/journal.pone.0290872. PMID: 37651403; PMCID: PMC10470882
20. Solca F, Dahl G, Zoephel A, Bader G, Sanderson M, Klein C, Kraemer O, Himmelsbach F, Haaksma E, Adolf GR. Target binding properties and cellular activity of afatinib (BIBW 2992), an irreversible ErbB family blocker. J Pharmacol Exp Ther. 2012;343(2):342–50. https://doi.org/10.1124/jpet.112.197756. Epub 2012 Aug 10
21. Sakamoto H, Tsukaguchi T, Hiroshima S, Kodama T, Kobayashi T, Fukami TA, Oikawa N, Tsukuda T, Ishii N, Aoki Y. CH5424802, a selective ALK inhibitor capable of blocking the resistant gatekeeper mutant. Cancer Cell. 2011;19(5):679–90. https://doi.org/10.1016/j.ccr.2011.04.004.
22. Furet P, Guagnano V, Fairhurst RA, Imbach-Weese P, Bruce I, Knapp M, Fritsch C, Blasco F, Blanz J, Aichholz R, Hamon J, Fabbro D, Caravatti G. Discovery of NVP-BYL719 a potent and selective phosphatidylinositol-3 kinase alpha inhibitor selected for clinical evaluation. Bioorg Med Chem Lett. 2013;23(13):3741–8. https://doi.org/10.1016/j.bmcl.2013.05.007. Epub 2013 May 14. Erratum in: Bioorg Med Chem Lett. 2013 23(16):4723
23. McTigue M, Murray BW, Chen JH, Deng YL, Solowiej J, Kania RS. Molecular conformations, interactions, and properties associated with drug efficiency and clinical performance among VEGFR TK inhibitors. Proc Natl Acad Sci U S A. 2012;109(45):18281–9. https://doi.org/10.1073/pnas.1207759109. Epub 2012 Sep 17. PMID: 22988103; PMCID: PMC3494931
24. Davis RR, Li B, Yun SY, Chan A, Nareddy P, Gunawan S, Ayaz M, Lawrence HR, Reuther GW, Lawrence NJ, Schönbrunn E. Structural insights into JAK2 inhibition by Ruxolitinib, Fedratinib, and derivatives thereof. J Med Chem. 2021;64(4):2228–41. https://doi.org/10.1021/acs.jmedchem.0c01952. Epub 2021 Feb 11. PMID: 33570945; PMCID: PMC8327781
25. Gonzalez-Del Pino GL, Li K, Park E, Schmoker AM, Ha BH, Eck MJ. Allosteric MEK inhibitors act on BRAF/MEK complexes to block MEK activation. Proc Natl Acad Sci USA. 2021;118(36):e2107207118. https://doi.org/10.1073/pnas.2107207118. PMID: 34470822; PMCID: PMC8433572
26. Huang WS, Liu S, Zou D, Thomas M, Wang Y, Zhou T, Romero J, Kohlmann A, Li F, Qi J, Cai L, Dwight TA, Xu Y, Xu R, Dodd R, Toms A, Parillon L, Lu X, Anjum R, Zhang S, Wang F, Keats J, Wardwell SD, Ning Y, Xu Q, Moran LE, Mohemmad QK, Jang HG, Clackson T, Narasimhan NI, Rivera VM, Zhu X, Dalgarno D, Shakespeare WC. Discovery of Brigatinib (AP26113), a phosphine oxide-containing, potent, orally active inhibitor of anaplastic lymphoma kinase. J Med Chem. 2016;59(10):4948–64. https://doi.org/10.1021/acs.jmedchem.6b00306. Epub 2016 May 12

27. Levinson NM, Boxer SG. Structural and spectroscopic analysis of the kinase inhibitor bosutinib and an isomer of bosutinib binding to the Abl tyrosine kinase domain. PLoS One. 2012;7(4):e29828. https://doi.org/10.1371/journal.pone.0029828. Epub 2012 Apr 6. PMID: 22493660; PMCID: PMC3320885
28. Zhan Z, Ai J, Liu Q, Ji Y, Chen T, Xu Y, Geng M, Duan W. Discovery of Anilinopyrimidines as dual inhibitors of c-met and VEGFR-2: synthesis, SAR, and cellular activity. ACS Med Chem Lett. 2014;5(6):673–8. https://doi.org/10.1021/ml500066m. PMID: 24944742; PMCID: PMC4060942
29. Friboulet L, Li N, Katayama R, Lee CC, Gainor JF, Crystal AS, Michellys PY, Awad MM, Yanagitani N, Kim S, Pferdekamper AC, Li J, Kasibhatla S, Sun F, Sun X, Hua S, McNamara P, Mahmood S, Lockerman EL, Fujita N, Nishio M, Harris JL, Shaw AT, Engelman JA. The ALK inhibitor ceritinib overcomes crizotinib resistance in non-small cell lung cancer. Cancer Discov. 2014;4(6):662–73. https://doi.org/10.1158/2159-8290.CD-13-0846. Epub 2014 Mar 27. PMID: 24675041; PMCID: PMC4068971
30. Rice KD, Aay N, Anand NK, Blazey CM, Bowles OJ, Bussenius J, Costanzo S, Curtis JK, Defina SC, Dubenko L, Engst S, Joshi AA, Kennedy AR, Kim AI, Koltun ES, Lougheed JC, Manalo JC, Martini JF, Nuss JM, Peto CJ, Tsang TH, Yu P, Johnston S. Novel Carboxamide-based allosteric MEK inhibitors: discovery and optimization efforts toward XL518 (GDC-0973). ACS Med Chem Lett. 2012;3(5):416–21. https://doi.org/10.1021/ml300049d. PMID: 24900486; PMCID: PMC4025802
31. Scott WJ, Hentemann MF, Rowley RB, Bull CO, Jenkins S, Bullion AM, Johnson J, Redman A, Robbins AH, Esler W, Fracasso RP, Garrison T, Hamilton M, Michels M, Wood JE, Wilkie DP, Xiao H, Levy J, Stasik E, Liu N, Schaefer M, Brands M, Lefranc J. Discovery and SAR of novel 2,3-Dihydroimidazo[1,2-c]quinazoline PI3K inhibitors: identification of Copanlisib (BAY 80-6946). ChemMedChem. 2016;11(14):1517–30. https://doi.org/10.1002/cmdc.201600148. Epub 2016 Jun 16. PMID: 27310202; PMCID: PMC5094563
32. Gerlinger M, Norton L, Swanton C. Acquired resistance to crizotinib from a mutation in CD74-ROS1. N Engl J Med. 2013;369(12):1172–3. https://doi.org/10.1056/NEJMc1309091.
33. Teplyakov A, Obmolova G, Wu SJ, Luo J, Kang J, O'Neil K, Gilliland GL. Epitope mapping of anti-interleukin-13 neutralizing antibody CNTO607. J Mol Biol. 2009;389(1):115–23. https://doi.org/10.1016/j.jmb.2009.03.076. Epub 2009 Apr 8
34. Waizenegger IC, Baum A, Steurer S, Stadtmüller H, Bader G, Schaaf O, Garin-Chesa P, Schlattl A, Schweifer N, Haslinger C, Colbatzky F, Mousa S, Kalkuhl A, Kraut N, Adolf GR. A novel RAF kinase inhibitor with DFG-out-binding mode: high efficacy in BRAF-mutant tumor xenograft models in the absence of normal tissue Hyperproliferation. Mol Cancer Ther. 2016;15(3):354–65. https://doi.org/10.1158/1535-7163.MCT-15-0617. Epub 2016 Feb 25
35. Gajiwala KS, Feng J, Ferre R, Ryan K, Brodsky O, Weinrich S, Kath JC, Stewart A. Insights into the aberrant activity of mutant EGFR kinase domain and drug recognition. Structure. 2013;21(2):209–19. https://doi.org/10.1016/j.str.2012.11.014. Epub 2012 Dec 27
36. Menichincheri M, Ardini E, Magnaghi P, Avanzi N, Banfi P, Bossi R, Buffa L, Canevari G, Ceriani L, Colombo M, Corti L, Donati D, Fasolini M, Felder E, Fiorelli C, Fiorentini F, Galvani A, Isacchi A, Borgia AL, Marchionni C, Nesi M, Orrenius C, Panzeri A, Pesenti E, Rusconi L, Saccardo MB, Vanotti E, Perrone E, Orsini P. Discovery of Entrectinib: a new 3-Aminoindazole as a potent anaplastic lymphoma kinase (ALK), c-ROS oncogene 1 kinase (ROS1), and Pan-tropomyosin receptor kinases (Pan-TRKs) inhibitor. J Med Chem. 2016;59(7):3392–408. https://doi.org/10.1021/acs.jmedchem.6b00064. Epub 2016 Mar 30. Erratum in: J Med Chem 2019 62(17):8364
37. Patani H, Bunney TD, Thiyagarajan N, Norman RA, Ogg D, Breed J, Ashford P, Potterton A, Edwards M, Williams SV, Thomson GS, Pang CS, Knowles MA, Breeze AL, Orengo C, Phillips C, Katan M. Landscape of activating cancer mutations in FGFR kinases and their differential responses to inhibitors in clinical use. Oncotarget. 2016;7(17):24252–68. https://doi.org/10.18632/oncotarget.8132. PMID: 26992226; PMCID: PMC5029699

38. Park JH, Liu Y, Lemmon MA, Radhakrishnan R. Erlotinib binds both inactive and active conformations of the EGFR tyrosine kinase domain. Biochem J. 2012;448(3):417–23. https://doi.org/10.1042/BJ20121513. PMID: 23101586; PMCID: PMC3507260
39. Kim V, Lim YR, Lee I, Lee JH, Han S, Pham TV, Kim H, Lee R, Kang LW, Kim D. Structural insights into CYP107G1 from rapamycin-producing Streptomyces rapamycinicus. Arch Biochem Biophys. 2020;692:108544. https://doi.org/10.1016/j.abb.2020.108544. Epub 2020 Aug 18
40. Ember SW, Zhu JY, Olesen SH, Martin MP, Becker A, Berndt N, Georg GI, Schönbrunn E. Acetyl-lysine binding site of bromodomain-containing protein 4 (BRD4) interacts with diverse kinase inhibitors. ACS Chem Biol. 2014;9(5):1160–71. https://doi.org/10.1021/cb500072z. Epub 2014 Mar 13. PMID: 24568369; PMCID: PMC4032195
41. Kalyukina M, Yosaatmadja Y, Middleditch MJ, Patterson AV, Smaill JB, Squire CJ. TAS-120 cancer target binding: defining reactivity and revealing the first fibroblast growth factor receptor 1 (FGFR1) irreversible structure. ChemMedChem. 2019;14(4):494–500. https://doi.org/10.1002/cmdc.201800719. Epub 2019 Jan 14
42. Ohbayashi N, Murayama K, Kato-Murayama M, Kukimoto-Niino M, Uejima T, Matsuda T, Ohsawa N, Yokoyoma S, Nojima H, Shirouzu M. Structural basis for the inhibition of cyclin G-associated kinase by Gefitinib. ChemistryOpen. 2018;7(9):721–7. https://doi.org/10.1002/open.201800177. PMID: 30214852; PMCID: PMC6129943
43. Kawase T, Nakazawa T, Eguchi T, Tsuzuki H, Ueno Y, Amano Y, Suzuki T, Mori M, Yoshida T. Effect of Fms-like tyrosine kinase 3 (FLT3) ligand (FL) on antitumor activity of gilteritinib, a FLT3 inhibitor, in mice xenografted with FL-overexpressing cells. Oncotarget. 2019;10(58):6111–23. https://doi.org/10.18632/oncotarget.27222. PMID: 31692922; PMCID: PMC6817455
44. Bender AT, Gardberg A, Pereira A, Johnson T, Wu Y, Grenningloh R, Head J, Morandi F, Haselmayer P, Liu-Bujalski L. Ability of Bruton's tyrosine kinase inhibitors to sequester Y551 and prevent phosphorylation determines potency for inhibition of fc receptor but not B-cell receptor signaling. Mol Pharmacol. 2017;91(3):208–19. https://doi.org/10.1124/mol.116.107037. Epub 2017 Jan 6
45. Somoza JR, Koditek D, Villaseñor AG, Novikov N, Wong MH, Liclican A, Xing W, Lagpacan L, Wang R, Schultz BE, Papalia GA, Samuel D, Lad L, McGrath ME. Structural, biochemical, and biophysical characterization of idelalisib binding to phosphoinositide 3-kinase δ. J Biol Chem. 2015;290(13):8439–46. https://doi.org/10.1074/jbc.M114.634683. Epub 2015 Jan 28. PMID: 25631052; PMCID: PMC4375495
46. Cowan-Jacob SW, Fendrich G, Floersheimer A, Furet P, Liebetanz J, Rummel G, Rheinberger P, Centeleghe M, Fabbro D, Manley PW. Structural biology contributions to the discovery of drugs to treat chronic myelogenous leukaemia. Acta Crystallogr D Biol Crystallogr. 2007;63(Pt 1):80–93. https://doi.org/10.1107/S0907444906047287. Epub 2006 Dec 13. PMID: 17164530; PMCID: PMC2483489
47. Guagnano V, Furet P, Spanka C, Bordas V, Le Douget M, Stamm C, Brueggen J, Jensen MR, Schnell C, Schmid H, Wartmann M, Berghausen J, Drueckes P, Zimmerlin A, Bussiere D, Murray J, Graus PD. Discovery of 3-(2,6-dichloro-3,5-dimethoxy-phenyl)-1-{6-[4-(4-ethyl-piperazin-1-yl)-phenylamino]-pyrimidin-4-yl}-1-methyl-urea (NVP-BGJ398), a potent and selective inhibitor of the fibroblast growth factor receptor family of receptor tyrosine kinase. J Med Chem. 2011;54(20):7066–83. https://doi.org/10.1021/jm2006222. Epub 2011 Sep 21
48. Qiu C, Tarrant MK, Choi SH, Sathyamurthy A, Bose R, Banjade S, Pal A, Bornmann WG, Lemmon MA, Cole PA, Leahy DJ. Mechanism of activation and inhibition of the HER4/ErbB4 kinase. Structure. 2008;16(3):460–7. https://doi.org/10.1016/j.str.2007.12.016. PMID: 18334220; PMCID: PMC2858219
49. Hoegenauer K, Soldermann N, Zécri F, Strang RS, Graveleau N, Wolf RM, Cooke NG, Smith AB, Hollingworth GJ, Blanz J, Gutmann S, Rummel G, Littlewood-Evans A, Burkhart C. Discovery of CDZ173 (Leniolisib), representing a structurally novel class of PI3K delta-selective inhibitors. ACS Med Chem Lett. 2017;8(9):975–80. https://doi.org/10.1021/acsmedchemlett.7b00293. PMID: 28947947; PMCID: PMC5601375

50. Okamoto K, Ikemori-Kawada M, Jestel A, von König K, Funahashi Y, Matsushima T, Tsuruoka A, Inoue A, Matsui J. Distinct binding mode of multikinase inhibitor lenvatinib revealed by biochemical characterization. ACS Med Chem Lett. 2014;6(1):89–94. https://doi.org/10.1021/ml500394m. PMID: 25589937; PMCID: PMC4291723
51. Zou HY, Li Q, Engstrom LD, West M, Appleman V, Wong KA, McTigue M, Deng YL, Liu W, Brooun A, Timofeevski S, McDonnell SR, Jiang P, Falk MD, Lappin PB, Affolter T, Nichols T, Hu W, Lam J, Johnson TW, Smeal T, Charest A, Fantin VR. PF-06463922 is a potent and selective next-generation ROS1/ALK inhibitor capable of blocking crizotinib-resistant ROS1 mutations. Proc Natl Acad Sci USA. 2015;112(11):3493–8. https://doi.org/10.1073/pnas.1420785112. Epub 2015 Mar 2. PMID: 25733882; PMCID: PMC4371934
52. Alexeeva M, Åberg E, Engh RA, Rothweiler U. The structure of a dual-specificity tyrosine phosphorylation-regulated kinase 1A-PKC412 complex reveals disulfide-bridge formation with the anomalous catalytic loop HRD(HCD) cysteine. Acta Crystallogr D Biol Crystallogr. 2015;71(Pt 5):1207–15. https://doi.org/10.1107/S1399004715005106. Epub 2015 Apr 25
53. Lategahn J, Tumbrink HL, Schultz-Fademrecht C, Heimsoeth A, Werr L, Niggenaber J, Keul M, Parmaksiz F, Baumann M, Menninger S, Zent E, Landel I, Weisner J, Jeyakumar K, Heyden L, Russ N, Müller F, Lorenz C, Brägelmann J, Spille I, Grabe T, Müller MP, Heuckmann JM, Klebl BM, Nussbaumer P, Sos ML, Rauh D. Insight into targeting Exon20 insertion mutations of the epidermal growth factor receptor with wild type-sparing inhibitors. J Med Chem. 2022;65(9):6643–55. https://doi.org/10.1021/acs.jmedchem.1c02080. Epub 2022 Apr 29
54. Yun CH, Mengwasser KE, Toms AV, Woo MS, Greulich H, Wong KK, Meyerson M, Eck MJ. The T790M mutation in EGFR kinase causes drug resistance by increasing the affinity for ATP. Proc Natl Acad Sci U S A. 2008;105(6):2070–5. https://doi.org/10.1073/pnas.0709662105. Epub 2008 Jan 28. PMID: 18227510; PMCID: PMC2538882
55. Weisberg E, Manley PW, Breitenstein W, Brüggen J, Cowan-Jacob SW, Ray A, Huntly B, Fabbro D, Fendrich G, Hall-Meyers E, Kung AL, Mestan J, Daley GQ, Callahan L, Catley L, Cavazza C, Azam M, Neuberg D, Wright RD, Gilliland DG, Griffin JD. Characterization of AMN107, a selective inhibitor of native and mutant Bcr-Abl. Cancer Cell. 2005;7(2):129–41. https://doi.org/10.1016/j.ccr.2005.01.007. Erratum in: Cancer Cell 2005 7(4):399. Mohammed, Azam [corrected to Azam, Mohammad]
56. Hilberg F, Roth GJ, Krssak M, Kautschitsch S, Sommergruber W, Tontsch-Grunt U, Garin-Chesa P, Bader G, Zoephel A, Quant J, Heckel A, Rettig WJ. BIBF 1120: triple angiokinase inhibitor with sustained receptor blockade and good antitumor efficacy. Cancer Res. 2008;68(12):4774–82. https://doi.org/10.1158/0008-5472.CAN-07-6307.
57. Yan XE, Ayaz P, Zhu SJ, Zhao P, Liang L, Zhang CH, Wu YC, Li JL, Choi HG, Huang X, Shan Y, Shaw DE, Yun CH. Structural basis of AZD9291 selectivity for EGFR T790M. J Med Chem. 2020;63(15):8502–11. https://doi.org/10.1021/acs.jmedchem.0c00891. Epub 2020 Aug 3
58. Klaeger S, Heinzlmeir S, Wilhelm M, Polzer H, Vick B, Koenig PA, Reinecke M, Ruprecht B, Petzoldt S, Meng C, Zecha J, Reiter K, Qiao H, Helm D, Koch H, Schoof M, Canevari G, Casale E, Depaolini SR, Feuchtinger A, Wu Z, Schmidt T, Rueckert L, Becker W, Huenges J, Garz AK, Gohlke BO, Zolg DP, Kayser G, Vooder T, Preissner R, Hahne H, Tõnisson N, Kramer K, Götze K, Bassermann F, Schlegl J, Ehrlich HC, Aiche S, Walch A, Greif PA, Schneider S, Felder ER, Ruland J, Médard G, Jeremias I, Spiekermann K, Kuster B. The target landscape of clinical kinase drugs. Science. 2017;358(6367):eaan4368. https://doi.org/10.1126/science.aan4368. PMID: 29191878; PMCID: PMC6542668
59. Lin Q, Chen X, Qu L, Guo M, Wei H, Dai S, Jiang L, Chen Y. Characterization of the cholangiocarcinoma drug pemigatinib against FGFR gatekeeper mutants. Commun Chem. 2022;5(1):100. https://doi.org/10.1038/s42004-022-00718-z. PMID: 36698015; PMCID: PMC9814635
60. Tap WD, Wainberg ZA, Anthony SP, Ibrahim PN, Zhang C, Healey JH, Chmielowski B, Staddon AP, Cohn AL, Shapiro GI, Keedy VL, Singh AS, Puzanov I, Kwak EL, Wagner AJ,

Von Hoff DD, Weiss GJ, Ramanathan RK, Zhang J, Habets G, Zhang Y, Burton EA, Visor G, Sanftner L, Severson P, Nguyen H, Kim MJ, Marimuthu A, Tsang G, Shellooe R, Gee C, West BL, Hirth P, Nolop K, van de Rijn M, Hsu HH, Peterfy C, Lin PS, Tong-Starksen S, Bollag G. Structure-guided blockade of CSF1R kinase in Tenosynovial Giant-cell tumor. N Engl J Med. 2015;373(5):428–37. https://doi.org/10.1056/NEJMoa1411366.

61. Gomez EB, Ebata K, Randeria HS, Rosendahl MS, Cedervall EP, Morales TH, Hanson LM, Brown NE, Gong X, Stephens J, Wu W, Lippincott I, Ku KS, Walgren RA, Abada PB, Ballard JA, Allerston CK, Brandhuber BJ. Preclinical characterization of pirtobrutinib, a highly selective, noncovalent (reversible) BTK inhibitor. Blood. 2023;142(1):62–72. https://doi.org/10.1182/blood.2022018674.
62. Zhou T, Commodore L, Huang WS, Wang Y, Thomas M, Keats J, Xu Q, Rivera VM, Shakespeare WC, Clackson T, Dalgarno DC, Zhu X. Structural mechanism of the Pan-BCR-ABL inhibitor ponatinib (AP24534): lessons for overcoming kinase inhibitor resistance. Chem Biol Drug Des. 2011;77(1):1–11. https://doi.org/10.1111/j.1747-0285.2010.01054.x. Epub 2010 Nov 30
63. Subbiah V, Shen T, Terzyan SS, Liu X, Hu X, Patel KP, Hu M, Cabanillas M, Behrang A, Meric-Bernstam F, Vo PTT, Mooers BHM, Wu J. Structural basis of acquired resistance to selpercatinib and pralsetinib mediated by non-gatekeeper RET mutations. Ann Oncol. 2021;32(2):261–8. https://doi.org/10.1016/j.annonc.2020.10.599. Epub 2020 Nov 5. PMID: 33161056; PMCID: PMC7883646
64. Gan N, Han Y, Zeng W, Wang Y, Xue J, Jiang Y. Structural mechanism of allosteric activation of TRPML1 by PI(3,5)P2 and rapamycin. Proc Natl Acad Sci USA. 2022;119(7):e2120404119. https://doi.org/10.1073/pnas.2120404119. PMID: 35131932; PMCID: PMC8851561
65. Smith BD, Kaufman MD, Lu WP, Gupta A, Leary CB, Wise SC, Rutkoski TJ, Ahn YM, Al-Ani G, Bulfer SL, Caldwell TM, Chun L, Ensinger CL, Hood MM, McKinley A, Patt WC, Ruiz-Soto R, Su Y, Telikepalli H, Town A, Turner BA, Vogeti L, Vogeti S, Yates K, Janku F, Abdul Razak AR, Rosen O, Heinrich MC, Flynn DL. Ripretinib (DCC-2618) is a switch control kinase inhibitor of a broad spectrum of oncogenic and drug-resistant KIT and PDGFRA variants. Cancer Cell. 2019;35(5):738–751.e9. https://doi.org/10.1016/j.ccell.2019.04.006.
66. Robarge KD, Lee W, Eigenbrot C, Ultsch M, Wiesmann C, Heald R, Price S, Hewitt J, Jackson P, Savy P, Burton B, Choo EF, Pang J, Boggs J, Yang A, Yang X, Baumgardner M. Structure based design of novel 6,5 heterobicyclic mitogen-activated protein kinase kinase (MEK) inhibitors leading to the discovery of imidazo[1,5-a] pyrazine G-479. Bioorg Med Chem Lett. 2014;24(19):4714–23. https://doi.org/10.1016/j.bmcl.2014.08.008. Epub 2014 Aug 15
67. Gajiwala KS, Wu JC, Christensen J, Deshmukh GD, Diehl W, DiNitto JP, English JM, Greig MJ, He YA, Jacques SL, Lunney EA, McTigue M, Molina D, Quenzer T, Wells PA, Yu X, Zhang Y, Zou A, Emmett MR, Marshall AG, Zhang HM, Demetri GD. KIT kinase mutants show unique mechanisms of drug resistance to imatinib and sunitinib in gastrointestinal stromal tumor patients. Proc Natl Acad Sci U S A. 2009;106(5):1542–7. https://doi.org/10.1073/pnas.0812413106. Epub 2009 Jan 21. PMID: 19164557; PMCID: PMC2635778
68. Dorsch D, Schadt O, Stieber F, Meyring M, Grädler U, Bladt F, Friese-Hamim M, Knühl C, Pehl U, Blaukat A. Identification and optimization of pyridazinones as potent and selective c-met kinase inhibitors. Bioorg Med Chem Lett. 2015;25(7):1597–602. https://doi.org/10.1016/j.bmcl.2015.02.002. Epub 2015 Feb 16
69. Chrencik JE, Patny A, Leung IK, Korniski B, Emmons TL, Hall T, Weinberg RA, Gormley JA, Williams JM, Day JE, Hirsch JL, Kiefer JR, Leone JW, Fischer HD, Sommers CD, Huang HC, Jacobsen EJ, Tenbrink RE, Tomasselli AG, Benson TE. Structural and thermodynamic characterization of the TYK2 and JAK3 kinase domains in complex with CP-690550 and CMP-6. J Mol Biol. 2010;400(3):413–33. https://doi.org/10.1016/j.jmb.2010.05.020. Epub 2010 May 15
70. Khan ZM, Real AM, Marsiglia WM, Chow A, Duffy ME, Yerabolu JR, Scopton AP, Dar AC. Structural basis for the action of the drug trametinib at KSR-bound MEK. Nature. 2020;588(7838):509–14. https://doi.org/10.1038/s41586-020-2760-4. Epub 2020 Sep 14. PMID: 32927473; PMCID: PMC7746607

71. Knowles PP, Murray-Rust J, Kjaer S, Scott RP, Hanrahan S, Santoro M, Ibáñez CF, McDonald NQ. Structure and chemical inhibition of the RET tyrosine kinase domain. J Biol Chem. 2006;281(44):33577–87. https://doi.org/10.1074/jbc.M605604200. Epub 2006 Aug 23
72. Bollag G, Hirth P, Tsai J, Zhang J, Ibrahim PN, Cho H, Spevak W, Zhang C, Zhang Y, Habets G, Burton EA, Wong B, Tsang G, West BL, Powell B, Shellooe R, Marimuthu A, Nguyen H, Zhang KY, Artis DR, Schlessinger J, Su F, Higgins B, Iyer R, D'Andrea K, Koehler A, Stumm M, Lin PS, Lee RJ, Grippo J, Puzanov I, Kim KB, Ribas A, McArthur GA, Sosman JA, Chapman PB, Flaherty KT, Xu X, Nathanson KL, Nolop K. Clinical efficacy of a RAF inhibitor needs broad target blockade in BRAF-mutant melanoma. Nature. 2010;467(7315):596–9. https://doi.org/10.1038/nature09454. PMID: 20823850; PMCID: PMC2948082
73. Rudolph D, Steegmaier M, Hoffmann M, Grauert M, Baum A, Quant J, Haslinger C, Garin-Chesa P, Adolf GR. BI 6727, a polo-like kinase inhibitor with improved pharmacokinetic profile and broad antitumor activity. Clin Cancer Res. 2009;15(9):3094–102. https://doi.org/10.1158/1078-0432.CCR-08-2445. Epub 2009 Apr 21
74. Guo Y, Liu Y, Hu N, Yu D, Zhou C, Shi G, Zhang B, Wei M, Liu J, Luo L, Tang Z, Song H, Guo Y, Liu X, Su D, Zhang S, Song X, Zhou X, Hong Y, Chen S, Cheng Z, Young S, Wei Q, Wang H, Wang Q, Lv L, Wang F, Xu H, Sun H, Xing H, Li N, Zhang W, Wang Z, Liu G, Sun Z, Zhou D, Li W, Liu L, Wang L, Wang Z. Discovery of Zanubrutinib (BGB-3111), a novel, potent, and selective covalent inhibitor of Bruton's tyrosine kinase. J Med Chem. 2019;62(17):7923–40. https://doi.org/10.1021/acs.jmedchem.9b00687. Epub 2019 Aug 19
75. Bahar ME, Kim HJ, Kim DR. Targeting the RAS/RAF/MAPK pathway for cancer therapy: from mechanism to clinical studies. Signal Transduct Target Ther. 2023;8(1):455. https://doi.org/10.1038/s41392-023-01705-z. PMID: 38105263; PMCID: PMC10725898
76. Crozet F, Levayer R. Emerging roles and mechanisms of ERK pathway mechanosensing. Cell Mol Life Sci. 2023;80(12):355. https://doi.org/10.1007/s00018-023-05007-z. PMID: 37947896; PMCID: PMC10638131
77. Hadwiger JA, Aranda RG, Fatima S. Atypical MAP kinases – new insights and directions from amoeba. J Cell Sci. 2023;136(20):jcs261447. https://doi.org/10.1242/jcs.261447. Epub 2023 Oct 16. PMID: 37850857; PMCID: PMC10617611
78. Burgermeister E. Mitogen-activated protein kinase and exploratory nuclear receptor crosstalk in cancer immunotherapy. Int J Mol Sci. 2023;24(19):14546. https://doi.org/10.3390/ijms241914546. PMID: 37833991; PMCID: PMC10572424
79. Ganguly P, Macleod T, Wong C, Harland M, McGonagle D. Revisiting p38 mitogen-activated protein kinases (MAPK) in inflammatory arthritis: a narrative of the emergence of MAPK-activated protein kinase inhibitors (MK2i). Pharmaceuticals (Basel). 2023;16(9):1286. https://doi.org/10.3390/ph16091286. PMID: 37765094; PMCID: PMC10537904
80. Deng Y, Adam V, Nepovimova E, Heger Z, Valko M, Wu Q, Wei W, Kuca K. c-Jun N-terminal kinase signaling in cellular senescence. Arch Toxicol. 2023;97(8):2089–109. https://doi.org/10.1007/s00204-023-03540-1. Epub 2023 Jun 19
81. Ma Y, Nicolet J. Specificity models in MAPK cascade signaling. FEBS Open Bio. 2023;13(7):1177–92. https://doi.org/10.1002/2211-5463.13619. Epub 2023 Jun 11. PMID: 37157227; PMCID: PMC10315774
82. Qi XM, Chen G. p38γ MAPK inflammatory and metabolic signaling in physiology and disease. Cells. 2023;12(13):1674. https://doi.org/10.3390/cells12131674. PMID: 37443708; PMCID: PMC10341180
83. Deschênes-Simard X, Malleshaiah M, Ferbeyre G. Extracellular signal-regulated kinases: one pathway, multiple fates. Cancers (Basel). 2023;16(1):95. https://doi.org/10.3390/cancers16010095. PMID: 38201521; PMCID: PMC10778234
84. Saleem S. Targeting MAPK signaling: a promising approach for treating inflammatory lung disease. Pathol Res Pract. 2024;254:155122. https://doi.org/10.1016/j.prp.2024.155122. Epub ahead of print

85. Hirashima T, Hino N, Aoki K, Matsuda M. Stretching the limits of extracellular signal-related kinase (ERK) signaling – cell mechanosensing to ERK activation. Curr Opin Cell Biol. 2023;84:102217. https://doi.org/10.1016/j.ceb.2023.102217. Epub 2023 Aug 11
86. Miller DC, Harnor SJ, Martin MP, Noble RA, Wedge SR, Cano C. Modulation of ERK5 activity as a therapeutic anti-cancer strategy. J Med Chem. 2023;66(7):4491–502. https://doi.org/10.1021/acs.jmedchem.3c00072. Epub 2023 Apr 1. PMID: 37002872; PMCID: PMC10108346
87. Hendrikse CSE, Theelen PMM, van der Ploeg P, Westgeest HM, Boere IA, Thijs AMJ, Ottevanger PB, van de Stolpe A, Lambrechts S, Bekkers RLM, Piek JMJ. The potential of RAS/RAF/MEK/ERK (MAPK) signaling pathway inhibitors in ovarian cancer: a systematic review and meta-analysis. Gynecol Oncol. 2023;171:83–94. https://doi.org/10.1016/j.ygyno.2023.01.038. Epub 2023 Feb 24
88. Latham SL, O'Donnell YEI, Croucher DR. Non-kinase targeting of oncogenic c-Jun N-terminal kinase (JNK) signaling: the future of clinically viable cancer treatments. Biochem Soc Trans. 2022;50(6):1823–36. https://doi.org/10.1042/BST20220808. PMID: 36454622; PMCID: PMC9788565
89. Wydra VR, Ditzinger RB, Seidler NJ, Hacker FW, Laufer SA. A patent review of MAPK inhibitors (2018–Present). Expert Opin Ther Pat. 2023;33(6):421–44. https://doi.org/10.1080/13543776.2023.2242584. Epub 2023 Aug 1
90. Khan MA, Tania M. Cordycepin and kinase inhibition in cancer. Drug Discov Today. 2023;28(3):103481. https://doi.org/10.1016/j.drudis.2022.103481. Epub 2022 Dec 27
91. Lee EH, Kim HT, Chun SY, Chung JW, Choi SH, Lee JN, Kim BS, Yoo ES, Kwon TG, Kim TH, Ha YS. Role of the JNK pathway in bladder cancer. Onco Targets Ther. 2022;15:963–71. https://doi.org/10.2147/OTT.S374908. PMID: 36091874; PMCID: PMC9462548
92. Min RWM, Aung FWM, Liu B, Arya A, Win S. Mechanism and therapeutic targets of c-Jun-N-terminal kinases activation in nonalcoholic fatty liver disease. Biomedicines. 2022;10(8):2035. https://doi.org/10.3390/biomedicines10082035. PMID: 36009582; PMCID: PMC9406172
93. Qin P, Ran Y, Liu Y, Wei C, Luan X, Niu H, Peng J, Sun J, Wu J. Recent advances of small molecule JNK3 inhibitors for Alzheimer's disease. Bioorg Chem. 2022;128:106090. https://doi.org/10.1016/j.bioorg.2022.106090. Epub 2022 Aug 9
94. Pua LJW, Mai CW, Chung FF, Khoo AS, Leong CO, Lim WM, Hii LW. Functional roles of JNK and p38 MAPK signaling in nasopharyngeal carcinoma. Int J Mol Sci. 2022;23(3):1108. https://doi.org/10.3390/ijms23031108. PMID: 35163030; PMCID: PMC8834850
95. Cho H, Hah JM. A perspective on the development of c-Jun N-terminal kinase inhibitors as therapeutics for Alzheimer's disease: investigating structure through docking studies. Biomedicines. 2021;9(10):1431. https://doi.org/10.3390/biomedicines9101431. PMID: 34680547; PMCID: PMC8533360
96. Chen J, Ye C, Wan C, Li G, Peng L, Peng Y, Fang R. The roles of c-Jun N-terminal kinase (JNK) in infectious diseases. Int J Mol Sci. 2021;22(17):9640. https://doi.org/10.3390/ijms22179640. PMID: 34502556; PMCID: PMC8431791
97. Le NT. The significance of ERK5 catalytic-independent functions in disease pathways. Front Cell Dev Biol. 2023;11:1235217. https://doi.org/10.3389/fcell.2023.1235217. PMID: 37601096; PMCID: PMC10436230
98. Chauhan RK, Sharma PK, Srivastava S. The role of signaling pathway in the biological cause of rheumatoid arthritis. Curr Drug Res Rev. 2021;13(2):130–9. https://doi.org/10.2174/2589977512999201109215004.
99. Chen YJ, Chan YJ, Chen WJ, Li YM, Zhang CY. Muramyl dipeptide promotes Aβ1-42 oligomer production via the NOD2/p-p38 MAPK/BACE1 signaling pathway in the SH-SY5Y cells. J Integr Neurosci. 2020;19(3):421–8. https://doi.org/10.31083/j.jin.2020.03.112.
100. Akamine T, Takaku S, Suzuki M, Niimi N, Yako H, Matoba K, Kawanami D, Utsunomiya K, Nishimura R, Sango K. Glycolaldehyde induces sensory neuron death through activation of the c-Jun N-terminal kinase and p-38 MAP kinase pathways. Histochem Cell Biol. 2020;153(2):111–9. https://doi.org/10.1007/s00418-019-01830-3. Epub 2019 Nov 16

101. Tejwani L, Jung Y, Kokubu H, Sowmithra S, Ni L, Lee C, Sanders B, Lee PJ, Xiang Y, Luttik K, Soriano A, Yoon J, Park J, Ro HH, Ju H, Liao C, Tieze SM, Rigo F, Jafar-Nejad P, Lim J. Reduction of nemo-like kinase increases lysosome biogenesis and ameliorates TDP-43-related neurodegeneration. J Clin Invest. 2023;133(16):e138207. https://doi.org/10.1172/JCI138207. PMID: 37384409; PMCID: PMC10425213
102. Yin X, Ren Y, Luo W, Liao M, Huang L, Zhuang X, Liu Y, Wang W. Nemo-like kinase (NLK) gene regulates apoptosis via the p53 signaling pathway in Litopenaeus vannamei under low-temperature stress. Dev Comp Immunol. 2022;131:104378. https://doi.org/10.1016/j.dci.2022.104378. Epub 2022 Feb 26
103. Li SZ, Zhang ZY, Chen J, Dong MY, Du XH, Gao J, Shu QP, Li C, Liang XY, Ding ZH, Du RL, Wang J, Zhang XD. NLK is required for Ras/ERK/SRF/ELK signaling to tune skeletal muscle development by phosphorylating SRF and antagonizing the SRF/MKL pathway. Cell Death Discov. 2022;8(1):4. https://doi.org/10.1038/s41420-021-00774-9. PMID: 35013153; PMCID: PMC8748963
104. Wilkes MC, Jung K, Lee BE, Saxena M, Sathianathen RS, Mercado JD, Perez C, Flygare J, Narla A, Glader B, Sakamoto KM. The active component of ginseng, ginsenoside Rb1, improves erythropoiesis in models of Diamond-Blackfan anemia by targeting Nemo-like kinase. J Biol Chem. 2021;297(3):100988. https://doi.org/10.1016/j.jbc.2021.100988. Epub 2021 Jul 21. PMID: 34298020; PMCID: PMC8379498
105. Ji YX, Wang Y, Li PL, Cai L, Wang XM, Bai L, Liu Z, Tian H, Tian S, Zhang P, Zhang XJ, Cheng X, Yuan Y, She ZG, Hu Y, Li H. A kinome screen reveals that nemo-like kinase is a key suppressor of hepatic gluconeogenesis. Cell Metab. 2021;33(6):1171–1186.e9. https://doi.org/10.1016/j.cmet.2021.04.006. Epub 2021 May 4
106. Wang X, Veeraraghavan J, Liu CC, Cao X, Qin L, Kim JA, Tan Y, Loo SK, Hu Y, Lin L, Lee S, Shea MJ, Mitchell T, Li S, Ellis MJ, Hilsenbeck SG, Schiff R, Wang XS. Therapeutic targeting of nemo-like kinase in primary and acquired endocrine-resistant breast cancer. Clin Cancer Res. 2021;27(9):2648–62. https://doi.org/10.1158/1078-0432.CCR-20-2961. Epub 2021 Feb 4. PMID: 33542078; PMCID: PMC8653766
107. Daams R, Sime W, Leandersson K, Sitnicka E, Massoumi R. Deletion of nemo-like kinase in T cells reduces single-positive CD8+ Thymocyte population. J Immunol. 2020;205(7):1830–41. https://doi.org/10.4049/jimmunol.2000109. Epub 2020 Aug 24
108. Li SZ, Shu QP, Song Y, Zhang HH, Liu Y, Jin BX, Liuyu TZ, Li C, Huang XC, Du RL, Song W, Zhong B, Zhang XD. Phosphorylation of MAVS/VISA by nemo-like kinase (NLK) for degradation regulates the antiviral innate immune response. Nat Commun. 2019;10(1):3233. https://doi.org/10.1038/s41467-019-11258-x. PMID: 31324787; PMCID: PMC6642205
109. Lei L, Wang Y, Zheng YW, Fei LR, Shen HY, Li ZH, Huang WJ, Yu JH, Xu HT. Overexpression of nemo-like kinase promotes the proliferation and invasion of lung cancer cells and indicates poor prognosis. Curr Cancer Drug Targets. 2019;19(8):674–80. https://doi.org/10.2174/1568009618666181119150521.
110. Li SZ, Zeng F, Li J, Shu QP, Zhang HH, Xu J, Ren JW, Zhang XD, Song XM, Du RL. Nemo-like kinase (NLK) primes colorectal cancer progression by releasing the E2F1 complex from HDAC1. Cancer Lett. 2018;431:43–53. https://doi.org/10.1016/j.canlet.2018.05.032. Epub 2018 May 25
111. Yang W, Gu L, Yang C, Liu T. Expression of nemo-like kinase in cervical squamous cell carcinoma: a clinicopathological study. Onco Targets Ther. 2018;11:743–9. https://doi.org/10.2147/OTT.S154188. PMID: 29445289; PMCID: PMC5810516
112. He XY, Tan ZL, Mou Q, Liu FJ, Liu S, Yu CW, Zhu J, Lv LY, Zhang J, Wang S, Bao LM, Peng B, Zhao H, Zou L. microRNA-221 enhances MYCN via targeting nemo-like kinase and functions as an oncogene related to poor prognosis in neuroblastoma. Clin Cancer Res. 2017;23(11):2905–18. https://doi.org/10.1158/1078-0432.CCR-16-1591. Epub 2016 Dec 21
113. Ke H, Masoumi KC, Ahlqvist K, Seckl MJ, Rydell-Törmänen K, Massoumi R. Nemo-like kinase regulates the expression of vascular endothelial growth factor (VEGF) in alveolar epithelial cells. Sci Rep. 2016;6:23987. https://doi.org/10.1038/srep23987. PMID: 27035511; PMCID: PMC4817507

114. Wilkes MC, Siva K, Varetti G, Mercado J, Wentworth EP, Perez CA, Saxena M, Kam S, Kapur S, Chen J, Narla A, Glader B, Lin S, Serrano M, Flygare J, Sakamoto KM. Metformin-induced suppression of Nemo-like kinase improves erythropoiesis in preclinical models of Diamond-Blackfan anemia through induction of miR-26a. Exp Hematol. 2020;91:65–77. https://doi.org/10.1016/j.exphem.2020.09.187. Epub 2020 Sep 12. PMID: 32926965; PMCID: PMC8963704
115. Wilkes MC, Siva K, Chen J, Varetti G, Youn MY, Chae H, Ek F, Olsson R, Lundbäck T, Dever DP, Nishimura T, Narla A, Glader B, Nakauchi H, Porteus MH, Repellin CE, Gazda HT, Lin S, Serrano M, Flygare J, Sakamoto KM. Diamond Blackfan anemia is mediated by hyperactive nemo-like kinase. Nat Commun. 2020;11(1):3344. https://doi.org/10.1038/s41467-020-17100-z. PMID: 32620751; PMCID: PMC7334220
116. Wang J, Yang ZH, Chen H, Li HH, Chen LY, Zhu Z, Zou Y, Ding CC, Yang J, He ZW. Nemo-like kinase as a negative regulator of nuclear receptor Nurr1 gene transcription in prostate cancer. BMC Cancer. 2016;16:257. https://doi.org/10.1186/s12885-016-2291-4. PMID: 27036119; PMCID: PMC4815267
117. Zhang HH, Li SZ, Zhang ZY, Hu XM, Hou PN, Gao L, Du RL, Zhang XD. Nemo-like kinase is critical for p53 stabilization and function in response to DNA damage. Cell Death Differ. 2014;21(10):1656–63. https://doi.org/10.1038/cdd.2014.78. Epub 2014 Jun 13. PMID: 24926618; PMCID: PMC4158690
118. Ota S, Ishitani S, Shimizu N, Matsumoto K, Itoh M, Ishitani T. NLK positively regulates Wnt/β-catenin signalling by phosphorylating LEF1 in neural progenitor cells. EMBO J. 2012;31(8):1904–15. https://doi.org/10.1038/emboj.2012.46. Epub 2012 Feb 28. PMID: 22373574; PMCID: PMC3343332
119. Li Z, Cui G, Wang J, Yu Z, Zhao L, Lv Z. Nemo-like kinase (NLK) involves in neuronal apoptosis after traumatic brain injury. Cell Mol Neurobiol. 2012;32(3):381–9. https://doi.org/10.1007/s10571-011-9766-2. Epub 2011 Nov 30
120. Nifuji A, Ideno H, Ohyama Y, Takanabe R, Araki R, Abe M, Noda M, Shibuya H. Nemo-like kinase (NLK) expression in osteoblastic cells and suppression of osteoblastic differentiation. Exp Cell Res. 2010;316(7):1127–36. https://doi.org/10.1016/j.yexcr.2010.01.023. Epub 2010 Jan 29
121. Kanei-Ishii C, Nomura T, Takagi T, Watanabe N, Nakayama KI, Ishii S. Fbxw7 acts as an E3 ubiquitin ligase that targets c-Myb for nemo-like kinase (NLK)-induced degradation. J Biol Chem. 2008;283(45):30540–8. https://doi.org/10.1074/jbc.M804340200. Epub 2008 Sep 2. PMID: 18765672; PMCID: PMC2662147
122. Thorpe CJ, Moon RT. Nemo-like kinase is an essential co-activator of Wnt signaling during early zebrafish development. Development. 2004;131(12):2899–909. https://doi.org/10.1242/dev.01171. Epub 2004 May 19
123. Yu T, Zeng Q, Mao H, Liu Y, Zhang H, Wang S, Hu C, Xu X. Grass carp (Ctenopharyngodon idella) NLK2 inhibits IFN I response through blocking MAVS-IRF3 axis. Fish Shellfish Immunol. 2022;131:206–17. https://doi.org/10.1016/j.fsi.2022.10.002. Epub 2022 Oct 8
124. Zhao Z, Liu W, Liu J, Wang J, Luo B. The effect of EBV on WIF1, NLK, and APC gene methylation and expression in gastric carcinoma and nasopharyngeal cancer. J Med Virol. 2017;89(10):1844–51. https://doi.org/10.1002/jmv.24863. Epub 2017 Jul 6
125. Yan X, Liu J, Wu H, Liu Y, Zheng S, Zhang C, Yang C. Impact of miR-208 and its target gene nemo-like kinase on the protective effect of Ginsenoside Rb1 in hypoxia/ischemia injuried cardiomyocytes. Cell Physiol Biochem. 2016;39(3):1187–95. https://doi.org/10.1159/000447825. Epub 2016 Sep 1
126. Tai J, Rao Y, Fang J, Huang Z, Yu Z, Chen X, Zhou W, Xiao X, Long T, Han Y, Liu Q, Li A, Ni X. Lentivirus-delivered nemo-like kinase small interfering RNA inhibits laryngeal cancer cell proliferation in vitro. Mol Med Rep. 2015;12(4):5619–24. https://doi.org/10.3892/mmr.2015.4189. Epub 2015 Aug 6. PMID: 26252054; PMCID: PMC4581764

127. Dong JR, Guo N, Zhao JP, Liu PD, Feng HH, Li Y. Inhibition of nemo-like kinase increases taxol sensitivity in laryngeal cancer. Asian Pac J Cancer Prev. 2013;14(12):7137–41. https://doi.org/10.7314/apjcp.2013.14.12.7137.
128. Gupta K, Kuznetsova I, Klimenkova O, Klimiankou M, Meyer J, Moore MA, Zeidler C, Welte K, Skokowa J. Bortezomib inhibits STAT5-dependent degradation of LEF-1, inducing granulocytic differentiation in congenital neutropenia CD34(+) cells. Blood. 2014;123(16):2550–61. https://doi.org/10.1182/blood-2012-09-456889. Epub 2014 Jan 6. PMID: 24394665; PMCID: PMC4968380
129. Shen Q, Bae HJ, Eun JW, Kim HS, Park SJ, Shin WC, Lee EK, Park S, Park WS, Lee JY, Nam SW. MiR-101 functions as a tumor suppressor by directly targeting nemo-like kinase in liver cancer. Cancer Lett. 2014;344(2):204–11. https://doi.org/10.1016/j.canlet.2013.10.030. Epub 2013 Nov 1
130. Wang K, Wang X, Zou J, Zhang A, Wan Y, Pu P, Song Z, Qian C, Chen Y, Yang S, Wang Y. miR-92b controls glioma proliferation and invasion through regulating Wnt/beta-catenin signaling via Nemo-like kinase. Neuro Oncol. 2013;15(5):578–88. https://doi.org/10.1093/neuonc/not004. Epub 2013 Feb 14. Erratum in: Neuro Oncol. 2013 15(7):970. PMID: 23416699; PMCID: PMC3635522
131. Cichocki F, Felices M, McCullar V, Presnell SR, Al-Attar A, Lutz CT, Miller JS. Cutting edge: microRNA-181 promotes human NK cell development by regulating Notch signaling. J Immunol. 2011;187(12):6171–5. https://doi.org/10.4049/jimmunol.1100835. Epub 2011 Nov 14. PMID: 22084432; PMCID: PMC3237765
132. Ji J, Yamashita T, Budhu A, Forgues M, Jia HL, Li C, Deng C, Wauthier E, Reid LM, Ye QH, Qin LX, Yang W, Wang HY, Tang ZY, Croce CM, Wang XW. Identification of microRNA-181 by genome-wide screening as a critical player in EpCAM-positive hepatic cancer stem cells. Hepatology. 2009;50(2):472–80. https://doi.org/10.1002/hep.22989. PMID: 19585654; PMCID: PMC2721019
133. Zeng YA, Rahnama M, Wang S, Sosu-Sedzorme W, Verheyen EM. Drosophila nemo antagonizes BMP signaling by phosphorylation of mad and inhibition of its nuclear accumulation. Development. 2007;134(11):2061–71. https://doi.org/10.1242/dev.02853.
134. Kurahashi T, Nomura T, Kanei-Ishii C, Shinkai Y, Ishii S. The Wnt-NLK signaling pathway inhibits A-Myb activity by inhibiting the association with coactivator CBP and methylating histone H3. Mol Biol Cell. 2005;16(10):4705–13. https://doi.org/10.1091/mbc.e05-05-0470. Epub 2005 Jul 29. PMID: 16055500; PMCID: PMC1237076
135. Kanei-Ishii C, Ninomiya-Tsuji J, Tanikawa J, Nomura T, Ishitani T, Kishida S, Kokura K, Kurahashi T, Ichikawa-Iwata E, Kim Y, Matsumoto K, Ishii S. Wnt-1 signal induces phosphorylation and degradation of c-Myb protein via TAK1, HIPK2, and NLK. Genes Dev. 2004;18(7):816–29. https://doi.org/10.1101/gad.1170604. PMID: 15082531; PMCID: PMC387421
136. Freeman M, Bienz M. EGF receptor/rolled MAP kinase signalling protects cells against activated armadillo in the drosophila eye. EMBO Rep. 2001;2(2):157–62. https://doi.org/10.1093/embo-reports/kve019. PMID: 11258709; PMCID: PMC1083815
137. Ishitani T, Ninomiya-Tsuji J, Nagai S, Nishita M, Meneghini M, Barker N, Waterman M, Bowerman B, Clevers H, Shibuya H, Matsumoto K. The TAK1-NLK-MAPK-related pathway antagonizes signalling between beta-catenin and transcription factor TCF. Nature. 1999;399(6738):798–802. https://doi.org/10.1038/21674.
138. Zheng Q, Li S, Wang A, Zhe M, Yang P, Wu Y, Zhao M, Zhu Y, Luo Y, Wang G, Ouyang L. MedComm Oncol. 2023;2:e53.
139. Pan X, Pei J, Wang A, Shuai W, Feng L, Bu F, Zhu Y, Zhang L, Wang G, Ouyang L. Development of small molecule extracellular signal-regulated kinases (ERKs) inhibitors for cancer therapy. Acta Pharm Sin B. 2022;12(5):2171–92. https://doi.org/10.1016/j.apsb.2021.12.022. Epub 2022 Jan 4. PMID: 35646548; PMCID: PMC9136582

Chapter 2
Biological Roles of MAP Kinases

Abstract The MAP kinases are serine/threonine kinases. They implicate in the pathogenesis of various disorders such as inflammation, metabolic disease, neurodegenerative disease, and cancer. The cell signaling network of MAP kinase is very complex and continuously evolving. They are activated by a large number of upstream kinases. Biology of MAP kinases and their roles in different diseases are described here. Structure of MAPK 8, 9, 10, interaction with other kinases, and localization of MAP kinases are summarized.

Keywords Cell signaling cascade · Roles of MAP kinases in diseases · Structure and interaction · Localization of MAP kinases

2.1 Introduction

The MAPK family comprises four major groups such as extracellular signal-regulated kinase 1/2 (ERK1/2), c-Jun N-terminal kinases (JNK1/2/3), p38 isoforms (p38 α/β/γ/δ), and ERK5. They share a Thr-Xaa-Tyr motif (TXY) in their activation segment. All MAPKs are activated by upstream MAPK kinases (MAPKKs). Several atypical MAPKs are ERK3, ERK4, ERK8, and nemo-like kinase (NLK). The biological roles ERK1/2 and p38 MAPK (α/β/γ/δ) in various diseases are well studied and published. Several small molecule ERK1/2 and p-38 MAPK inhibitors have entered into clinical trials (described in the next chapter). On the other hand, JNKs are less explored. Therefore, this chapter describes the biological roles of JNK in detail.

The c-Jun N-terminal kinases (JNKs, also termed stress-activated protein kinase, SAPK) were initially described in the early 1990s as a family of serine/threonine protein kinases, activated by a range of stress stimuli and able to phosphorylate the N-terminal transactivation domain of the c-Jun transcription factor [1]. Here biological roles of MAPK 8, 9, and 10 are described. Three distinct genes encoding *jnk*s have been identified such as *jnk1* (*MAPK8*), *jnk2* (*MAPK9*), and *jnk3* (*MAPK10*), and at least 10 different splicing isoforms remain in mammalian cells with

S. K. De, *Drug Design and Discovery*,
https://doi.org/10.1007/978-3-031-75682-5_2

molecular weights 46–55 kDa [2–4]. JNK1 has 4 splice isoforms or variants (JNK1-α1, JNK1-α2, JNK 1-β1, JNK1-β2), JNK2 has four splice variants (JNK2-α1, JNK2-α2, JNK2-β1, JNK2-β2), and JNK3 has two splice isoforms (JNK3-α1, JNK3-α2). Three main JNK isoforms share more than 90% amino acid sequence similarity. The ATP pocket is >98% homologous or highly conserved. They are also closely related to other MAP kinase families particularly p38α (MAPK14), p38β (MAPK11), and extracellular-regulated kinases ERK1 (MAPK3) and ERK2 (MAPK1). JNK-1 and the other JNK enzymes, JNK-2 and JNK-3, are activated in response to cellular stresses such as heat shock, irradiation, hypoxia or oxidative stress, osmotic, chemotoxins, hormones, growth factors, and peroxides. By phosphorylation, JNKs regulate the activity of numerous proteins that reside in the mitochondria or act in the nucleus. JNK activates downstream molecules such as ATF2, ELK1, SMAD4, p53, c-Jun, and HSF1. JNK inhibits downstream molecules including NFAT4, NFATC1, and STAT3. They are also activated in response to various cytokines such as tumor necrosis factor (TNF-α) and participate in the onset of apoptosis [5–7]. They phosphorylate different substrates in response to TNF-α, interleukin 1β (IL-1β), and other cellular stresses. It is reported that excessive up-regulation of JNK activity is associated with a number of disease states [8–108] such as CNS [40, 42, 76, 108–122], type 2 diabetes and obesity [10, 11, 15, 25, 49, 50, 123–129], arthritis [49, 75, 129–133], asthma [23], cancer [18, 21, 30, 44, 51, 66, 70, 72, 74, 80, 85–103, 134–150], age-related macular degeneration [106, 154, 155], inflammation [24, 60], fibrosis [43, 58, 91, 99, 153], ischemia [52, 53, 81], cardiovascular [59], malaria [64], and other diseases [151, 153–160]. Therefore, JNK inhibitors are expected to be effective therapeutic agents against a variety of these diseases [8–160]. A number of reviews focus on the biology of JNK [1–107] and a few of them focus on the medicinal chemistry aspect of JNK inhibitor development [see Chaps. 3, 4, and 5]. The reader will be provided with a background of the biology of JNKs as well as the recent research findings that have validated the important role of JNK inhibitors in different diseases.

2.2 MAPK Pathway: The Basics

The signaling cascade of JNK is very complex and continuously evolving [1–3]. JNKs are activated by MKK4 and MKK7 via dual phosphorylation on the Thr-183 and Tyr-185 residues in response to environmental stress and mitogenic factors (Fig. 2.1). The activation of MKKs is performed by a large number of upstream kinases including ASK1, TAK1, MLKs, and MKKKs such as Tpl2 and others. JNKs are deactivated by mitogen-activated protein kinase phosphatase 1 (MKP1) and mitogen-activated protein kinase phosphatase 5 (MKP5), which often function in a negative feedback loop. In addition to c-Jun, JNK also phosphorylates ATF2 and other Jun-family proteins that function as components of activator protein 1 (AP-1) transcriptional factor complex as well as non-transcription factors such as insulin receptor substrate-1 (IRS-1) and Bcl-2. Homodimers and heterodimers of the

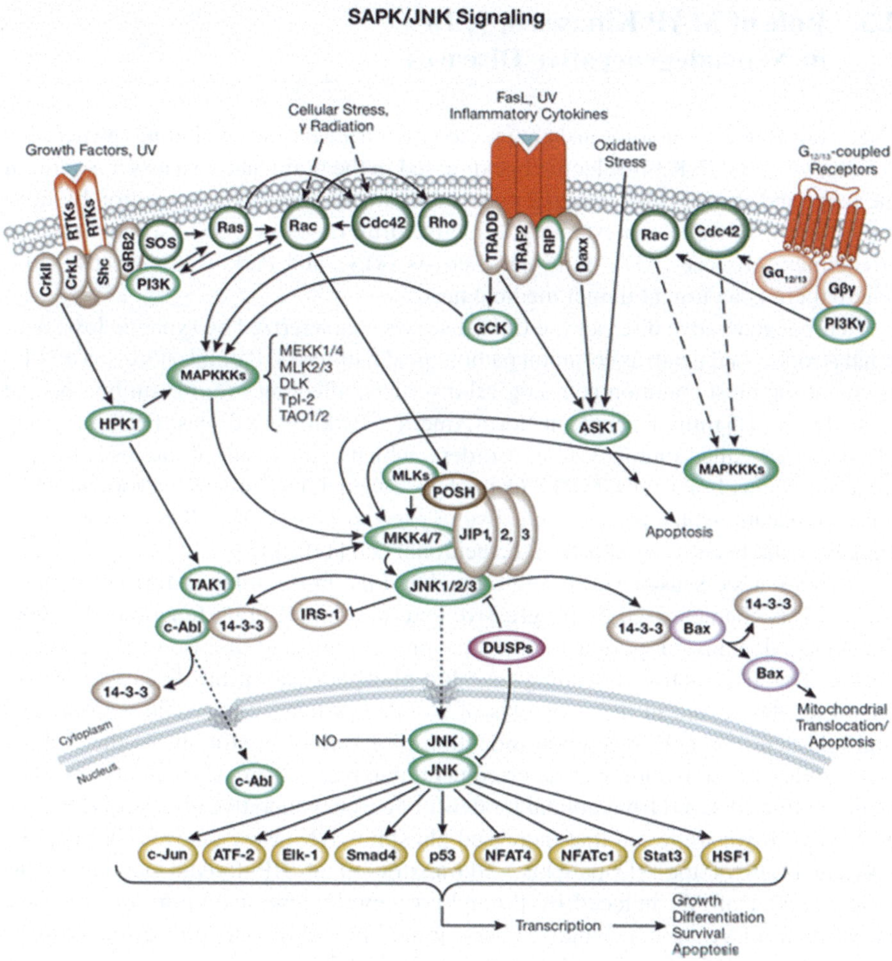

Fig. 2.1 JNK pathway network (illustration reproduced courtesy of Cell Signaling Technology, Inc)

products of the c-Jun and c-Fos proto-oncogene families constitute AP-1. At least three mammalian Jun proteins (c-Jun, Jun-B, and Jun-D) and four Fos family members (c-Fos, Fra-1, Fra-2, and Fos –B) have been characterized. The phosphorylation of these transcriptional factors by the JNKs causes increased AP-1 activity in response to cytokines, environmental stress, radiation, and growth factors [1–6].

JNK signaling specificity may be mediated through the formation of specific protein-protein interaction. Scaffolding proteins for the JNKs are JIPs (JNK-interacting proteins). Over-expression of JIPs causes the cytoplasmic retention of the JNK and thereby inhibits gene expression mediated by the JNK signaling pathway. JIP-1 contains a phosphotyrosine binding domain (PTB) in its C terminus and a JNK binding domain (JBD) at the N terminus. It binds to JNK, MKK7, MLK, and HPK1.

2.3 Role of MAP Kinases 8, 9, 10 in Neurodegenerative Diseases

JNK1 and JNK2 are ubiquitously expressed in murine tissues, including the nervous system, whereas JNK3 is selectively expressed in the brain and, to a lesser extent, in the heart and testis. This restricted expression, together with the activation by noxious stimuli, JNK3 becomes a very attractive drug target in CNS disorders such as Alzheimer's disease (AD), multiple sclerosis (MS), and Parkinson's disease (PD) where there is an urgent unmet medical need.

Neurodegenerative diseases are CNS diseases characterized by synaptic loss, neuronal atrophy, and death as common pathological hallmarks. Parkinson's disease (PD) is one of the most common neurological disorders, affecting over 6.3 million people globally, and 1.5 million people in North America. Parkinson's disease (PD) is a group of conditions called motor system disorders, which is the result of the loss of dopamine-producing brain cells. MPTP (1-methyl-4-phenyl-1,2,3,4-tetrahydropyridine) is a neurotoxicant, which produces reactive oxygen species (ROS). ROS activates JNK signaling that leads to apoptosis in the neuronal cell [110, 118].

Alzheimer's disease (AD) is age-related and the most common form of dementia, and it is characterized by progressive deterioration of brain functions. In 1906, Dr. Alois Alzheimer reported for the first time an unusual disease of the cerebral cortex. The JNK pathway leading to c-Jun phosphorylation has been well established to play a role in the apoptosis of isolated primary embryonic neurons and multiple neuronal cell lines upon induction of a variety of stimuli. At the cellular level, JNK over-activation was reported in human brains from AD patients or rodent brain sections derived from animal models of neurodegenerative diseases [108, 109, 112, 115]. It was reported that increased phospho-JNKs were detected in the postmortem brain of the AD patients. Administration of JIP-derived peptide in the rodent model of AD induced by β-amyloid peptide administration prevented the impairment of synaptic plasticity. In the animal model of ischemic stroke induced by glutamate excitotoxicity, mice deficient in JNK3 were resistant to kainic acid. These data indicated that JNK3, but not JNK1 or JNK2 involved in glutamate excitotoxicity, an important component in ischemic death, and is expressed in a brain region that is vulnerable to global ischemic conditions. Therefore, JNK inhibition emerged as an attractive drug target for multiple CNS diseases associated with neuronal cell death [108–120].

2.4 Role of MAP Kinases 8, 9, 10 in Obesity and Type-2 Diabetes

Obesity and type-2 diabetes are the most prevalent and serious metabolic diseases characterized by insulin resistance, affecting more than 50% of adults in the USA. It has been described that JNK activity is abnormally elevated in various diabetic target tissues under obese and diabetic conditions. JNK is activated during obesity. JNK phosphorylates the adapter protein insulin receptor substrate 1 (IRS-1) on Ser 307 and inhibits signaling by the insulin receptor which is one of the causes of diabetes. The physiological response to these diseases involves the production of pro-inflammatory cytokines, such as TNF-α that activates various serine kinases, via phosphorylation of insulin receptor substrate (IRS-1) at Ser 307. During obesity, adipocytes produce TNF-α, which promotes insulin resistance and serine phosphorylation of IRS-1, whereas disruption of TNF receptor-1 partially restores insulin signaling and glucose tolerance in obese mice. Many protein kinases phosphorylate IRS1 in vitro including JNK. IRS1 contains a binding site for the docking domain of JNK. Thus, activation of the JNK pathway by pro-inflammatory cytokines inhibits insulin signaling via phosphorylation of IRS1 at Ser 307. These studies demonstrated the potential use of JNK inhibitors [121–128] in the treatment of obesity and type-2 diabetes.

2.5 Role of MAP Kinases 8, 9, 10 in Rheumatoid Arthritis

Rheumatoid arthritis (RA) is a very complex systemic inflammatory disorder. It affects many tissues and organs but mainly attacks the joints, producing an inflammatory synovitis. Moreover, the joint swelling and pain caused by the inflammatory process, many RA patients ultimately suffer debilitating joint damage and deformation. Approximately 1% of the worldwide population is afflicted by RA. Several lines of pharmaceutical and genetic evidence in cellular and animal models strongly suggest the relevance and importance of the activated JNK in the pathogenesis of RA. An abnormal activation of JNK was detected in both human arthritis joints from RA patients and rodent arthritis joints from animal models of arthritis. The selective JNK inhibitors may block pro-inflammatory cytokines in particular IL-β and TNF-α as well as MMP production in human synovitis, macrophages, and lymphocytes. The use of selective JNK inhibitors in rats or mice with adjuvant arthritis and collagen-induced arthritis have been reported that protected joints from destruction and significantly reduced paw swelling by inhibition of cytokine and collagenase expression [49, 129–131].

2.6 Role of MAP Kinases 8, 9, 10 in Bronchial Asthma

Bronchial asthma is a complex, chronic inflammatory disease of the lungs or the airways, recognized by the presence of a cellular inflammatory process and by bronchial hyper-responsiveness associated with structural changes of the airways. An estimated 300 million people worldwide suffer from asthma, with 500,000 hospitalizations each year and 250,000 annual deaths attributed to this disease. It is predicted that the number of people with asthma will increase by more than 100 million by 2025. The annual economic cost of asthma is $19.8 billion (American Academy of Allergy Asthma and Immunology, www.aaaai.org). This type of disorder has been shown to be driven by many cell types in the airways, including T lymphocytes, eosinophils, mast cells, neutrophils, and epithelial cells. It has been reported that JNK inhibitors significantly inhibited RANTES production in activated human airway smooth cells. Moreover, the JNK inhibitors showed good efficacy in chronic rat and mouse models for their abilities to reduce cellular infiltration, inflammation, hyper-responsiveness, and smooth muscle proliferation. These studies suggested important roles of JNKs in allergic inflammation, and airway remodeling processes associated with hyper-responsiveness. Hence, stopping excessive JNK activities is expected to be beneficial for the treatment of asthma [23].

2.7 Role of MAP Kinases 8, 9, 10 in Oncology

The causal routes by which signal transduction pathways contribute to cellular transformation and tumorigenesis are now well established [132–148]. A substantial body of evidence suggests that JNK activation and c-Jun phosphorylation are required for the transformation of induced RAS, an oncogene that is mutationally activated in almost 30% of human cancers. RAS induces phosphorylation of c-Jun on the same serine residues phosphorylated by JNK, and acts cooperatively with c-Jun to enhance cellular transformation. Uncontrolled cellular growth, proliferation, and migration along with de-regulated angiogenesis lead to the formation of malignant tumors. The JNK signal transduction may not act exclusively in apoptosis; sustained JNK activation leading to AP-1 activation has recently been implicated to contribute to the cellular survival of specific cancer types such as glial tumors and BCL-ABL transformed B lymphoblasts. In the case of glial tumors, enhanced JNK/AP-1 activity was seen in most of the primary brain tumor samples. For the transformed B lymphoblasts, BCL-ABL was shown to activate the JNK pathway with up-regulated expression of the anti-apoptotic Bcl-2 gene.

Recently, Shibata et al. described JNK1 inhibition as a critical regulator for gastric cancer in mice [132]. In this study, authors explored the role of JNK in gastric carcinogenesis using JNK1 knockout mice as an *N*-methyl-*N*-nitrosourea (MNU)-mediated gastric cancer model. Mice lacking JNK1 demonstrated a marked decrease in gastric carcinogenesis due to MNU, relative to wild-type that indicated the role of JNK, which is associated with ROS production, and decreased cell proliferation. Another study by Potapova et al. showed JNK2-mediated tumor growth regulation in human prostate carcinoma PC3 cells [133]. El-Najjar et al. described thymoquinone, a component of black seed essential oil, induced apoptosis through JNK signaling pathway. This result showed that the thymoquinone inhibited the proliferation of a panel of human colon cancer cells (Caco-2, HCT-116, LoVo, DLD-1, and HT-29) without exhibiting cytotoxicity to normal human intestinal cells [134]. Recent findings showed that JNK activity is essential for survival and proliferation of T-cell acute lymphoblastic leukemia cells [135]. Subsequently, Chen and co-workers reported the role of JNK in skin tumorigenesis [136]. They demonstrated that in JNK2-deficient mice, the multiplicity of papillomas induced 12-O-tetradecanoylphorbol-13-acetate (TPA) was lower than that in wild-type mice. These data further supported that JNK2 is critical in the tumor promotion process. Antonyla et al. have shown that JNK activation also contributes to the pathogenesis of human brain tumors [137]. Most recently, Kaoud et al. reported that JNK2 selective peptide inhibits breast cancer cell migration. In this report, they extensively studied JIP peptide as a JNK inhibitor that inhibits both JNK activation and c-Jun phosphorylation in HEK293 breast cancer cells [138].

Thus, the central role of JNK in several physiological processes makes it an attractive target in drug discovery, establishing multiple opportunities for the design of small molecule inhibitors that might modulate specific components of the JNK signaling pathways [1–182].

2.8 Structures of MAPK 8, 9, 10

MAPK 8, 9, 10 (JNKs) have small N-terminal lobe consisting of β-strands. They have large C-terminal lobe with α-helices [183–186]. The N-terminal and C-terminal are connected by a flexible hinge region. ATP site stays between the interface of N and C lobes. MAPK 8, 9, 10 (JNKs) contain 11 subdomains (I-XI, Fig. 2.2). They have large C-terminal lobe with α-helices. ATP site stays between the interface of N and C lobes. In subdomain I, there is a glycine-rich sequence (G-loop) Gly-X-Gly-X-X-Gly-X-X (residues Gly71 to Val78). In subdomain II, there is Lys43 which forms a salt bridge with a glutamate from subdomain III. In subdomain IV, there is a β4-strand that provides the structure of the N-lobe. In subdomain V, there is a hydrophobic β5-strand. The gatekeeper residue Met146 stays at subdomain V that forms hydrogen bonding interactions with the ATP [186]. The kinase activation loop is located between subdomains VII and VIII. This kinase activation loop contains the conserved amino acid motifs DFG (Asp207-Phe208-Gly209) and APE (Ala231-Pro232-Glu233). The DFG motif activates the protein. If Phe208 is inside the hydrophobic pocket, the conformation is called DFG-in, and the enzyme is activated. If Phe208 is outside this hydrophobic pocket, the conformation is called DFG-out, the enzyme is inactivated as Asp207 cannot properly coordinate Mg^{2+} in the active site, therefore, transference of phosphate group from ATP to the substrate is hampered. In subdomain VIII, there are threonine and tyrosine residues (TPY motif, Thre221-Pro222-Tyr223) that phosphorylation activates JNK protein. In the subdomain IX, there is hydrophobic αF-helix and aspartate. The subdomains X and XI contain mostly α-helices for binding substrate proteins. The molecular weights of the three JNK isoforms (JNK1, JNK2, and JNK3) are 46 kDa or 55 kDa as there is a C-terminal extension in some JNK isoforms (Fig. 2.2). The amino acids sequence of JNK3 is 45% identical in amino acid sequence to ERK2 and 51% identical to p38. JNK3 shares 77% amino acid sequence identity with JNK2 and 75% identity with JNK1. The ATP binding pocket is 98% identical with all isoforms of

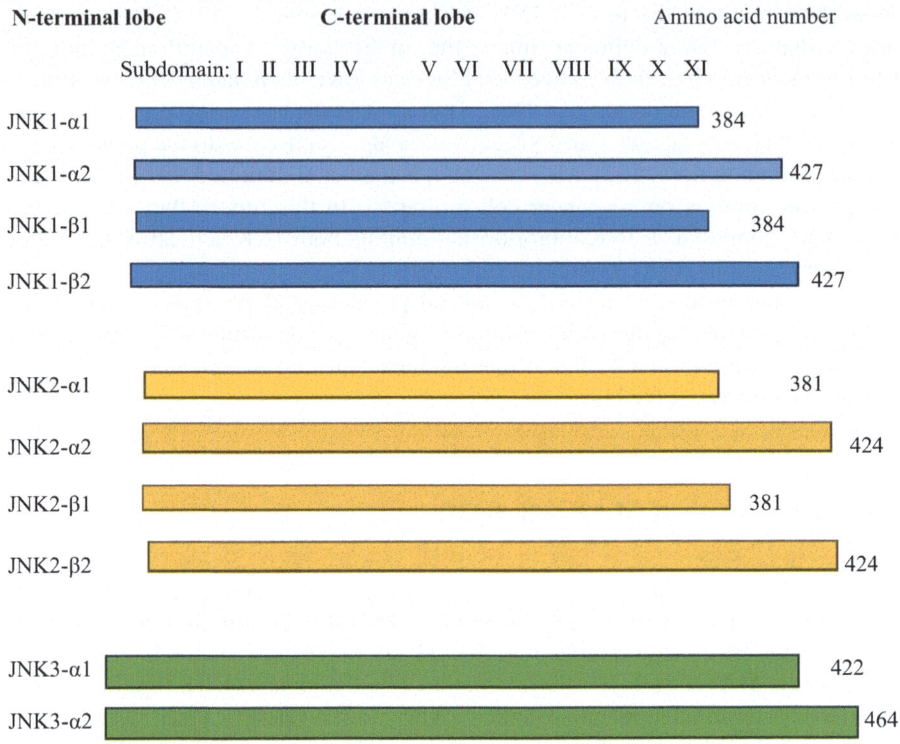

Fig. 2.2 Protein structural diagram of ten different isoforms of JNK

JNKs. However, there is a difference in residues in the N-lobe-1, N-lobe 2, and C-helix as shown in Fig. 2.3. In the N-lobe 1, there is Val54 for JNK2, whereas the same position is Ile54 for JNK1, and Ile92 for JNK3. In the N-lobe 2, Ile106 is for JNK1, whereas the same position is leucine for JNK2 and JNK3. In the C-helix, Leu77 is for JKN2, whereas the same position is methionine for JNK1 and JNK3. The hinge-region, C-lobe, and DFG motif are identical for all isoforms of JNKs.

2.9 Interactions of MAPK8, 9, 10 with Other Proteins

MAPK8 has been described to interact with activating transcription factor 2 [187–190], c-Jun [191–197], CRK [198], DUSP1 [199], DUSP10 [200], GSTP1 [201], IRS1 [202, 203], ITCH [204, 205], MAP2K4 [196, 197, 208–210], MAP2K7 [197, 206], MAP3K1 [207], MAP3K2 [206], MAPK8IP1 [208, 209], MAPK8IP3 [210, 211], Myc [212], REL [192], SH3BP5 [213], and SPIB [214].

MAPK9 interacts with Grb2 [215, 216], MAPK8IP1 [217, 218], MAPK8IP2 [217], MAPK8IP3 [210, 211], p53 [219, 220], and TOB1 [221].

MAPK10 interacts with MAPK8IP3 [210, 211, 213].

Glycine rich-loop

JNK1	I32	G33	S34	G35	A36	Q37	G38	I39
JNK2	I32	G33	S34	G35	A36	Q37	G38	I39
JNK3	I70	G71	S72	G73	A74	Q75	G76	I77

N-lobe 1

JNK1	A53	I54	K55
JNK2	A53	**V54**	K55
JNK3	A91	I92	K93

N-lobe 2

JNK1	I86	L88	**I106**	V107
JNK2	I86	L88	L106	V107
JNK3	I124	L126	L144	V145

C-helix

JNK1	R69	E73	L74	M77
JNK2	R69	E73	L74	**L77**
JNK3	R107	E111	L112	M115

Hinge-region

JNK1	M108	E109	L110	M111	D112	A113	N114
JNK2	M108	E109	L110	M111	D112	A113	N114
JNK3	M146	E147	L148	M149	D150	A151	N152

C-lobe

JNK1	C116	D151	K153	P154	S155	N156	I157	V158
JNK2	C116	D151	K153	P154	S155	N156	I157	V158
JNK3	C154	D189	K191	P192	S193	N194	I195	V196

DFG-motif

JNK1	L168	D169	F170	G171
JNK2	L168	D169	F170	G171
JNK3	L206	D207	F208	G209

Fig. 2.3 Sequence alignment in different areas and differences in amino acids are highlighted in red color

2.10 Localization of MAP Kinases

ERK1 and ERK2 are located in cytoplasmic, nuclear, and membrane-associated compartments as shown in Fig. 2.4. MAPK 8, 9, 10 are found in both cytoplasm and nucleus [222–224]. Nevertheless, JNK1 and JNK2 are mainly located in the cytoplasm and nucleus whereas JNK3 is found in cytoplasm, nucleus, membrane, and mitochondrion. Similarly, p38 α/β/γ/δ can occur in all cell compartments. Moreover, activated MAP kinases can translocate to the nucleus. But localization is dynamic and it depends on the duration and nature of stimulation. Since MAPKs can occur in different cell compartments, isoform-specific MAPK inhibitors may play important roles in different cellular diseases.

ERK1/2 are located in cytoplasmic-, nuclear-, and membrane-associated compartments. MAPK 8, 9, 10 are found in both cytoplasm and nucleus. Nevertheless, JNK1 and JNK2 are mainly located in the cytoplasm and nucleus whereas JNK3 is found in cytoplasm, nucleus, membrane, and mitochondrion. Similarly, p38 α/β/γ/δ can occur in all cell compartments. Moreover, activated MAP kinases can translocate to the nucleus. But localization is dynamic and it depends on the duration and nature of stimulation.

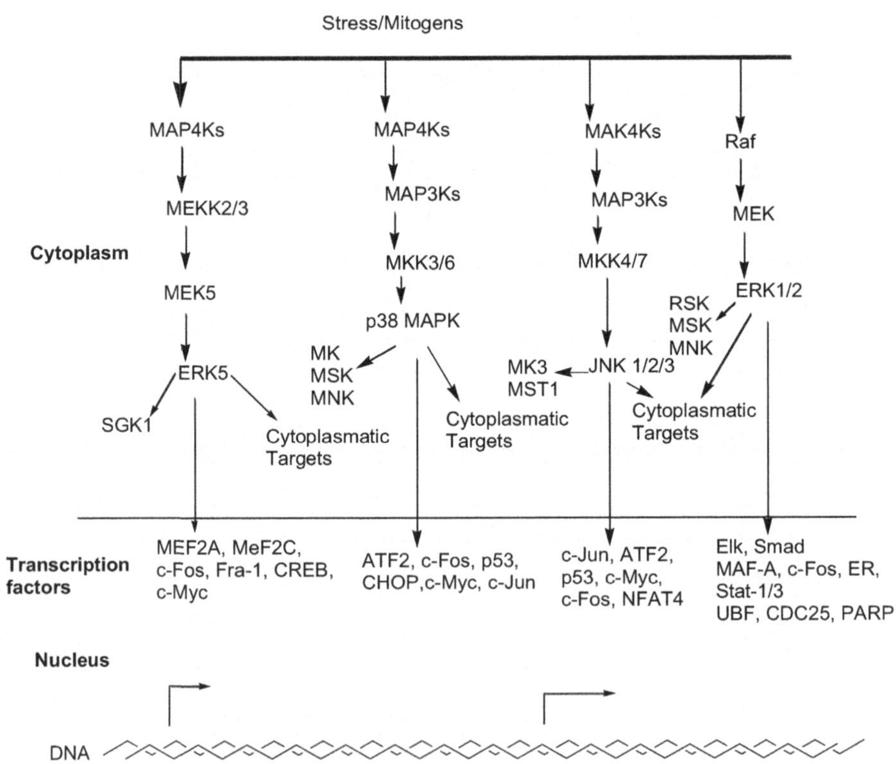

Fig. 2.4 Localization of MAPKs

2.11 Landmark of MAP Kinases 8, 9, 10

Table 2.1 summarizes the MAPK 8, 9, 10 discovery area.

Some key properties should be considered for the drug design [159] as shown in Fig. 2.5.

Table 2.1 Landmark of MAP kinases 8, 9, 10 research area

Year	Discovery
1990–1994	JNK, stress activated Protein kinase activities
1995	Cloning of JNK1, JNK2, JNK3
1997	JNK scaffold protein, JIP1
1998	JNK3 crystal structure
2001	First small molecule JNK inhibitor
2003	Cell permeable JIP1 peptide in neuroprotection
2004	JNK1 crystal structure in complex with JNK inhibitor
2008	JNK2 crystal structure
2008	Small molecule inhibitor targeting JNK-JIP interaction site
2011	First JNK bi-dentate inhibitor
2012	First JNK covalent inhibitor
2012	Small molecule JNK inhibitor Tanzisertib (CC-930) Clinical trials for idiopathic pulmonary fibrosis
2017	Brimapitide, a peptide (XG-102) Clinical trials for inflammatory eye diseases
2023	CC-90001 (BMS-986360) Clinical trial on pulmonary fibrosis
2023	JNK covalent inhibitors

Peripherally –Acting Drugs	Centrally-acting Drugs
1. MW <550	MW<450
2. LogP<5	LogP <4 and >1
3. PSA <140A^2	PSA <80 A^2
4. H-bond donors <6	H-bond donors <4
5. H-bond acceptors <8	H-bond acceptors <7
6. Freely rotating bonds <12	Freely rotating bonds <8

Blood-brain Barrier (BBB) for CNS drug

Other important parameters are LE, ADME, DMPK, PK/PD.

Fig. 2.5 Drug-like: These are the key properties to be considered

References

1. Manning AM, Davis RJ. Targeting JNK for therapeutic benefit: from junk to gold. Nat Rev Drug Discov. 2003;2:554–65.
2. Manning G, Whyte DB, Martinez R, Hunter T, Sudarsanam S. The protein kinase complement of human genome. Science. 2002;298:1912–34.
3. Bogoyevitch MA, Ngoei KRW, Zhao TT, Yeap YYC, Ng DCH. c-Jun N-terminal kinase signaling: recent advances and challenges. Biochim Biophys Acta. 2010;1804:463–75.
4. Siddiqui MA, Reddy PA. Small molecule JNK inhibitors. J Med Chem. 2010;53:3005–12.
5. Peng J, Andersen JK. The role of c-Jun N-terminal kinase in Parkinson's disease. IUBMB Life. 2003;55:267–71.
6. LoGrasso P, Kamenecka T. Inhibitors of c-Jun N-terminal kinase (JNK). Mini Rev Med Chem. 2008;8:755–66.
7. Kallunki T, Deng T, Hibi M, Karin M. c-Jun can recruit JNK to phosphorylate dimerization partners via specific docking interactions. Cell. 1996;87:929–39.
8. Yang DD, Kuan C-Y, Whitmarsh AJ, Rincon M, Zheng TS, Davis RJ, Rakic P, Flavell RA. Absence of excitotoxicity-induced apoptosis in the hippocampus of mice lacking the JNK3 gene. Nature. 1997;389:865–70.
9. Kamata H, Honda SI, Maeda S, Chang L, Hirata H, Karin M. Reactive oxygen species promote TNF alpha induced death and sustained JNK activation by inhibiting MAP kinase phosphatases. Cell. 2005;120:649–61.
10. Kaneto H, Nakatani Y, Miyatsuka T, Kawamori D, Matsuoka T, Matsuhisa M, Kajimoto Y, Ichijo H, Yamasaki Y, Hori M. Possible novel therapy for diabetes with cell-permeable JNK-inhibitory peptide. Nat Med. 2004;10:1128–32.
11. Bonny C, Oberson A, Negri S, Sauser C, Schorderet DF. Cell-permeable peptide inhibitors of JNK: novel blockers of beta-cell death. Diabetes. 2001;50:77–82.
12. Bogoyevitch MA, Arthur PG. Inhibitors of c-Jun N-terminal kinases-JuNK no more. Biochim Biophys Acta. 2008;1784:76–93.
13. Johnson GL, Lapadat R. Mitogen-activated protein kinase pathways mediated by ERK, JNK and p38 protein kinase. Science. 2002;298:1911–2.
14. Pulverer BJ, Kyriakis JM, Avruch J, Nikolakaki E, Woodgett JR. Phosphorylation of c-Jun mediated by MAP kinases. Nature. 1991;353:670–3.
15. Hirosumi J, Tuncman G, Chang L, Gorgun CZ, Uysal KT, Maeda K, Karin M, Hotamisligil GS. A central role for JNK in obesity and insulin resistance. Nature. 2002;420:333–6.
16. Kahn SE, Hull RL, Utzschneider KM. Mechanism linking obesity to insulin resistance and type 2 diabetes. Nature. 2006;444:840–6.
17. Bennett BL, Sasaki DT, Murray BW, O'Leary EC, Sakata ST, Xu W, Leisten JC, Motiwala A, Pierce S, Satoh Y, Bhagwat SS, Manning AM, Anderson DW. SP600125, an anthrapyrazolone inhibitor of JNK. Proc Natl Acad Sci U S A. 2001;98:13681–6.
18. Wagner EF, Nebreda AR. Signal integration by JNK and p38 MAPK pathways in cancer development. Nat Rev Cancer. 2009;9:537–49.
19. Morphy R. Selectively nonselective kinase inhibition: striking the right balance. J Med Chem. 2010;53:1413–37.
20. Bogoyevitch MA, Fairlie DP. A new paradign for protein kinase inhibition: blocking phosphorylation without directly targeting ATP binding. Drug Discov Today. 2007;12:622–33.
21. Kennedy NJ, Davis RJ. Role of JNK in tumor development. Cell Cycle. 2003;2:199–201.
22. Koch P, Gehringer M, Laufer SA. Inhibitors of c-Jun N-terminal kinases: an update. J Med Chem. 2015;59:72–95.
23. Kumar A, Singh UK, Kini SG, Garg V, Agrawal S, Tomar PK, Pathak P, Chaudhary A, Gupta P, Malik A. JNK pathway signaling: a novel and smarter therapeutic targets for various biological diseases. Future Med Chem. 2015;7:2065–86.
24. Graczyk PP. JNK inhibitors as anti-inflammatory and neuroprotective agents. Future Med Chem. 2013;5:539–51.

25. Li H, Yu X. Emerging role of JNK in insulin resistance. Curr Diabetes Rev. 2013;9:422–8.
26. Barr RK, Kendrick TS, Bogoyevitch MA. Identification of the critical features of a small peptide inhibitor of JNK activity. J Biol Chem. 2002;277:10987–97.
27. Gao S, Howard S, LoGrasso PV. Pharmacological inhibition of c-Jun N-terminal kinase reduces food intake and sensitizes Leptin's anorectic signaling actions. Sci Rep. 2017;7:41795.
28. Ngoei KR, Catimel B, Church N, Lio DS, Dogovski C, Perugini MA, Watt PM, Cheng HCH, Ng DC, Bogoyevitch MA. Characterization of a novel JNK (c-Jun N-terminal kinase) inhibitory peptide. Biochem J. 2011;434:399–413.
29. Chambers JW, Cherry L, Laughlin JD, Figuera-Losada M, Lograsso PV. Selective inhibition of mitochondrial JNK signaling achieved using peptide mimicry of the sab kinase interacting motif-1 (KIM1). ACS Chem Biol. 2011;6:808–18.
30. Messoussi A, Feneyrolles C, Bros A, Deroide A, Dayde-Cazals B, Cheve G, Van N, Hijfte NV, Fauvel B, Bougrin K, Yasri A. Recent Progress in the design, study, and development of c-Jun N-terminal kinase inhibitors as anticancer agents. Chem Biol. 2014;21:1433–43.
31. Seki E, Brenner DA, Karin M. A liver full of JNK: signaling in regulation of cell function and disease pathogenesis, and clinical approaches. Gastroenterology. 2012;143:307–20.
32. Davis RJ. Signal transduction by the JNK group of MAP kinases. Cell. 2000;103:239–52.
33. Davies C, Tournier C. Exploring the function of the JNK (c-Jun N-terminal kinase) signalling pathway in physiological and pathological processes to design novel therapeutic strategies. Biochem Soc Trans. 2012;40:85–9.
34. Temkin V, Karin M. From death receptor to reactive oxygen species and c-Jun N-terminal protein kinase: the receptor-interacting protein 1 odyssey. Immunol Rev. 2007;220:8–21.
35. Karin M, Gallagher E. From JNK to pay dirt: Jun kinases, their biochemistry, physiology and clinical importance. IUBMB Life. 2005;57:283–95.
36. Solinas G, Naugler W, Galimi F, Lee MS, Karin M. Saturated fatty acids inhibit induction of insulin gene transcription by JNK-mediated phosphorylation of insulin-receptor substrates. Proc Natl Acad Sci USA. 2006;103:16454–9.
37. Chen CY, Del Gatto-Konczak F, Wu Z, Karin M. Stabilization of interleukin-2 mRNA by the c-Jun NH_2-terminal kinase pathway. Science. 1998;280:1945–9.
38. Tang G, Minemoto Y, Dibling B, Purcell NH, Li Z, Karin M, Lin A. Inhibition of JNK activation through NF-kappaB target genes. Nature. 2001;414:313–7.
39. Gao M, Labuda T, Xia Y, Gallagher E, Fang D, Liu YC, Karin M. Jun turnover is controlled through JNK-dependent phosphorylation of the E3 ligase Itch. Science. 2004;306:271–5.
40. Ghosh AS, Wang B, Pozniak CD, Chen M, Watts R, Lewcock JW. DLK-induces developmental neuronal degeneration via selective regulation of proapoptotic JNK activity. J Cell Biol. 2011;194:751–64.
41. Laughlin JD, Nwachukwu JC, Figuera-Losada M, Cherry L, Nettles KW, LoGrasso PV. Structural mechanism of allostery and autoinhibiton in JNK family kinases. Structure. 2012;20:2174–84.
42. Tournier C. The 2 faces of JNK signaling in cancer. Genes Cancer. 2013;4:397–400.
43. Grynberg K, Ma FY, Nikolic-Paterson DJ. The JNK signaling pathway in renal fibrosis. Front Physiol. 2017;8:829.
44. Bubici C, Papa S. JNK signalling in cancer: in need of new, smarter therapeutic targets. Br J Pharmacol. 2014;171:24–37.
45. Gaestel M, Kracht M. Peptides as signaling inhibitors for mammalian MAP kinase cascades. Curr Pharm Des. 2009;15:2471–80.
46. Lombard CK, Davis AL, Inukai T, Maly DJ. Allosteric modulation of JNK docking site interactions with ATP-competitive inhibitors. Biochemistry. 2018;57:5897–909.
47. Konson A, Pradeep S, D'Acunto CW, Seger R. Pigment epithelium-derived factor and its phosphomimetic mutant induce JNK-dependent apoptosis and p38-mediated migration arrest. Cell Physiol Biochem. 2018;49:512–29.
48. Ansideri F, Macedo JT, Eitel M, El-Gokha A, Zinad DS, Scarpellini C, Kudolo M, Schollmeyer D, Boeckler FM, Blaum BS, Laufer SA, Koch P. Structural optimization of

a Pyridinylimidazole scaffold: shifting the selectivity from p38α mitogen-activated protein kinase to c-Jun N-terminal kinase 3. ACS Omega. 2018;3:7809–31.
49. Ge HX, Zou FM, Li Y, Liu AM, Tu M. JNK pathway in osteoarthritis: pathological and therapeutic aspects. J Recept Signal Transduct Res. 2017;37:431–6.
50. Solinas G, Becattini B. JNK at the crossroad of obesity, insulin resistance, and cell stress response. Mol Metab. 2016;6:174–84.
51. Li Z, Lim SK, Liang X, Lim YP. The transcriptional coactivator WBP2 primes triple-negative breast cancer cells for responses to Wnt signaling via the JNK/Jun kinase pathway. J Biol Chem. 2018;293:20014–28.
52. King LA, Toledo AH, Rivera-Chavez FA, Toledo-Pereyra LH. Role of p38 and JNK in liver ischemia and reperfusion. J Hepatobiliary Pancreat Surg. 2009;16:763–70.
53. Shvedova M, Anfinogenova Y, Atochina-Vasserman EN, Schepetkin IA, Atochin DN. c-Jun N-terminal kinases (JNKs) in myocardial and cerebral ischemia/reperfusion injury. Front Pharmacol. 2018;9:975.
54. Gehringer M, Muth F, Koch P, Laufer SA. c-Jun N-terminal kinase inhibitors: a patent review (2010-2014). Expert Opin Ther Pat. 2015;25:849–72.
55. Sabapathy K. Role of the JNK pathway in human diseases. Prog Mol Biol Transl Sci. 2012;106:145–69.
56. Vallerie SN, Hotamisligil GS. The role of JNK proteins in metabolism. Sci Transl Med. 2010;2:60rv5.
57. Noguchi H, Miyagi-Shiohira C, Nakashima Y, Ebi N, Hamada E, Tamaki Y, Kuwae K, Kobayashi N, Saitoh I, Watanabe M. Modified cell-permeable JNK inhibitors efficiently prevents islet apoptosis and improves the outcome of islet transplantation. Sci Rep. 2018;8:11082.
58. Grande MT, López-Novoa JM. Therapeutical relevance of MAP-kinase inhibitors in renal diseases: current knowledge and future clinical perspectives. Curr Med Chem. 2008;15:2054–70.
59. Muslin AJ. MAPK signalling in cardiovascular health and disease: molecular mechanisms and therapeutic targets. Clin Sci. 2008;115:203.
60. Roy PK, Rashid F, Bragg J, Ibdah JA. Role of the JNK signal transduction pathway in inflammatory bowel disease. World J Gastroenterol. 2008;14:200–2.
61. Salh B. c-Jun N-terminal kinases as potential therapeutic targets. Expert Opin Ther Tar. 2007;11:1339–53.
62. Weston CR, Davis RJ. The JNK signal transduction pathway. Curr Opin Genet Dev. 2002;12:14–21.
63. Rosette C, Karin M. Ultraviolet light and osmotic stress: activation of the JNK cascade through multiple growth factor and cytokine receptors. Science. 1996;274:1194–7.
64. Souvannaseng L, Hun LV, Baker H, Klyver JM, Wang B, Pakpour N, Bridgewater JM, Napoli E, Giulivi C, Riehle MA, Luckhart S. Inhibition of JNK signaling in the Asian malaria vector Anopheles stephensi extends mosquito longevity and improves resistance to plasmodium falciparum infection. PLoS Pathog. 2018;14:e1007418.
65. Khan MA, Farahvash A, Douda DN, Licht JC, Grasemann H, Sweezey N, Palaniyar N. JNK activation turns on LPS- and gram-negative bacteria-induced NADPH oxidase-dependent suicidal NETosis. Sci Rep. 2017;7:3409.
66. Feng J, Yan PF, Zhao HY, Zhang FC, Zhao WH, Feng M. Inhibitor of nicotinamide phosphoribosyltransferase sensitizes glioblastoma cells to temozolomide via activating ROS/JNK signaling pathway. Biomed Res Int. 2016;2016:1450843.
67. Dai M, Yang J, Xie M, Lin J, Luo M, Hua H, Xu G, Lin H, Song D, Cheng Y, Guo B, Zhao J, Gonzalez FJ, Liu A. Inhibition of JNK signalling mediates PPARα-dependent protection against intrahepatic cholestasis by fenofibrate. Br J Pharmacol. 2017;174:3000–17.
68. He Y, Cai C, Sun S, Wang X, Li W, Li H. Effect of JNK inhibitor SP600125 on hair cell regeneration in zebrafish (Danio rerio) larvae. Oncotarget. 2016;7:51640–50.
69. Palmer SS, Altan M, Denis D, Tos EG, Gotteland JP, Osteen KG, Bruner-Tran KL, Nataraja SG. Bentamapimod (JNK Inhibitor AS602801) induces regression of endometriotic lesions in animal models. Reprod Sci. 2016;23:11–23.

70. Phelps-Polirer K, Abt MA, Smith D, Yeh ES. Co-targeting of JNK and HUNK in resistant HER2-positive breast cancer. PLoS One. 2016;11:e0153025.
71. Zhou J, Edgar BA, Boutros M. ATF3 acts as a rheostat to control JNK signalling during intestinal regeneration. Nat Commun. 2017;8:14289.
72. Jin HO, Hong SE, Park JA, Chang YH, Hong YJ, Park IC, Lee JK. Inhibition of JNK-mediated autophagy enhances NSCLC cell sensitivity to mTORC1/2 inhibitors. Sci Rep. 2016;6:28945.
73. Shylaja R, Kavitha R, Meganathan C. Atom based 3D-QSAR, molecular docking and density functional theory approaches to identify novel JNK-1 inhibitor. Comb Chem High Throughput Screen. 2016;19:771–97.
74. Liu X, Zhang J, Li J, Volk A, Breslin P, Zhang J, Zhang Z. The synergistic repressive effect of NF-κB and JNK inhibitor on the clonogenic capacity of Jurkat leukemia cells. PLoS One. 2014;9:e115490.
75. Kondo A, Otsuka T, Matsushima-Nishiwaki R, Kuroyanagi G, Mizutani J, Wada I, Kozawa O, Tokuda H. Inhibition of SAPK/JNK leads to enhanced IL-1-induced IL-6 synthesis in osteoblasts. Arch Biochem Biophys. 2013;535:227–33.
76. Zhuo ZH, Sun YZ, Jin PN, Li FY, Zhang YL, Wang HL. Selective targeting of MAPK family kinases JNK over p38 by rationally designed peptides as potential therapeutics for neurological disorders and epilepsy. Mol BioSyst. 2016;12:2532–40.
77. Yabu T, Shiba H, Shibasaki Y, Nakanishi T, Imamura S, Touhata K, Yamashita M. Stress-induced ceramide generation and apoptosis via the phosphorylation and activation of nSMase1 by JNK signaling. Cell Death Differ. 2015;22:258–73.
78. Park J, Chai JS, Kim SW, Paick JS, Cho MC. Inhibition of Jun N-terminal kinase improves erectile function by alleviation of Cavernosal apoptosis in a rat model of cavernous nerve injury. Urology. 2018;113:e9–253.e16.
79. Prasad KD, Trinath J, Biswas A, Sekar K, Balaji KN, Guru Row TN. Alkyl chain substituted 1,9-pyrazoloanthrones exhibit prominent inhibitory effect on c-Jun N-terminal kinase (JNK). Org Biomol Chem. 2014;12:4656–62.
80. Ge D, Gao J, Han L, Li Y, Liu HH, Yang WC, Chang F, Liu J, Yu M, Zhao J. Novel effects of sphingosylphosphorylcholine on the apoptosis of breast cancer via autophagy/AKT/p38 and JNK signaling. J Cell Physiol. 2019;234:11451–62.
81. Xu P, Zhang G, Sha L, Hou S. DUSP1 alleviates cerebral ischaemia reperfusion injury via inactivating JNK-Mff pathways and repressing mitochondrial fission. Life Sci. 2018;210:251–62.
82. Kim MS, Han JY, Kim SH, Kim HY, Jeon D, Lee K. Polyhexamethylene guanidine phosphate induces IL-6 and TNF-α expression through JNK-dependent pathway in human lung epithelial cells. J Toxicol Sci. 2018;43:485–92.
83. Ng PY, McIntosh KA, Hargrave G, Ho KH, Paul A, Plevin R. Inhibition of cytokine-mediated JNK signalling by purinergic $P2Y_{11}$ receptors, a novel protective mechanism in endothelial cells. Cell Signal. 2018;51:59–71.
84. Lessard SJ, MacDonald TL, Pathak P, Han MS, Coffey VG, Edge J, Rivas DA, Hirshman MF, Davis RJ, Goodyear LJ. JNK regulates muscle remodeling via myostatin/SMAD inhibition. Nat Commun. 2018;9:3030.
85. Almasi S, Kennedy BE, El-Aghil M, Sterea AM, Gujar S, Partida-Sánchez S, El Hiani Y. TRPM2 channel-mediated regulation of autophagy maintains mitochondrial function and promotes gastric cancer cell survival via the JNK-signaling pathway. J Biol Chem. 2018;293:3637–50.
86. Cicenas J, Zalyte E, Rimkus A, Dapkus D, Noreika R, Urbonavicius SJNK. p38, ERK, and SGK1 inhibitors in cancer. Cancers (Basel). 2017;10(1)
87. Park WH. MAPK inhibitors, particularly the JNK inhibitor, increase cell death effects in H_2O_2-treated lung cancer cells via increased superoxide anion and glutathione depletion. Oncol Rep. 2018;39:860–70.
88. Hanigan TW, Aboukhatwa SM, Taha TY, Frasor J, Petukhov PA. Divergent JNK phosphorylation of HDAC3 in triple-negative breast cancer cells determines HDAC inhibitor binding and selectivity. Cell Chem Biol. 2017;24:1356–67.

89. Zhao X, Guo Y, Jiang C, Chang Q, Zhang S, Luo T, Zhang B, Jia X, Hung MC, Dong C, Lin X. JNK1 negatively controls antifungal innate immunity by suppressing CD23 expression. Nat Med. 2017;23:337–46.
90. Caly L, Li HM, Bogoyevitch MA, Jans DA. c-Jun N-terminal kinase activity is required for efficient respiratory syncytial virus production. Biochem Biophys Res Commun. 2017;483:64–8.
91. Kitanaka T, Nakano R, Kitanaka N, Kimura T, Okabayashi K, Narita T, Sugiya H. JNK activation is essential for activation of MEK/ERK signaling in IL-1β-induced COX-2 expression in synovial fibroblasts. Sci Rep. 2017;7:39914.
92. Konishi H, Fujiya M, Tanaka H, Ueno N, Moriichi K, Sasajima J, Ikuta K, Akutsu H, Tanabe H, Kohgo Y. Probiotic-derived ferrichrome inhibits colon cancer progression via JNK-mediated apoptosis. Nat Commun. 2016;7:12365.
93. Sayeed A, Lu H, Liu Q, Deming D, Duffy A, McCue P, Dicker AP, Davis RJ, Gabrilovich D, Rodeck U, Altieri DC, Languino LR. β1 integrin- and JNK-dependent tumor growth upon hypofractionated radiation. Oncotarget. 2016;7:52618–30.
94. Lu ZY, Chen WC, Li YH, Li L, Zhang H, Pang Y, Xiao ZF, Xiao HW, Xiao Y. TNF-α enhances vascular cell adhesion molecule-1 expression in human bone marrow mesenchymal stem cells via the NF-κB, ERK and JNK signaling pathways. Mol Med Rep. 2016;14:643–8.
95. Pathria G, Garg B, Garg K, Wagner C, Wagner SN. Dual c-Jun N-terminal kinase-cyclin D1 and extracellular signal-related kinase-c-Jun disjunction in human melanoma. Br J Dermatol. 2016;175:1221–31.
96. Jang WY, Lee JY, Lee ST, Jun DY, Kim YH. Inhibition of JNK2 and JNK3 by JNK inhibitor IX induces prometaphase arrest-dependent apoptotic cell death in human Jurkat T cells. Biochem Biophys Res Commun. 2014;452:845–51.
97. Wang F, Ke ZF, Wang R, Wang YF, Huang LL, Wang LT. Astrocyte elevated gene-1 (AEG-1) promotes osteosarcoma cell invasion through the JNK/c-Jun/MMP-2 pathway. Biochem Biophys Res Commun. 2014;452:933–9.
98. Chen J, Fu H, Wang Z, Yin F, Li J, Hua Y, Cai Z. A new synthetic ursolic acid derivative IUA with anti-tumor efficacy against osteosarcoma cells via inhibition of JNK signaling pathway. Cell Physiol Biochem. 2014;34:724–33.
99. Li D, Chen D, Zhang X, Wang H, Song Z, Xu W, He Y, Yin Y, Cao J. c-Jun N-terminal kinase and Akt signalling pathways regulating tumour necrosis factor-α-induced interleukin-32 expression in human lung fibroblasts: implications in airway inflammation. Immunology. 2015;144:282–90.
100. Zhong Y, Naito Y, Cope L, Naranjo-Suarez S, Saunders T, Hong SM, Goggins MG, Herman JM, Wolfgang CL, Iacobuzio-Donahue CA. Functional p38 MAPK identified by biomarker profiling of pancreatic cancer restrains growth through JNK inhibition and correlates with improved survival. Clin Cancer Res. 2014;20:6200–11.
101. Lu CC, Chen HP, Chiang JH, Jin YA, Kuo SC, Wu TS, Hour MJ, Yang JS, Chiu YJ. Quinazoline analog HMJ-30 inhibits angiogenesis: involvement of endothelial cell apoptosis through ROS-JNK-mediated death receptor 5 signaling. Oncol Rep. 2014;32:597–606.
102. Gutierrez GJ, Tsuji T, Chen M, Jiang W, Ronai ZA. Interplay between Cdh1 and JNK activity during the cell cycle. Nat Cell Biol. 2010;12:686–95.
103. Gutierrez GJ, Tsuji T, Cross JV, Davis RJ, Templeton DJ, Jiang W, Ronai ZA. JNK-mediated phosphorylation of Cdc25C regulates cell cycle entry and G(2)/M DNA damage checkpoint. J Biol Chem. 2010;285:14217–28.
104. Topisirovic I, Gutierrez GJ, Chen M, Appella E, Borden KL, Ronai ZA. Control of p53 multimerization by Ubc13 is JNK-regulated. Proc Natl Acad Sci USA. 2009;106:12676–81.
105. Lanna A, Gomes DC, Muller-Durovic B, McDonnell T, Escors D, Gilroy DW, Lee JH, Karin M, Akbar AN. A sestrin-dependent Erk-Jnk-p38 MAPK activation complex inhibits immunity during aging. Nat Immunol. 2017;18:354–63.
106. Du H, Sun X, Guma M, Luo J, Ouyang H, Zhang X, Zeng J, Quach J, Nguyen DH, Shaw PX, Karin M, Zhang K. JNK inhibition reduces apoptosis and neovascularization in a murine model of age-related macular degeneration. Proc Natl Acad Sci USA. 2013;110:2377–82.

107. Holzer RG, Park EJ, Li N, Tran H, Chen M, Choi C, Solinas G, Karin M. Saturated fatty acids induce c-Src clustering within membrane subdomains, leading to JNK activation. Cell. 2011;147:173–84.
108. Antoniou X, Falconi M, Di Marino D, Borsello T. JNK3 as a therapeutic target for neurodegenerative diseases. J Alzheimers Dis. 2011;24:633–42.
109. Yarza R, Vela S, Solas M, Ramirez MJ. c-Jun N-terminal kinase (JNK) signaling as a therapeutic target for Alzheimer's disease. Front Pharmacol. 2016;6:321.
110. Wang G, Pan J, Chen SD. Kinases and kinase signaling pathways: potential therapeutic targets in Parkinson's disease. Prog Neurobiol. 2012;98:207–21.
111. Haeusgen W, Boehm R, Zhao Y, Herdegen T, Waetzig V. Specific activities of individual c-Jun N-terminal kinases in the brain. Neuroscience. 2009;161:951–9.
112. Rodriguez-Perdigon M, Solas M, Ramirez MJ. JNK: a putative link between insulin signaling and VGLUT1 in Alzheimer's disease. J Alzheimers Dis. 2016;50:963–7.
113. Jia C, Keasey MP, Lovins C, Hagg T. Inhibition of astrocyte FAK-JNK signaling promotes subventricular zone neurogenesis through CNTF. Glia. 2018;66:2456–69.
114. Guha I, Slamova I, Chun S, Clegg A, Golos M, Thrasivoulou C, Simons JP, Al-Shawi R. The effects of short-term JNK inhibition on the survival and growth of aged sympathetic neurons. Neurobiol Aging. 2016;46:138–48.
115. Abdullah M, Takase H, Nunome M, Enomoto H, Ito J, Gong JS, Michikawa M. Amyloid-β reduces exosome release from astrocytes by enhancing JNK phosphorylation. J Alzheimers Dis. 2016;53:1433–41.
116. Wityak J, McGee KF, Conlon MP, Song RH, Duffy BC, Clayton B, Lynch M, Wang G, Freeman E, Haber J, Kitchen DB, Manning DD, Ismail J, Khmelnitsky Y, Michels P, Webster J, Irigoyen M, Luche M, Hultman M, Bai M, Kuok ID, Newell R, Lamers M, Leonard P, Yates D, Matthews K, Ongeri L, Clifton S, Mead T, Deupree S, Wheelan P, Lyons K, Wilson C, Kiselyov A, Toledo-Sherman L, Beconi M, Muñoz-Sanjuan I, Bard J, Dominguez C. Lead optimization toward proof-of-concept tools for Huntington's disease within a 4-(1H-pyrazol-4-yl)pyrimidine class of pan-JNK inhibitors. J Med Chem. 2015;58:2967–87.
117. Cheng YL, Choi Y, Seow WL, Manzanero S, Sobey CG, Jo DG, Arumugam TV. Evidence that neuronal Notch-1 promotes JNK/c-Jun activation and cell death following ischemic stress. Brain Res. 2014;1586:193–202.
118. Han XH, Cheng MN, Chen L, Fang H, Wang LJ, Li XT, Qu ZQ. 7,8-dihydroxyflavone protects PC12 cells against 6-hydroxydopamine induced cell death through modulating PI3K/Akt and JNK pathways. Neurosci Lett. 2014;581:85–8.
119. Saporito MS, Hudkins RL, Maroney AC. Discovery of CEP-1347/KT-7515, an inhibitor of the JNK/SAPK pathway for the treatment of neurodegenerative diseases. Prog Med Chem. 2002;40:23–62.
120. Chen M, Sun J, Lu C, Chen X, Ba H, Lin Q, Cai J, Dai J. The impact of neuronal Notch-1/JNK pathway on intracerebral hemorrhage-induced neuronal injury of rat model. Oncotarget. 2016;7:73903–11.
121. Kaneto H, Matsuoka TA, Nakatani Y, Kawamori D, Matsuhisa M, Yamasaki Y. Oxidative stress and the JNK pathway in diabetes. Curr Diabetes Rev. 2005;1:65–72.
122. Stebbins JL, De SK, Machleidt T, Becattini B, Vazquez J, Kuntzen C, Chen LH, Cellitti JF, Riel-Mehan M, Emdadi A, Solinas G, Karin M, Pellecchia M. Identification of a new JNK inhibitor targeting the JNK-JIP interaction site. Proc Natl Acad Sci. 2008;105:16809–13.
123. Vazquez J, De SK, Chen LH, Riel-Mehan M, Emdadi A, Cellitti J, Stebbins JL, Rega MF, Pellecchia M. Development of paramagnetic probes for molecular recognition studies in protein kinases. J Med Chem. 2008;51:3460–5.
124. De SK, Stebbins JL, Chen L-H, Riel-Mehan M, Machleidt T, Dahl R, Yuan H, Emdadi A, Barile E, Chen V, Murphy R, Pellecchia M. Design, synthesis, and structure-activity relationship of substrate competitive, selective, and *in vivo* active triazole and thiadiazole inhibitors of the c-Jun-N-terminal kinase. J Med Chem. 2009;52:1943–52.

125. De SK, Chen V, Stebbins JL, Chen L-H, Cellitti JF, Machleidt T, Barile E, Riel-Mehan M, Dahl R, Yang L, Emdaddi A, Murphy M, Pellecchia M. Synthesis and optimization of thiadiazole derivatives as a novel class of substrate competitive c-Jun N-terminal kinase inhibitors. Bioorg Med Chem. 2010;18:590–6.
126. De SK, Barile E, Chen V, Stebbins JL, Cellitti JF, Machleidt T, Carlson CB, Yang L, Dahl R, Pellecchia M. Design, synthesis, and structure-activity relationship studies of thiophene-3-carboxamide derivatives as dual inhibitors of the c-Jun-N-terminal kinase. Bioorg Med Chem. 2011;19:2582–8.
127. Stebbins JL, De SK, Pavlickova P, Chen V, Machleidt T, Chen L-H, Kuntzen C, Kitada S, Karin M, Pellecchia M. Design and characterization of a potent and selective dual ATP and substrate-competitive subnanomolar bidentate c-Jun N-terminal kinase inhibitor. J Med Chem. 2011;54:6206–14.
128. De SK, Chen L-H, Stebbins JL, Machleidt T, Riel-Mehan M, Dahl R, Chen V, Yuan H, Barile E, Emdadi A, Murphy R, Pellecchia M. Discovery of 2-(5-nitrothiazol-2yl)thio)benzo[*d*]thiazoles as novel JNK inhibitors. Bioorg Med Chem. 2009;17:2712–2717
129. Zhang X, Zhao G, Zhang Y, Wang J, Wang Y, Cheng L, Sun M, Rui Y. Activation of JNK signaling in osteoblasts is inversely correlated with collagen synthesis in age-related osteoporosis. Biochem Biophys Res Commun. 2018;504:771–6.
130. Guma M, Ronacher LM, Firestein GS, Karin M, Corr M. JNK-1 deficiency limits macrophage-mediated antigen-induced arthritis. Arthritis Rheum. 2011;63:1603–12.
131. Guma M, Kashiwakura J, Crain B, Kawakami Y, Beutler B, Firestein GS, Kawakami T, Karin M, Corr M. JNK1 controls mast cell degranulation and IL-1{beta} production in inflammatory arthritis. Proc Natl Acad Sci USA. 2010;107:22122–7.
132. Shibata W, Maeda S, Hikiba Y, Yanai A, Sakamoto K, Nakahawa H, Ogura K, Karin M, Omata M. JNK1 is a critical regulator for the development of gastric cancer in mice. Cancer Res. 2008;68:5031–9.
133. Potapova O, Anisimov SV, Gorospe M, Dougherty RH, Gaarde WA, Boheler KR, Holbrook NJ. Targets of JNK2 – mediated tumor growth regulation revealed by serial analysis of gene expression. Cancer Res. 2002;62:3257–63.
134. El-Najjar N, Chatila M, Moukadem H, Vuorela H, Ocker M, Gandesiri M, Schneider-Stock R, Gali-Mustasib H. Reactive oxygen species mediate thymoquinone-induced apoptosis and activate ERK and JNK signaling. Apoptosis. 2010;15:183–95.
135. Cui J, Wang Q, Wang J, Lu M, Zhu N, Li Y, Feng J, Shen B, Zhang J. Basal JNK activity is essential for survival and proliferation of T-cell acute lymphoblastic leukemia cells. Mol Cancer Thera. 2009;8:3214–22.
136. Chen P, O'Neal JF, Ebelt ND, Cantrell MA, Mitra S, Nasrazadani A, Vandenbroek TL, Heasley LE, Van Den Berg CL. JNK2 effects on tumor development, genetic instability and replicative stress in an oncogene-driven mouse mammary tumor model. PLoS One. 2010;5:e10443.
137. Antonyak MA, Kenyon LC, Godwin AK, James DC, Emlet DR, Okamoto I, Tnani M, Holgado-Madruga M, Moscatello DK, Wong AJ. Elevated JNK activation contributes to the pathogenesis of human brain tumors. Oncogene. 2002;21:5038–46.
138. Kaoud TS, Mitra S, Lee S, Taliaferro J, Cantrell M, Linse KD, Van Den Berg CL, Dalby KN. Development of JNK2-selective peptide inhibitors that inhibit breast cancer cell migration. ACS Chem Biol. 2011;6:658–66.
139. Wang X, Chao L, Zhen J, Chen L, Ma G, Li X. Phosphorylated JNK is over expressed in human papillary thyroid carcinomas and associated with lymph node metastasis. Cancer Lett. 2010;293:175–80.
140. Vo VA, Lee JW, Lee HJ, Chun W, Lim SY, Kim SS. Inhibition of JNK potentiates temozolomide-induced cytotoxicity in U87MG glioblastoma cells via suppression of Akt phosphorylation. Anticancer Res. 2014;34:5509–15.
141. Chen X, Chen Y, Lin X, Su S, Hou X, Zhang Q, Tian Y. The drug combination of SB202190 and SP600125 significantly inhibit the growth and metastasis of Olaparib-resistant ovarian cancer cell. Curr Pharm Biotechnol. 2018;19:506–13.

142. Zhang Y, Chen F. Reactive oxygen species (ROS), troublemakers between nuclear factor-kappaB (NF-kappaB) and c-Jun NH(2)-terminal kinase (JNK). Cancer Res. 2004;64:1902–5.
143. Reno EM, Haughian JM, Jackson TA, Thorne AM, Bradford AP. c-Jun N-terminal kinase regulates apoptosis in endometrial cancer cells. Apoptosis. 2009;14:809–20.
144. Celluale C, Girnius N, Jiang F, Cavangh-Kyros J, Lu S, Garlick DS, Mercurio AM, Davis RJ. Role of JNK in mammary gland development and breast cancer. Cancer Res. 2012;72:472–81.
145. Lopez-Bergami P. The role of mitogen- and stress-activated protein kinase pathways in melanoma. Pigm Cell Melanoma Res. 2011;24:902–21.
146. Okada M, Kuramoto K, Takeda H, Watarai H, Sakaki H, Seino S, Seino M, Suzuki S, Kitanaka C. The novel JNK inhibitor AS602801 inhibits cancer stem cells in vitro and in vivo. Oncotarget. 2016;7:27021–32.
147. Zhang JY, Selim MA. The role of the c-Jun N-terminal kinase signaling pathway in skin cancer. Am J Cancer. 2012;2:691–8.
148. Seino M, Okada M, Shibuya K, Seino S, Suzuki S, Ohta T, Kurachi H, Kitanaka C. Requirement of JNK signaling for self-renewal and tumor-initiating capacity of ovarian cancer stem cells. Anticancer Res. 2014;34:4723–31.
149. Ceballos-Olvera I, Chavez-Salinas S, Medina F, Ludert JE, Del-Angel RM. JNK phosphorylation, induced during dengue virus infection, is important for viral infection and requires the presence of cholesterol. Virology. 2010;396:30–6.
150. Kaoud TS, Park H, Mitra S, Yan C, Tseng CC, Shi Y, Jose J, Taliaferro JM, Lee K, Ren P, Hong J, Dalby KN. Manipulating JNK signaling with (−)-zuonin A. ACS Chem Biol. 2012;7:1873–83.
151. Walker NM, Mazzoni SM, Vittal R, Fingar DC, Lama VN. c-Jun N-terminal kinase (JNK)-mediated induction of mSin1 expression and mTORC2 activation in mesenchymal cells during fibrosis. J Biol Chem. 2018;293:17229–39.
152. Okada Y, Saika S, Shirai K, Yamanaka O, Kitano A, Wang Z, Yang H, Reinach P. JNK MAPK signaling contributes in vivo to injury-induced corneal epithelial migration. Ophthalmic Res. 2009;42:185–92.
153. Kyosseva SV. Targeting MAPK signaling in age-related macular degeneration. Ophthalmol Eye Dis. 2016;8:23–30.
154. Borghi A, Verstrepen L, Beyaert R. TRAF2 multitasking in TNF receptor-induced signaling to NF-κB, MAP kinases and cell death. Biochem Pharmacol. 2016;116:1–10.
155. Atsriku C, Hoffmann M, Ye Y, Kumar G, Surapaneni S. Metabolism and disposition of a potent and selective JNK inhibitor [14C]tanzisertib following oral administration to rats, dogs and humans. Xenobiotica. 2015;45:428–41.
156. Atsriku C, Hoffmann M, Ye Y, Kumar G, Surapaneni S. In vitro metabolism of a novel JNK inhibitor tanzisertib: interspecies differences in oxido-reduction and characterization of enzymes involved in metabolism. Xenobiotica. 2016;45:465–80.
157. Guma M, Rius J, Duong-Polk KX, Haddad GG, Lindsey JD, Karin M. Genetic and pharmacological inhibition of JNK ameliorates hypoxia-induced retinopathy through interference with VEGF expression. Proc Natl Acad Sci USA. 2009;106:8760–5.
158. Nakagawa H, Maeda S, Hikiba Y, Ohmae T, Shibata W, Yanai A, Sakamoto K, Ogura K, Noguchi T, Karin M, Ichijo H, Omata M. Deletion of apoptosis signal-regulating kinase 1 attenuates acetaminophen-induced liver injury by inhibiting c-Jun N-terminal kinase activation. Gastroenterology. 2008;135:1311–21.
159. Lipinski CA. Lead- and drug-like compounds: the rule-of-five revolution. Drug Discov Today Technol. 2004;1:337–41.
160. Tanaka KI, Shimoda M, Kasai M, Ikeda M, Ishima Y, Kawahara M. Involvement of SAPK/JNK signaling pathway in copper enhanced zinc-induced neuronal cell death. Toxicol Sci. 2019;169(1):293–302. https://doi.org/10.1093/toxsci/kfz043. PMID: 30768131
161. Mohammad J, Singh RR, Riggle C, Haugrud B, Abdalla MY, Reindl KM. JNK inhibition blocks piperlongumine-induced cell death and transcriptional activation of heme oxygen-

ase-1 in pancreatic cancer cells. Apoptosis. 2019;24(9–10):730–44. https://doi.org/10.1007/s10495-019-01553-9. PMID: 31243599; PMCID: PMC6713602
162. Wang Y, Zhao M, He S, Luo Y, Zhao Y, Cheng J, Gong Y, Xie J, Wang Y, Hu B, Tian L, Liu X, Li C, Huang Q. Necroptosis regulates tumor repopulation after radiotherapy via RIP1/RIP3/MLKL/JNK/IL8 pathway. J Exp Clin Cancer Res. 2019;38(1):461. https://doi.org/10.1186/s13046-019-1423-5. PMID: 31706322; PMCID: PMC6842489
163. Wu Q, Wu W, Jacevic V, Franca TCC, Wang X, Kuca K. Selective inhibitors for JNK signalling: a potential targeted therapy in cancer. J Enzyme Inhib Med Chem. 2020;35(1):574–83. https://doi.org/10.1080/14756366.2020.1720013. PMID: 31994958; PMCID: PMC7034130
164. Tanaka Y, Uchi H, Ito T, Furue M. Indirubin-pregnane X receptor-JNK axis accelerates skin wound healing. Sci Rep. 2019;9(1):18174. https://doi.org/10.1038/s41598-019-54754-2. PMID: 31796845; PMCID: PMC6890704
165. Jiang M, Fan J, Qu X, Li S, Nilsson SK, Sun YBY, Chen Y, Yu D, Liu D, Liu BC, Tang M, Chen W, Ren Y, Nikolic-Paterson DJ, Jiang X, Li J, Yu X. Combined blockade of Smad3 and JNK pathways ameliorates progressive fibrosis in folic acid nephropathy. Front Pharmacol. 2019;10:880. https://doi.org/10.3389/fphar.2019.00880. PMID: 31447676; PMCID: PMC6695473
166. Xiao Z, Xu F, Zhu X, Bai B, Guo L, Liang G, Shan X, Zhang Y, Zhao Y, Zhang B. Inhibition of JNK phosphorylation by curcumin analog C66 protects LPS-induced acute lung injury. Drug Des Devel Ther. 2019;13:4161–71. https://doi.org/10.2147/DDDT.S215712.
167. Li G, Qi W, Li X, Zhao J, Luo M, Chen J. Recent advances in c-Jun N-terminal kinase (JNK) inhibitors. Curr Med Chem. 2021;28(3):607–27. https://doi.org/10.2174/0929867327666200210144114. PMID: 32039671
168. Qin P, Ran Y, Liu Y, Wei C, Luan X, Niu H, Peng J, Sun J, Wu J. Recent advances of small molecule JNK3 inhibitors for Alzheimer's disease. Bioorg Chem. 2022;128:106090. https://doi.org/10.1016/j.bioorg.2022.106090. Epub 2022 Aug 9. PMID: 35964505
169. Wu L, Zeng S, Cao Y, Huang Z, Liu S, Peng H, Zhi C, Ma S, Hu K, Yuan Z. Inhibition of HDAC4 attenuated JNK/c-Jun-dependent neuronal apoptosis and early brain injury following subarachnoid hemorrhage by transcriptionally suppressing MKK7. Front Cell Neurosci. 2019;13:468. https://doi.org/10.3389/fncel.2019.00468. PMID: 31708743; PMCID: PMC6823346
170. Busquets O, Ettcheto M, Cano A, Manzine P, Sanchez-Lopez E, Espinosa-Jimenez T, Verdaguer E, Dario Castro-Torres R, Beas-Zarate C, Sureda F, Olloquequi J. Role of c-Jun N-terminal kinases (JNKs) in epilepsy and metabolic cognitive impairment. Int J Mol Sci. 2019;21(1):255. https://doi.org/10.3390/ijms21010255. PMID: 31905931; PMCID: PMC6981493
171. Hao MM, Bergner AJ, Nguyen HTH, Dissanayake P, Burnett LE, Hopkins CD, Zeng K, Young HM, Stamp LA. Role of JNK, MEK and adenylyl cyclase signalling in speed and directionality of enteric neural crest-derived cells. Dev Biol. 2019;455(2):362–8. https://doi.org/10.1016/j.ydbio.2019.07.007. Epub 2019 Jul 12. PMID: 31306639
172. Chou AH, Liao CC, Lee HC, Liou JT, Liu FC. The MAP2K4/JNK/c-Jun signaling pathway plays a key role in dexmedetomidine protection against acetaminophen-induced liver toxicity. Drug Des Devel Ther. 2019;13:3887–98. https://doi.org/10.2147/DDDT.S215473. PMID: 31814709; PMCID: PMC6861533
173. Zheng J, Dai Q, Han K, Hong W, Jia D, Mo Y, Lv Y, Tang H, Fu H, Geng W. JNK-IN-8, a c-Jun N-terminal kinase inhibitor, improves functional recovery through suppressing neuroinflammation in ischemic stroke. J Cell Physiol. 2020;235(3):2792–9. https://doi.org/10.1002/jcp.29183. Epub 2019 Sep 20. PMID: 31541462; PMCID: PMC6916328
174. Shi X, Yang W, Wang N, Zhu J. Circulating JNK pathway-associated phosphatase level correlates with decreased risk, activity, inflammation level and reduced clinical response to tumor necrosis factor-α inhibitor in Crohn disease patients. Medicine (Baltimore). 2019;98(33):e16622. https://doi.org/10.1097/MD.0000000000016622. PMID: 31415355; PMCID: PMC6831387

175. Al-Kandari N, Fadel F, Al-Saleh F, Khashab F, Al-Maghrebi M. The Thioredoxin system is regulated by the ASK-1/JNK/p38/Survivin pathway during germ cell apoptosis. Molecules. 2019;24(18):3333. https://doi.org/10.3390/molecules24183333. PMID: 31547465; PMCID: PMC6767173
176. Kaoud TS, Johnson WH, Ebelt ND, Piserchio A, Zamora-Olivares D, Van Ravenstein SX, Pridgen JR, Edupuganti R, Sammons R, Cano M, Warthaka M, Harger M, Tavares CDJ, Park J, Radwan MF, Ren P, Anslyn EV, Tsai KY, Ghose R, Dalby KN. Modulating multi-functional ERK complexes by covalent targeting of a recruitment site in vivo. Nat Commun. 2019;10(1):5232. https://doi.org/10.1038/s41467-019-12996-8. PMID: 31745079; PMCID: PMC6863825
177. Duong MTH, Lee JH, Ahn HC. C-Jun N-terminal kinase inhibitors: structural insight into kinase-inhibitor complexes. Comput Struct Biotechnol J. 2020;18:1440–57. https://doi.org/10.1016/j.csbj.2020.06.013. PMID: 32637042; PMCID: PMC7327381
178. Moore C, Palau VE, Mahboob R, Lightner J, Stone W, Krishnan K. Upregulation of pERK and c-JUN by γ-tocotrienol and not α-tocopherol are essential to the differential effect on apoptosis in prostate cancer cells. BMC Cancer. 2020;20(1):428. https://doi.org/10.1186/s12885-020-06947-6. PMID: 32414345; PMCID: PMC7227069
179. Kim KS, Park KS. XRP44X enhances the cytotoxic activity of natural killer cells by activating the c-JUN N-terminal kinase signaling pathway. Dev Reprod. 2020;24(1):53–62. https://doi.org/10.12717/DR.2020.24.1.53. Epub 2020 Mar 31. PMID: 32411918; PMCID: PMC7201060
180. Moustardas P, Aberdam D, Lagali N. MAPK pathways in ocular pathophysiology: potential therapeutic drugs and challenges. Cells. 2023;12(4):617. https://doi.org/10.3390/cells12040617. PMID: 36831285; PMCID: PMC9954064
181. De SK. Novel 2,4- diaminopyrimidine 5-carboxamides as c-Jun *N*-terminal kinase inhibitors for treating the liver fibrotic disorder. Endocr Metab Immune Disord Drug Targets. 2023;23(8):1118–23. https://doi.org/10.2174/1871530323666230220123228. PMID: 36803749
182. Gu L, He X, Zhang Y, Li S, Tang J, Ma R, Yang X, Huang H, Peng Y, Xie Y, Peng Z, Meng J, Hu G, Tao L, Liu X, Yang H. Fluorofenidone protects against acute liver failure in mice by regulating MKK4/JNK pathway. Biomed Pharmacother. 2023;164:114844. https://doi.org/10.1016/j.biopha.2023.114844. Epub 2023 May 22. PMID: 37224750
183. Hepp Rehfeldt SC, Majolo F, Goettert MI, Laufer S. c-Jun N-terminal kinase inhibitors as potential leads for new therapeutics for Alzheimer's diseases. Int J Mol Sci. 2020;21(24):9677. https://doi.org/10.3390/ijms21249677. PMID: 33352989; PMCID: PMC7765872
184. Yan H, He L, Lv D, Yang J, Yuan Z. The role of the dysregulated JNK signaling pathway in the pathogenesis of human diseases and its potential therapeutic strategies: a comprehensive review. Biomol Ther. 2024;14(2):243. https://doi.org/10.3390/biom14020243. PMID: 38397480; PMCID: PMC10887252
185. Davis RJ. Signal transduction by the JNK group of MAP kinases. Cell. 2000;103(2):239–52. https://doi.org/10.1016/s0092-8674(00)00116-1. PMID: 11057897
186. Zuccotto F, Ardini E, Casale E, Angiolini M. Through the "gatekeeper door": exploiting the active kinase conformation. J Med Chem. 2010;53(7):2681–94. https://doi.org/10.1021/jm901443h. PMID: 20000735
187. Raingeaud J, Gupta S, Rogers JS, Dickens M, Han J, Ulevitch RJ, Davis RJ. Pro-inflammatory cytokines and environmental stress cause p38 mitogen-activated protein kinase activation by dual phosphorylation on tyrosine and threonine. J Biol Chem. 1995;270(13):7420–6. https://doi.org/10.1074/jbc.270.13.7420. PMID: 7535770
188. Fuchs SY, Xie B, Adler V, Fried VA, Davis RJ, Ronai Z. c-Jun NH2-terminal kinases target the ubiquitination of their associated transcription factors. J Biol Chem. 1997;272(51):32163–8. https://doi.org/10.1074/jbc.272.51.32163. PMID: 9405416

189. Chen Z, Cobb MH. Regulation of stress-responsive mitogen-activated protein (MAP) kinase pathways by TAO2. J Biol Chem. 2001;276(19):16070–5. https://doi.org/10.1074/jbc.M100681200. Epub 2001 Mar 8. PMID: 11279118
190. Tournier C, Whitmarsh AJ, Cavanagh J, Barrett T, Davis RJ. Mitogen-activated protein kinase kinase 7 is an activator of the c-Jun NH2-terminal kinase. Proc Natl Acad Sci USA. 1997;94(14):7337–42. https://doi.org/10.1073/pnas.94.14.7337. PMID: 9207092; PMCID: PMC23822
191. Dérijard B, Hibi M, Wu IH, Barrett T, Su B, Deng T, Karin M, Davis RJ. JNK1: a protein kinase stimulated by UV light and ha-Ras that binds and phosphorylates the c-Jun activation domain. Cell. 1994;76(6):1025–37. https://doi.org/10.1016/0092-8674(94)90380-8. PMID: 8137421
192. Meyer CF, Wang X, Chang C, Templeton D, Tan TH. Interaction between c-Rel and the mitogen-activated protein kinase kinase kinase 1 signaling cascade in mediating kappaB enhancer activation. J Biol Chem. 1996;271(15):8971–6. https://doi.org/10.1074/jbc.271.15.8971. PMID: 8621542
193. Ishitani T, Takaesu G, Ninomiya-Tsuji J, Shibuya H, Gaynor RB, Matsumoto K. Role of the TAB2-related protein TAB3 in IL-1 and TNF signaling. EMBO J. 2003;22(23):6277–88. https://doi.org/10.1093/emboj/cdg605. PMID: 14633987; PMCID: PMC291846
194. Nishitoh H, Saitoh M, Mochida Y, Takeda K, Nakano H, Rothe M, Miyazono K, Ichijo H. ASK1 is essential for JNK/SAPK activation by TRAF2. Mol Cell. 1998;2(3):389–95. https://doi.org/10.1016/s1097-2765(00)80283-x. PMID: 9774977
195. Yazgan O, Pfarr CM. Regulation of two JunD isoforms by Jun N-terminal kinases. J Biol Chem. 2002;277(33):29710–8. https://doi.org/10.1074/jbc.M204552200. Epub 2002 Jun 6. PMID: 12052834
196. Tada K, Okazaki T, Sakon S, Kobarai T, Kurosawa K, Yamaoka S, Hashimoto H, Mak TW, Yagita H, Okumura K, Yeh WC, Nakano H. Critical roles of TRAF2 and TRAF5 in tumor necrosis factor-induced NF-kappa B activation and protection from cell death. J Biol Chem. 2001;276(39):36530–4. https://doi.org/10.1074/jbc.M104837200. Epub 2001 Jul 30. PMID: 11479302
197. Cano E, Hazzalin CA, Kardalinou E, Buckle RS, Mahadevan LC. Neither ERK nor JNK/SAPK MAP kinase subtypes are essential for histone H3/HMG-14 phosphorylation or c-fos and c-Jun induction. J Cell Sci. 1995;108(Pt 11):3599–609. https://doi.org/10.1242/jcs.108.11.3599. PMID: 8586671
198. Girardin SE, Yaniv M. A direct interaction between JNK1 and CrkII is critical for Rac1-induced JNK activation. EMBO J. 2001;20(13):3437–46. https://doi.org/10.1093/emboj/20.13.3437. PMID: 11432831; PMCID: PMC125507
199. Slack DN, Seternes OM, Gabrielsen M, Keyse SM. Distinct binding determinants for ERK2/p38alpha and JNK map kinases mediate catalytic activation and substrate selectivity of map kinase phosphatase-1. J Biol Chem. 2001;276(19):16491–500. https://doi.org/10.1074/jbc.M010966200. Epub 2001 Jan 30. PMID: 11278799
200. Tanoue T, Moriguchi T, Nishida E. Molecular cloning and characterization of a novel dual specificity phosphatase, MKP-5. J Biol Chem. 1999;274(28):19949–56. https://doi.org/10.1074/jbc.274.28.19949. PMID: 10391943
201. Wang T, Arifoglu P, Ronai Z, Tew KD. Glutathione S-transferase P1-1 (GSTP1-1) inhibits c-Jun N-terminal kinase (JNK1) signaling through interaction with the C terminus. J Biol Chem. 2001;276(24):20999–1003. https://doi.org/10.1074/jbc.M101355200. Epub 2001 Mar 9. PMID: 11279197
202. Aguirre V, Werner ED, Giraud J, Lee YH, Shoelson SE, White MF. Phosphorylation of Ser307 in insulin receptor substrate-1 blocks interactions with the insulin receptor and inhibits insulin action. J Biol Chem. 2002;277(2):1531–7. https://doi.org/10.1074/jbc.M101521200. Epub 2001 Oct 17. PMID: 11606564
203. Aguirre V, Uchida T, Yenush L, Davis R, White MF. The c-Jun NH(2)-terminal kinase promotes insulin resistance during association with insulin receptor substrate-1 and phos-

phorylation of Ser(307). J Biol Chem. 2000;275(12):9047–54. https://doi.org/10.1074/jbc.275.12.9047. PMID: 10722755

204. Gao M, Labuda T, Xia Y, Gallagher E, Fang D, Liu YC, Karin M. Jun turnover is controlled through JNK-dependent phosphorylation of the E3 ligase Itch. Science. 2004;306(5694):271–5. https://doi.org/10.1126/science.1099414. Epub 2004 Sep 9. PMID: 15358865
205. Gallagher E, Gao M, Liu YC, Karin M. Activation of the E3 ubiquitin ligase Itch through a phosphorylation-induced conformational change. Proc Natl Acad Sci USA. 2006;103(6):1717–22. https://doi.org/10.1073/pnas.0510664103. Epub 2006 Jan 30. Erratum in: Proc Natl Acad Sci U S A. 2010 107(16):7616. PMID: 16446428; PMCID: PMC1413664
206. Cheng J, Yang J, Xia Y, Karin M, Su B. Synergistic interaction of MEK kinase 2, c-Jun N-terminal kinase (JNK) kinase 2, and JNK1 results in efficient and specific JNK1 activation. Mol Cell Biol. 2000;20(7):2334–42. https://doi.org/10.1128/MCB.20.7.2334-2342.2000. PMID: 10713157; PMCID: PMC85399
207. Xu S, Cobb MH. MEKK1 binds directly to the c-Jun N-terminal kinases/stress-activated protein kinases. J Biol Chem. 1997;272(51):32056–60. https://doi.org/10.1074/jbc.272.51.32056. PMID: 9405400
208. Elion EA. Routing MAP kinase cascades. Science. 1998;281(5383):1625–6. https://doi.org/10.1126/science.281.5383.1625. PMID: 9767029
209. Cai Y, Lechner MS, Nihalani D, Prindle MJ, Holzman LB, Dressler GR. Phosphorylation of Pax2 by the c-Jun N-terminal kinase and enhanced Pax2-dependent transcription activation. J Biol Chem. 2002;277(2):1217–22. https://doi.org/10.1074/jbc.M109663200. Epub 2001 Nov 7. PMID: 11700324
210. Ito M, Yoshioka K, Akechi M, Yamashita S, Takamatsu N, Sugiyama K, Hibi M, Nakabeppu Y, Shiba T, Yamamoto KI. JSAP1, a novel Jun N-terminal protein kinase (JNK)-binding protein that functions as a scaffold factor in the JNK signaling pathway. Mol Cell Biol. 1999;19(11):7539–48. https://doi.org/10.1128/MCB.19.11.7539. PMID: 10523642; PMCID: PMC84763
211. Kelkar N, Gupta S, Dickens M, Davis RJ. Interaction of a mitogen-activated protein kinase signaling module with the neuronal protein JIP3. Mol Cell Biol. 2000;20(3):1030–43. https://doi.org/10.1128/MCB.20.3.1030-1043.2000. PMID: 10629060; PMCID: PMC85220
212. Noguchi K, Kitanaka C, Yamana H, Kokubu A, Mochizuki T, Kuchino Y. Regulation of c-Myc through phosphorylation at Ser-62 and Ser-71 by c-Jun N-terminal kinase. J Biol Chem. 1999;274(46):32580–7. https://doi.org/10.1074/jbc.274.46.32580. PMID: 10551811
213. Wiltshire C, Matsushita M, Tsukada S, Gillespie DA, May GH. A new c-Jun N-terminal kinase (JNK)-interacting protein, sab (SH3BP5), associates with mitochondria. Biochem J. 2002;367(Pt 3):577–85. https://doi.org/10.1042/BJ20020553. PMID: 12167088; PMCID: PMC1222945
214. Mao C, Ray-Gallet D, Tavitian A, Moreau-Gachelin F. Differential phosphorylations of Spi-B and Spi-1 transcription factors. Oncogene. 1996;12(4):863–73. PMID: 8632909
215. Saleem A, Datta R, Yuan ZM, Kharbanda S, Kufe D. Involvement of stress-activated protein kinase in the cellular response to 1-beta-D-arabinofuranosylcytosine and other DNA-damaging agents. Cell Growth Differ. 1995;6(12):1651–8. PMID: 9019171
216. Kharbanda S, Saleem A, Shafman T, Emoto Y, Taneja N, Rubin E, Weichselbaum R, Woodgett J, Avruch J, Kyriakis J, et al. Ionizing radiation stimulates a Grb2-mediated association of the stress-activated protein kinase with phosphatidylinositol 3-kinase. J Biol Chem. 1995;270(32):18871–4. https://doi.org/10.1074/jbc.270.32.18871. PMID: 7642542
217. Yasuda J, Whitmarsh AJ, Cavanagh J, Sharma M, Davis RJ. The JIP group of mitogen-activated protein kinase scaffold proteins. Mol Cell Biol. 1999;19(10):7245–54. https://doi.org/10.1128/MCB.19.10.7245. PMID: 10490659; PMCID: PMC84717

218. Whitmarsh AJ, Cavanagh J, Tournier C, Yasuda J, Davis RJ. A mammalian scaffold complex that selectively mediates MAP kinase activation. Science. 1998;281(5383):1671–4. https://doi.org/10.1126/science.281.5383.1671. PMID: 9733513
219. Hu MC, Qiu WR, Wang YP. JNK1, JNK2 and JNK3 are p53 N-terminal serine 34 kinases. Oncogene. 1997;15(19):2277–87. https://doi.org/10.1038/sj.onc.1201401. PMID: 9393873
220. Lin Y, Khokhlatchev A, Figeys D, Avruch J. Death-associated protein 4 binds MST1 and augments MST1-induced apoptosis. J Biol Chem. 2002;277(50):47991–8001. https://doi.org/10.1074/jbc.M202630200. Epub 2002 Oct 15. PMID: 12384512
221. Maekawa M, Nishida E, Tanoue T. Identification of the anti-proliferative protein Tob as a MAPK substrate. J Biol Chem. 2002;277(40):37783–7. https://doi.org/10.1074/jbc.M204506200. Epub 2002 Jul 31. PMID: 12151396
222. Martin-Vega A, Cobb MH. Navigating the ERK1/2 MAPK Cascade. Biomol Ther. 2023;13(10):1555. https://doi.org/10.3390/biom13101555. PMID: 37892237; PMCID: PMC10605237
223. Plotnikov A, Zehorai E, Procaccia S, Seger R. The MAPK cascades: signaling components, nuclear roles and mechanisms of nuclear translocation. Biochim Biophys Acta. 2011;1813(9):1619–33. https://doi.org/10.1016/j.bbamcr.2010.12.012. Epub 2010 Dec 16. PMID: 21167873
224. Cargnello M, Roux PP. Activation and function of the MAPKs and their substrates, the MAPK-activated protein kinases. Microbiol Mol Biol Rev. 2011;75(1):50–83. https://doi.org/10.1128/MMBR.00031-10. Erratum in: Microbiol Mol Biol Rev. 2012 76(2):496. PMID: 21372320; PMCID: PMC3063353

Chapter 3
ATP-Competitive Inhibitors of MAP Kinases

Abstract Inhibitors binding in the ATP site are a very promising and successful drug design strategy in the field of kinase. Several ATP-competitive inhibitors are in the market for the treatment of cancer and other diseases. Many research institutes discovered preclinical ATP-competitive inhibitors of MAP kinases using fragment-based drug design or optimization of their initial hit from high-throughput screening to improve in vitro potency only but lack PK profile. Some investigators from pharmaceutical companies discovered lead compounds using X-ray structure-based drug design strategy. A few industries disclosed clinical candidates from hit to lead with improved potency and favorable PK profile. Readers should gain vast information about how an initial hit becomes a clinical candidate.

Keywords ATP-site binder · Preclinical and clinical candidate · CC-90001 · Ulixertinib · AZD0364 · LY3214996 · ASTX029 · ASN007 · KO-947 · ONC201 · LTT462 · GDC-0994 · MK-8353 · MW150 · RO-3201195 · Pamapimod · Adezmapimod · SB-220025 · Dilmapimod · Losmapimod · Neflamapimod · AMG-548 · PH-797804 · Talmapimod · TAK-715 · AZD6703 · AZD7624 · and Ralimetinib

3.1 Introduction

Most known kinase inhibitors bind to the ATP site of the kinase catalytic domain. MAPK inhibitors could be classified into four categories such as ATP-competitive inhibitor, ATP-non-competitive inhibitor (allosteric), bi-dentate inhibitor, and covalent inhibitor. ATP-competitive inhibitors are a class of compounds that target the JNK, ERK, and p38 which are members of the mitogen-activated protein kinase (MAPK) family. These inhibitors compete with ATP (adenosine triphosphate) for binding to the ATP-binding site of the MAPKs, thereby blocking the kinase activity of MAPKs. By inhibiting MAPK activity, these compounds can modulate various cellular processes and have potential therapeutic applications in diseases where

S. K. De, *Drug Design and Discovery*,
https://doi.org/10.1007/978-3-031-75682-5_3

dysregulated MAPK signaling is implicated, such as cancer, neurodegenerative disorders, and inflammatory diseases.

3.2 ATP-Competitive Inhibitors of JNK

ATP-competitive inhibitors of JNK1, JNK2, and JNK3 from industries and research institutes are summarized here.

3.2.1 *Inhibitors from Serono*

The first small molecule ATP competitive JNK inhibitor coded **SP600125** (**1**, Fig. 3.1) was reported by Celgene [1, 2] (formerly Signal Pharmaceutical). **SP600125** is an ATP competitive inhibitor and inhibits the phosphorylation of c-Jun. Scientists at Serono identified compound **2** from their HTS campaign [3]. Extensive SAR studies led to optimized compound **3** (Fig. 3.1) as a first potent and selective JNK inhibitor of this series that demonstrated inhibition of neuronal cell

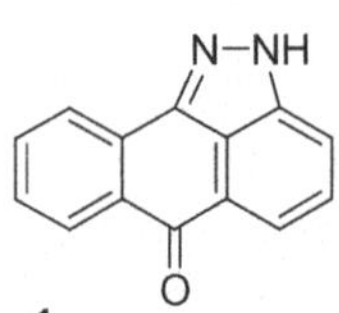

1

SP600125

IC_{50} (JNK1) = 40 nM
IC_{50} (JNK2) = 40 nM
IC_{50} (JNK3) = 90 nM

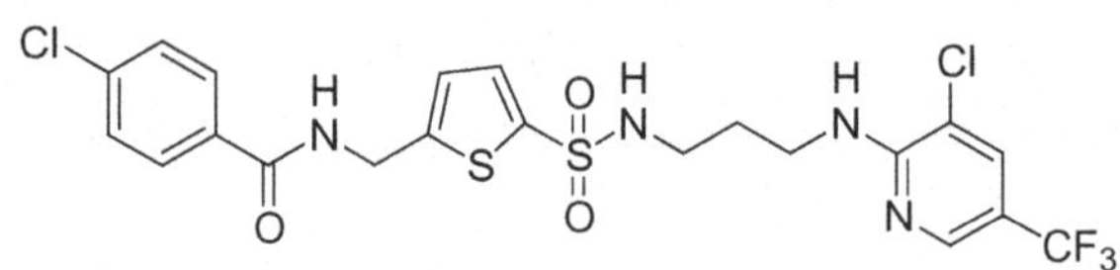

2

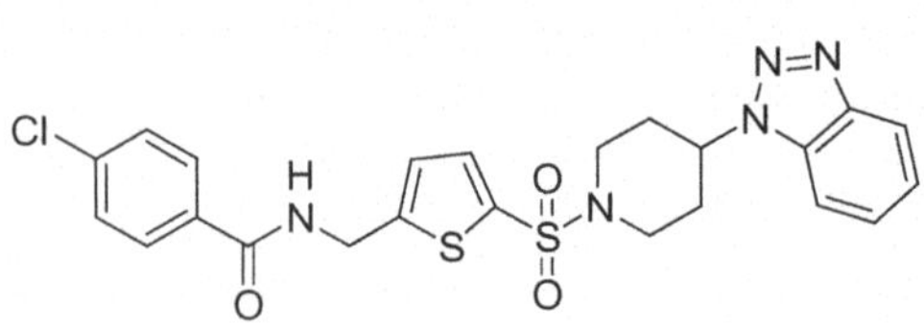

3

IC_{50} (JNK1) = 650 nM

IC_{50} (JNK3) = 150 nM

Fig. 3.1 Inhibitors from Serono

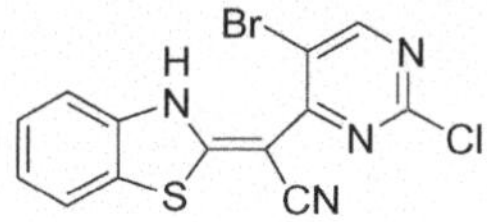

4

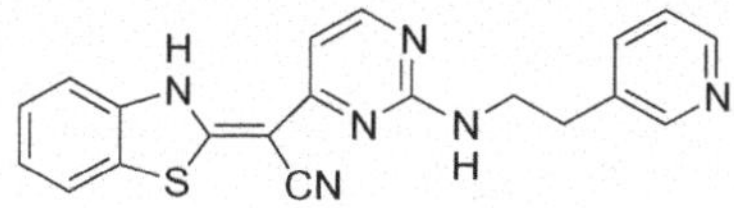

5

IC_{50} (JNAK1) = 150 nM

IC_{50} (JNK2) = 220 nM

IC_{50} (JNK3) = 70 nM

IC_{50} (p38α) = 4.8 μM

IC_{50} (ERK1) > 10 μM

Fig. 3.2 Benzothiazol-2-yl-acetonitrile derivatives

death induced by NGF-deprived superior cervical ganglia cells (SCG cells) with an IC_{50} value of 1.7 μM. Compound **3** is selective over 80 kinases tested.

Investigators from Serono identified (benzothiazol-2-yl)acetonitrile derivatives such as compound **4** as potent and selective JNK inhibitors [4]. Extensive SAR studies resulted in compound **5**. Pyrimidine ring is crucial for potency and selectivity (Fig. 3.2). It is also selective over closely related MAP kinases such as 68 times against p38α and 142 times against ERK1. It is a potent JNK3 inhibitor and selective over MKK4, MKK7, MEK, AKT, PI3K, PDK1, CHK1, PKC, EGFR, IKK, CDK2, and other kinases. Compound **5** has 38% bioavailability in rats. It demonstrates a dose-dependent decrease of TNFα plasma level with an ED_{50} of 3 mg/kg via oral route indicating anti-inflammatory effects in a mouse model of collagen-induced adjuvant (CIA) rheumatoid arthritis.

3.3 Inhibitors from Eisai

Scientists at Eisai tried to obtain a JNK3 inhibitor from a known imidazole scaffold of p38 inhibitor (compound **6,** Fig. 3.3). They modified compound **6** to lead compound **7** as a selective JNK3 inhibitor and 12 times selective over p38 [5]. Docking simulations indicate compound **7** binds at the ATP site and hydroxyl group forms a hydrogen bonding interaction with Asn152 and/or Ser193 of JNK3. Compound **7** inhibits c-Jun phosphorylation and at 1 micromolar concentration it shows strong neuroprotective properties.

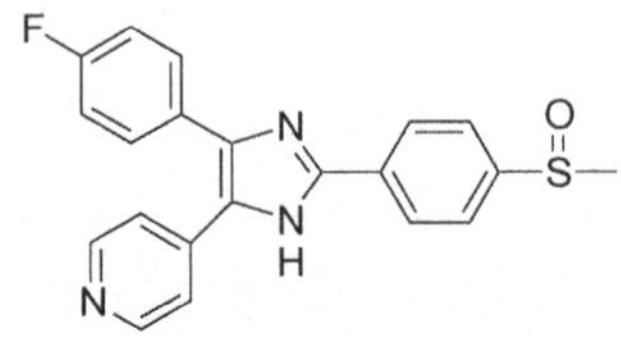

6

IC_{50} (p38) = 42 nM

IC_{50} (JNK3) = 10 μM

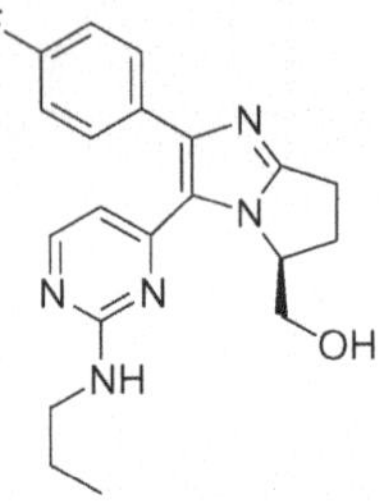

7

IC_{50} (p38) = 30 nM

IC_{50} (JNK3) = 2.5 nM

Fig. 3.3 Imidazole derivatives

3.4 Inhibitors from Pfizer

Investigators at Pfizer identified 4-substituted-2-aminopyrimidine such as compound **8** (Fig. 3.4) from their compound library collection [6]. It is a novel ATP-competitive pan-JNK inhibitor with K_i value of 26 nM (JNK1). It has moderate cellular activity but high clearance. They optimized this compound resulting in compound **9** with the K_i values of 12–17 nM in vitro assays and 98 nM in cell-based assay respectively. Compound **9** shows low clearance, high rat microsomal stability, and 100 folds selective over other MAP kinases (MKK1, MKK4, MKK7, and p38 α, γ, δ) and 27–65 times selective against GSK-3β and CDK2, and/or other kinases tested.

3.5 Inhibitors from Abbott Lab

Investigators at Abbott lab identified compound **10** from their HTS as a JNK1 inhibitor [7]. It is selective over other MAP kinases such as p38α and ERK2 (Fig. 3.5). The crystal structure of compound **10** in complex with JNK1 (PDB code: 2G0I) revealed that the chlorine group stayed toward the small pocket and oxygen of N-OH formed a hydrogen bonding interaction with the backbone amide of Met111.

Abbott lab reported a series of 4-anilinopyrimidine derivatives as potent and selective JNK inhibitors. Compound **11** was identified from HTS and it was selective over p38 and ERK2 [8]. Extensive SAR studies resulted in compound **12** with a good cellular potency. The replacement of the ether linkage with a substituted amino group forms a hydrogen bond with the backbone of carbonyl of Met111. Replacing acid to amide improves cell permeability. The crystal structure of compound **12** in complex with JNK1 (PDB code: 2NO3) showed that the pyrimidine N1 and 2-anilino formed 2 important hydrogen bonds with Met111 in the hinge region of the ATP binding site.

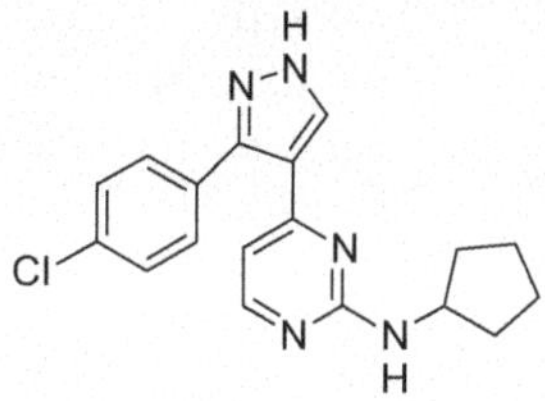

8

K_i (JNK1) = 26 nM

LE = 0.44

ELogD = 4.42

HML Cl: 116 μL/min/kg

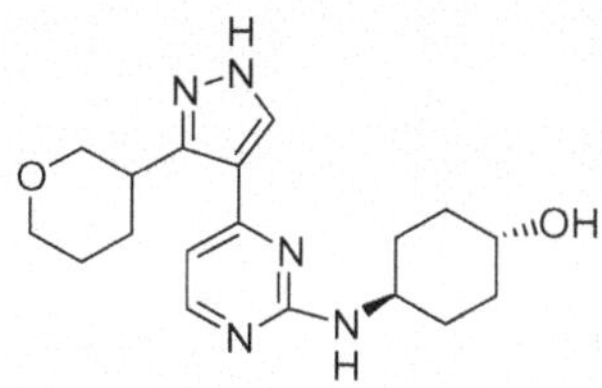

9

K_i (JNK1) = 12 nM

K_i (JNK2) = 17

K_i (JNK3) = 16 nM

HML Cl: 15 μL/min/Kg

Fig. 3.4 4-Pyrazolo-2-aminopyrimidine derivatives

Fig. 3.5 4-Anilinopyrimidine derivatives

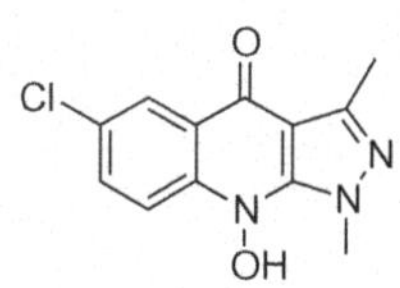

10

IC_{50} ((JNK1) = 1 μM

IC_{50} (p38α) > 50 μM

IC_{50} (ERK2) > 50 μM

Ref2

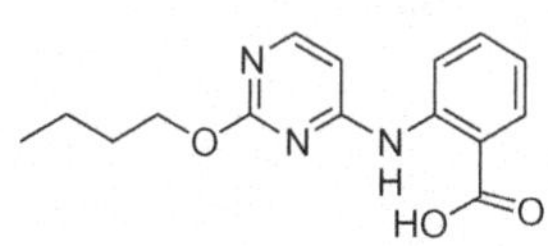

11

IC_{50} (JNK1) = 1.9 μM

IC_{50} (p38) > 100 μM

IC_{50} (ERK2) > 100 μM

IC_{50} (c-Jun) = 22.3 μM

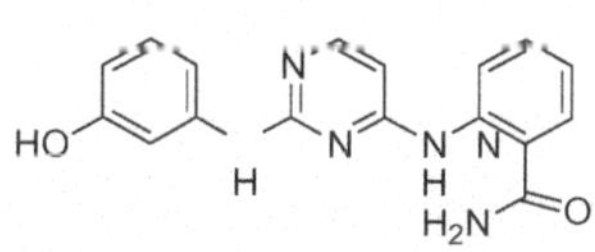

12

IC_{50} (JNK1) = 9 nM

IC_{50} (p38) > 50 μM

IC_{50} (ERK2) = 25 μM

IC_{50} (c-Jun) = 1.22 μM

Abbott lab reported a class of 4-aminopyridine derivatives as JNK inhibitors [9]. They identified compound **13** from HTS as a selective JNK inhibitor over p38 and ERK2 (Fig. 3.6). Extensive SAR studies resulted in compound **14** with improved potency. The crystal structure of a compound in this series with JNK1 reveals that

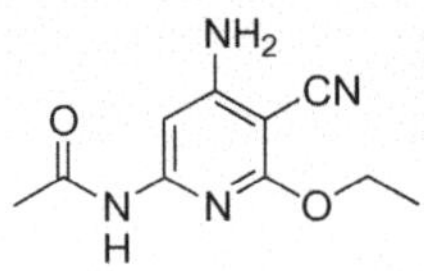

13

IC_{50} (JNK1) = 0.75 μM

IC_{50} (JNK2) = 1.1 μM

IC_{50} (p38) > 50 μM

IC_{50} (ERK2) = 25 μM

EC_{50} (c-Jun) = 6.8 μM

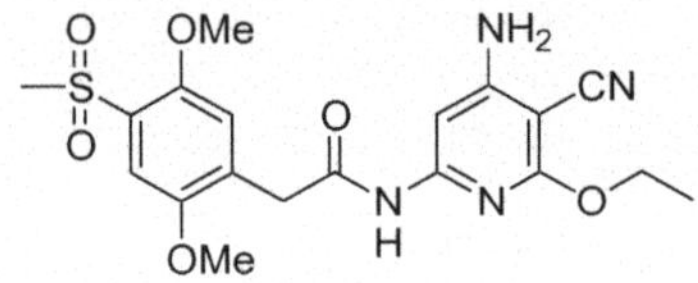

14

IC_{50} (JNK1) = 38 nM

IC_{50} (JNK3) = 150 nM

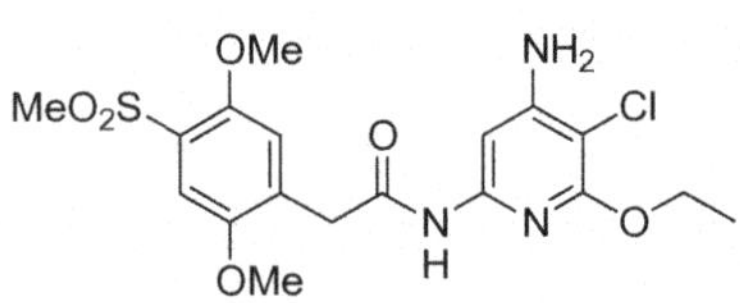

15

IC_{50} (JNK1) = 36 nM

IC_{50} (JNK2) = 70 nM

EC_{50} (c-Jun) = 1.75 μM

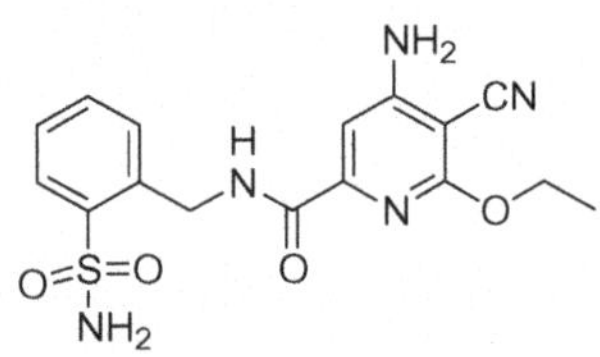

16

IC_{50} (JNK1) = 12 nM

IC_{50} (JNK2) = 53 nM

IC_{50} (JNK3) = 37 nM

EC_{50} (c-Jun) = 110 nM

Oral bioavailability: 54.7% (rat)

Fig. 3.6 4-Aminopyridine derivatives

the pyridine ring stays within the ATP pocket (PDB code: 2GMX). The C-4 amine and C-6 amide carbonyl form two weak hydrogen bonding interactions with Glu109 and Met111 in the hinge region. The cyano group forms a weak bond with Lys55. Further optimization of this series results in compound **15** with improved potency in vitro and cell-based assays. Reverse amide bond results in compound **16** with improved enzymatic potency and cellular activity [10]. Compound **16** provides better metabolic stability and favorable oral bioavailability (54.7%) in rats.

3.6 Inhibitors from Elan Pharmaceuticals

A team of scientists from Elan Pharmaceuticals reported [11] disubstituted thiophene and thiazole-based inhibitors of JNK. From their HTS campaign compound **17** (Fig. 3.7) was identified as an initial hit with good solubility, excellent passive permeability, and modest potency against JNK3, which are all important features

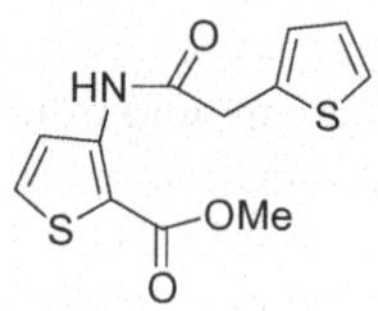

17

IC_{50} (JNK1) = 1.6 μM

IC_{50} (JNK2) = 23 μM

IC_{50} (JNK3) = 2.2 μM

IC_{50} (p38α) >50μM

IC_{50} (ERK2) > 50 μM

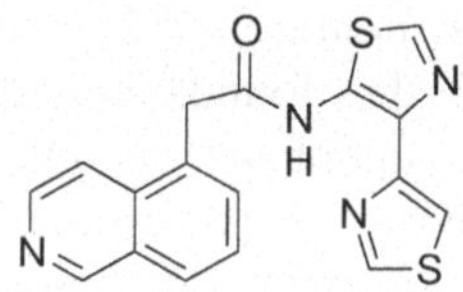

18

IC_{50} (JNK1) = 48 nM

IC_{50} (JNK2) = 470 nM

IC_{50} (JNK3) = 77 nM

IC_{50}(p38α) >50 μM

IC_{50} (ERK2) >50 μM

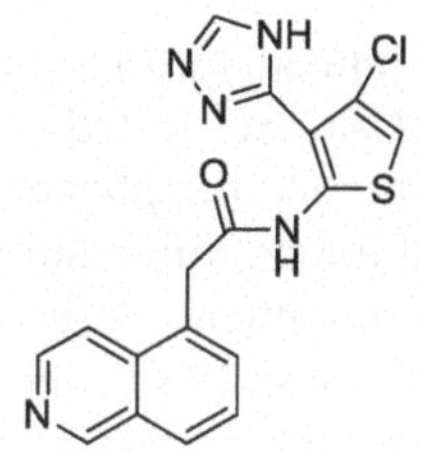

19

IC_{50} (JNK1) = 14 nM

IC_{50} (JNK2) = 43 nM

IC_{50} (JNK3) = 11 nM

IC_{50} (p38α) >50 μM

IC_{50} (ERK2) >50 μM

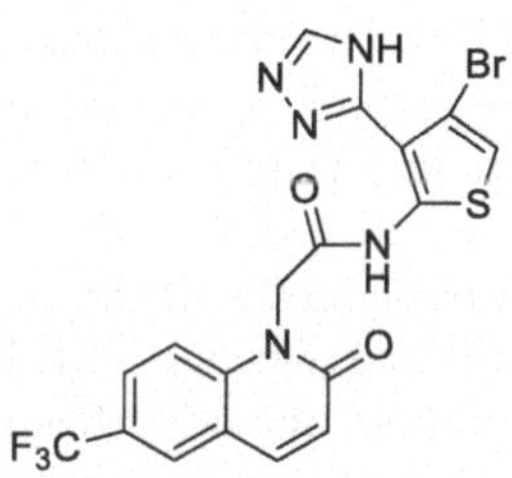

20

IC_{50} (JNK1) = 2 nM

IC_{50} (JNK2) = 5 nM

IC_{50} (JNK3) = 9 nM

IC_{50} (p38α) >50 μM

IC_{50} (ERK2) > 50 μM

Fig. 3.7 Thiophene derivatives

for the development of CNS drugs. The crystal structure of compound **17** in complex with JNK3 (PDB code: 3OXI) reveals that this is an ATP-competitive kinase inhibitor where critical hydrogen bond forms between carbonyl of **17** and the NH of Met149 in the hinge region and gatekeeper residue Met146 is within proximity to the inhibitor **17**. Based on these factors, they optimized compound **17** and found compound **18** as a lead compound from this series where thiophene was replaced by thiazole with the IC_{50} values of 48 nM, 470 nM, and 77 nM against JNK1, JNK2, and JNK3 respectively. Compound **18** showed good microsomal stability in mouse and human. Furthermore, compound **18** showed excellent selectivity against a panel of 36 kinases tested so far. Along the same line, Elan pharmaceutical described tri-substituted thiophenes as JNK inhibitors [12] starting from the same screening hit **17**. Further optimization of **17**, they found compound **19** as the best compound from this thiophene series where thiophene occupied tri-substituted functionalities. Compound **19** showed selectivity over p38α and ERK2. A crystal structure of JNK3 (PDB code: 3PTG) in complex with a compound of this series shows that the 4-position chlorine of thiophene is close to the Met 146 and interacts with the sulfur of Met149. Thiophene ring makes a hydrophobic interaction with the gatekeeper residue Met146. Compound **19** showed the best potency, metabolic stability, permeability, and was chosen for in vivo study in the kainic acid mouse model. Pretreatment of compound **19**, a single oral dose, 300 mg/Kg showed a significant 51% reduction of phospho-c-Jun in the hippocampus of FVB mice 4 h after IP injection of 25 mg/Kg kainic acid. This result suggests that these compounds are orally active, brain penetrant, and potent JNK inhibitors. Further optimization of this series results in compound **20** with lower clearance, greater brain exposure, and longer half-life [13].

Elan pharmaceuticals identified compound **21** from their in-house compound screening campaign [14]. Optimization of compound **21** by replacing thiol to oxo and other modifications obtained compound **22** and its analogs (**23**, **24**) which showed in vitro CNS-like pharmacokinetic properties (Fig. 3.8). The crystal structure of **21** in complex with JNK3 reveals that the naphthalene group just occupies a deep selectivity pocket. Since other MAP kinases do not have the capacity to place the big naphthalene group, thus representing this series very selective against JNK3 only. The isoform selectivity can be explained as well: JNK1 has a big Ile106 in the selectivity pocket that cannot accommodate the bulky naphthalene, whereas JNK2 and JNK3 have the less bulky residue Leu144 at the same position. Scientists from Elan pharmaceuticals tried further optimization of their previous compound **21** where the central core was replaced from triazole to pyrazole that resulted in compound **25** [15]. Compound **25** exhibited good microsomal stability both in mouse and human such as 100% and 73% remaining after 30 min incubation, high brain exposure, and low clearance. From these data, we may conclude that this series of JNK inhibitors possibly are viable neuroprotective agents.

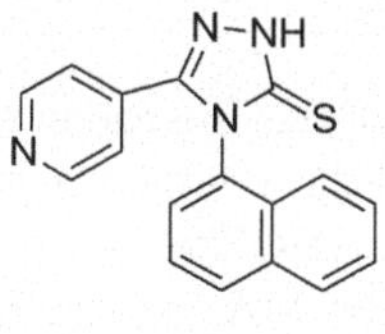

21

IC$_{50}$ (JNK1) = 2.95 μM

IC$_{50}$ (JNK2) = 7.28 μM

IC$_{50}$ (JNK3) = 1.07 μM

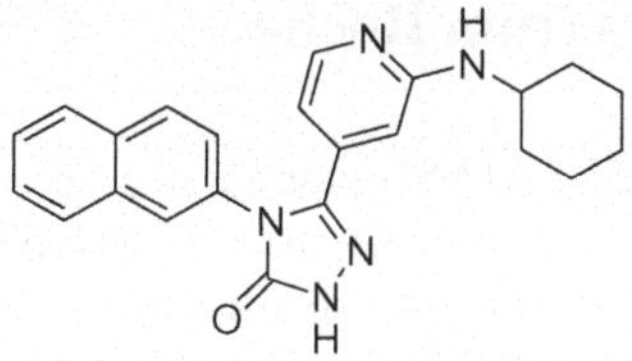

22

IC50 (JNK1) = 1 μM

IC50 (JNK2) = 0.069 μM

IC$_{50}$ (JNK3) = 0.024 μM

IC$_{50}$ (p38α) >50 μM

IC$_{50}$ (ERK2) > 50 μM

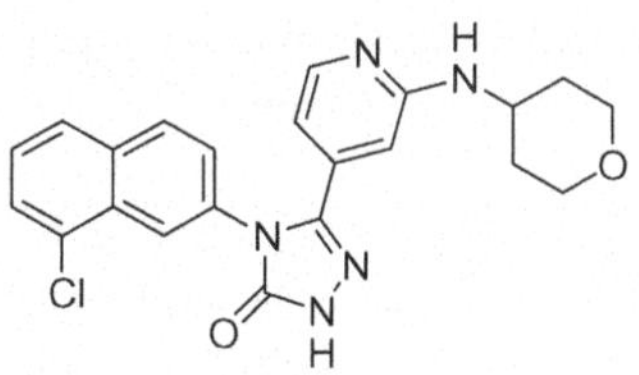

23

IC$_{50}$ (JNK1) = 0.11 μM

IC$_{50}$ (JNK2) = 0.15 μM

IC$_{50}$ (JNK3) = 0.014 μM

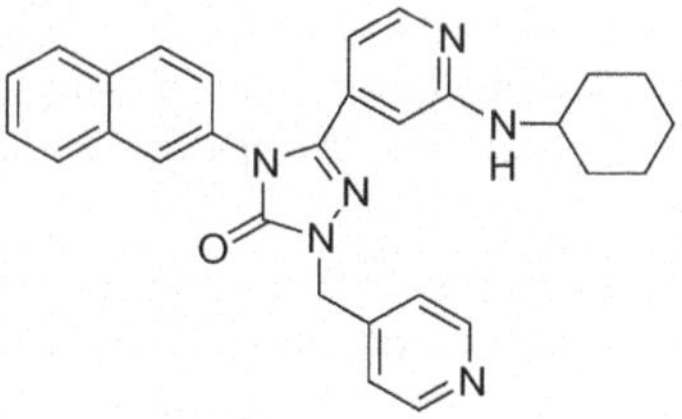

24

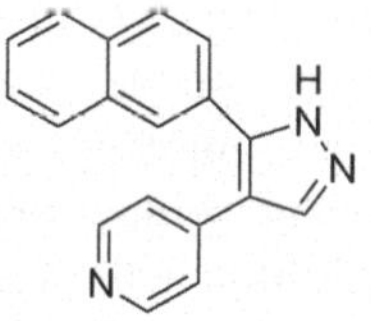

25

IC$_{50}$ (JNK1) = 7.15 μM

IC$_{50}$ (JNK2) = 0.86 μM

IC$_{50}$ (JNK3) = 0.245 μM

Fig. 3.8 JNK inhibitors from Elan Pharmaceuticals

3.7 Inhibitors from Roche

The p-38α is another MAP kinase and is closely related to JNK. Researchers at Roche provided the crystal structure of **BIRB796** (compound **26**) complexed with JNK2 (PDB code: 3E7O) [16]. It was reported previously that **BIRB796** is a p-38α inhibitor. This crystal structure revealed further insights into the MAP kinase family. This compound binds with JNK2 in a DFG-out conformation and inhibits JNK2 activation by stabilizing the activation loop in a conformation. Based on the structure they suggested that the design of selective DFG-out binding JNK inhibitors may have great potential targets.

Roche identified compound **27** from their CDK2 program. Structure-based drug design guided the optimization of initial hit led to compound **28** (Fig. 3.9) as potent and selective JNK1 and JNK2 inhibitors [17]. Replacing 1,2,3-benzotriazole with indole provided compound **28** with favorable bioavailability in rats. Further optimization resulted in compound **29** with improvement of potency [18]. Compound **29** has an IC_{50} value of 3 nM for JNK1 and 20 nM for JNK2, and 40-fold selectivity against other kinases with good physicochemical and pharmacokinetic properties. Indeed, compound **29** has oral bioavailability 100% in rats and 15% in dogs at the dose of 2 mg/kg.

Roche discovered a series of 4-quinolone derivatives as selective JNK inhibitors such as compound **30** [19]. Structure-based drug design was applied to improve potency and physicochemical properties of the series resulting in compound **31** (Fig. 3.10). Compound **31** shows excellent kinase selectivity and impressive efficacy in a rodent asthma model. The crystal structure of a compound of this series in complex with JNK1 (PDB code: 4G1W) reveals that the oxygen atom of carbonyl of quinolone forms a hydrogen bonding interaction with Met111 in the hinge region. However, compound **31** potency in human kidney (HK-2) cells and microsomal clearance values did not appear good to further study. Roche identified 8-azaquinolone such as compound **32** as JNK inhibitor [20]. Further optimization leads to a trans-hydroxyadamantyl azaquinolone compound **33** which shows improvement in JNK biochemical potency, cellular potency, solubility, and acceptable rodent PK. Introducing *trans*-hydroxyadamantyl moiety into the azaquinolone series resulting in compound **33** shows improved potency for the acute kidney injury (AKI) indication, and confirms proof of mechanism for JNK inhibition in the TNF-α challenge model. The crystal structure of **34** (*trans*-aminoadamantyl azaquinolone) in complex with JNK1β1-JIP (PDB code: 4E73) shows that the backbone NH from Met111 forms a hydrogen bond with the carbonyl of the azaquinolone. The phenyl ring stays between the side chain of Ile32 and the Asp112/Ala113 plane from the end of the hinge region. The aryl chain sits away from the ATP pocket. The amide-linked trans-adamantyl moiety stays in the shallow pocket outside of the N-terminal lobe of the kinase active site and makes a hydrophobic interaction with the protein.

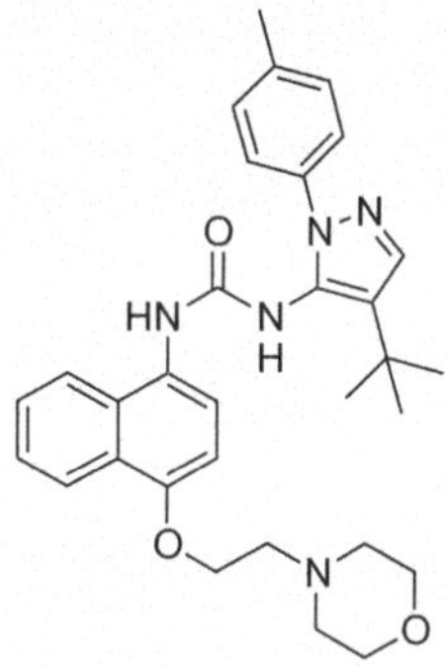

26 (BIRB796)

K_d (JNK2) = 4.6 nM

K_d (JNK3) = 62 nM

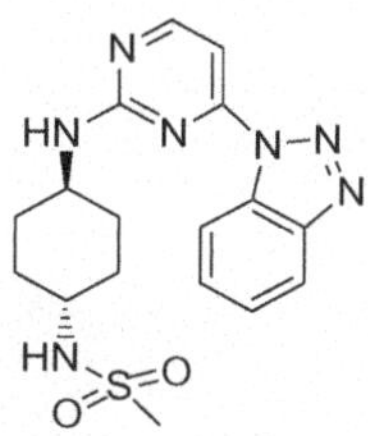

27

IC_{50} (CDK2) = 300 nM

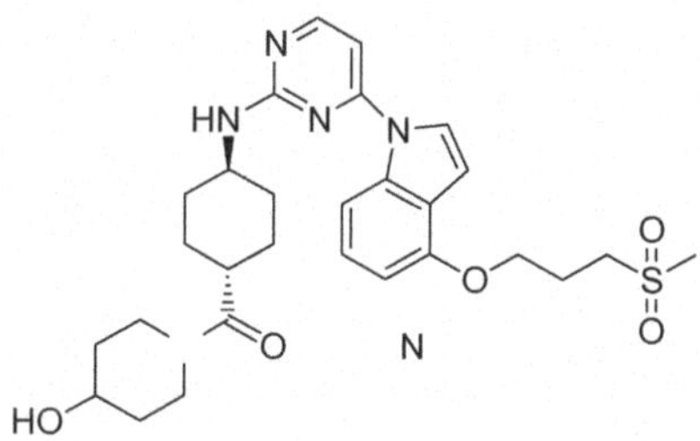

28

IC_{50} (JNK1) = 16 nM

IC_{50} (JNK2) = 66 nM

IC_{50} (c-Jun) = 1.1 μM

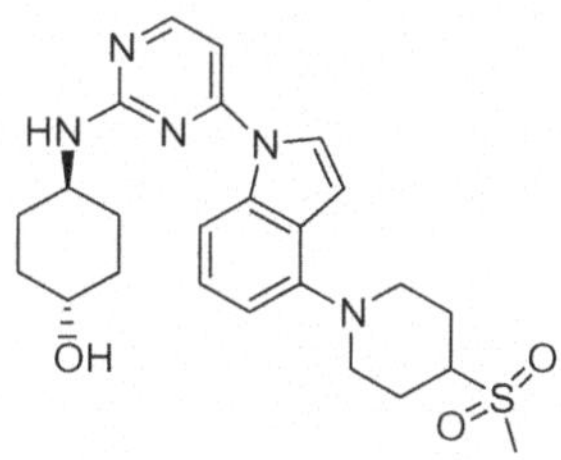

29

IC_{50} (JNK1) = 3 nM

IC_{50} (JNK2) = 20 nM

IC_{50} (c-Jun) = 0.24 μM

Fig. 3.9 Pyrazoles and 2-aminopyrimidine derivatives

3.8 Inhibitors from Takeda

Investigators at Takeda discovered a novel series of 4-phenylisoquinolones as JNK inhibitors for the treatment of heart failure [21]. Several reports have suggested that JNK may play crucial roles in the progression of heart failure. Takeda identified compound **35** (Fig. 3.11) from HTS as a JNK1 inhibitor and it was selective over other MAK kinases such as p38 and ERK1. SAR studies led to the lead compound **36**. The docking study of an analog of this series reveals that the compound binds at the ATP site. It forms a hydrogen bond between the carbonyl oxygen of the isoquinolone group and the backbone NH group of Met111 in the hinge region. Asp169 is

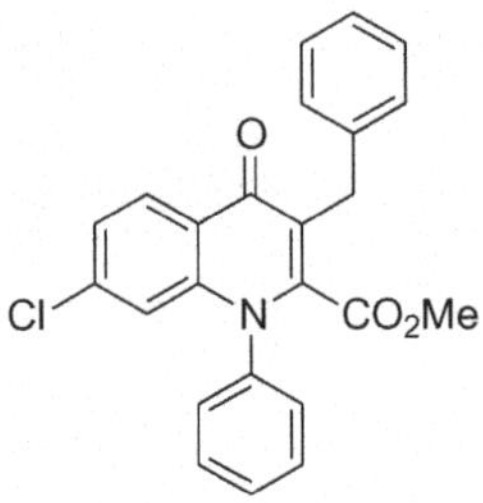

30

IC_{50} (JNK1) = 0.37 μM

IC_{50} (JNK2) = 0.96 μM

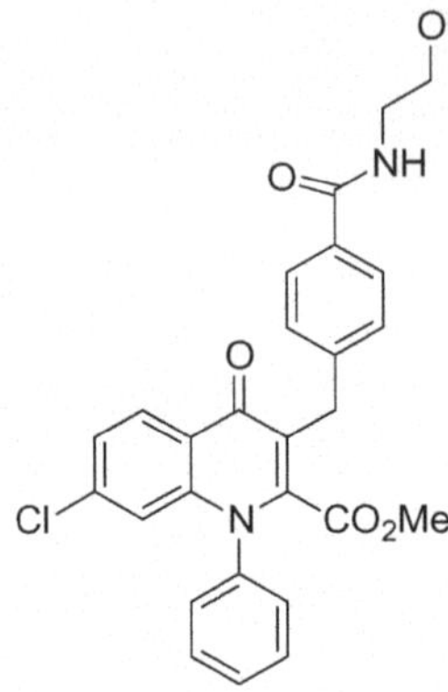

31

IC_{50} (JNK1) = 62 nM

IC_{50} (JNK2) = 170 nM

IC_{50} (p38γ) > 10 μM

EC_{50} (HK-2) = 10 μM

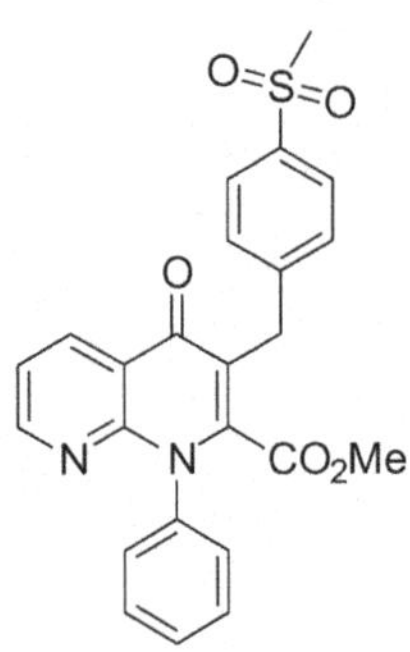

32

IC_{50} (JNK1) = 91 nM

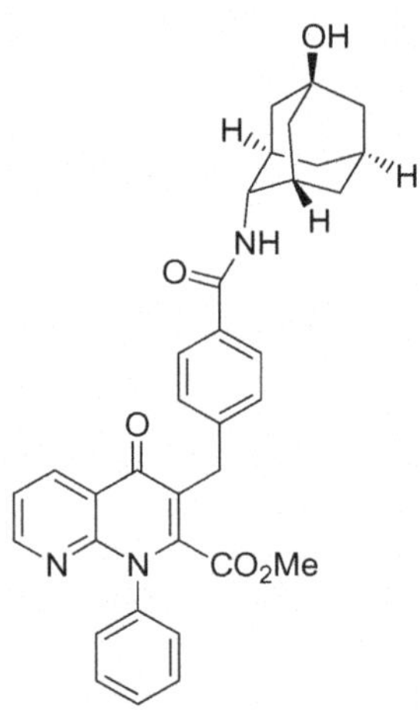

33

IC_{50} (JNK1) = 12 nM

EC_{50} (HK-2) = 4.1 μM

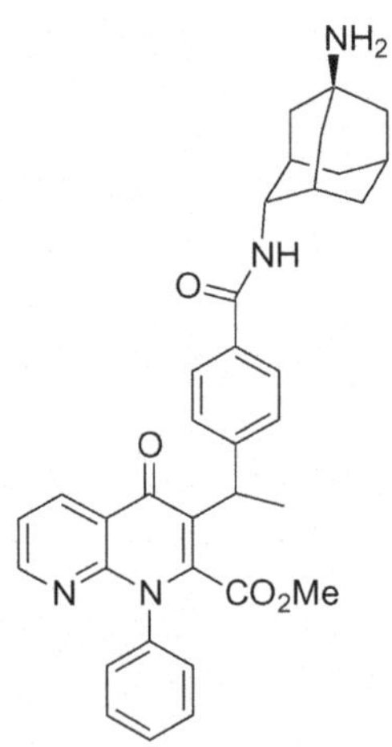

34

IC_{50} (JNK1) = 49 nM

EC_{50} (KH-2) = 21.9 μM

Fig. 3.10 Quinolone and azaquinolone derivatives

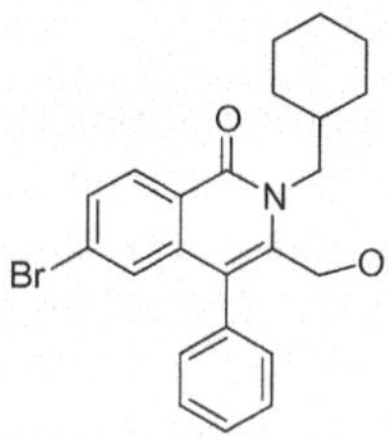

35

IC_{50} (JNK1) = 1.4 μM

IC_{50} (p38α) > 10 μM

IC_{50} (ERK1) > 10 μM

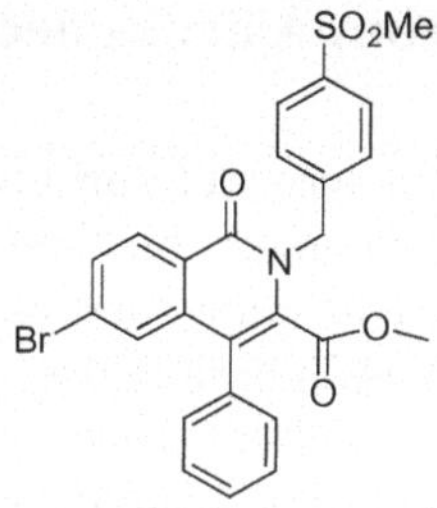

36

IC_{50} (JNK1) = 0.03 μM

IC_{50} (JNK2) = 0.034 μM

IC_{50} (p38) > 10 μM

IC_{50} (ERK1) >10 μM

EC_{50} (H9c2 cell) = 2.6 μM

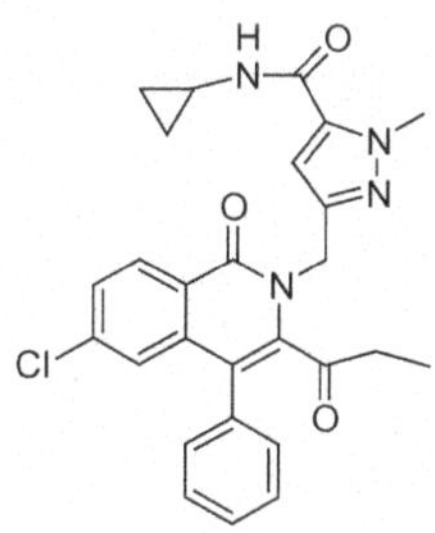

37

IC_{50} (JNK1) = 6.4 nM

IC_{50} (AP-1) = 0.72 μM

Fig. 3.11 4-phenylisoquinolone derivatives

near to the methoxycarbonyl group at the 3-position and the phenyl group at the 4-position. Compound **36** inhibits cardiac hypertrophy at a dose of 10 mg/kg in rat pressure-overload models without affecting blood pressure. Further optimization of this series results in compound **36** with improved potency in both enzymatic and cell-based assays [22]. Compound **37** shows a remarkable anti-hypertrophic activity at doses of more than 1 mg/kg (po) in a pressure-overload model. The crystal structure of an analog of this series in complex with JNK3 reveals that the compound binds in the ATP binding site (PDB code: 2ZDU). The carbonyl group of the isoquinolone ring forms a hydrogen bonding interaction with the NH group of Met149 in the hinge region.

3.9 Inhibitors from AstraZeneca

AstraZeneca reported a series of substituted 2′-anilino-4,4′-bipyridines as JNK3 inhibitors. From HTS campaign hits and structure-based guided resulted in compound **38** (Fig. 3.12). This compound binds in the JNK3 ATP binding active site [23]. Investigators at AstraZeneca also reported a series of 6-anilinoindazole derivatives as selective JNK3 inhibitors such as compounds **39** and **40**. Further optimization results in compound **41**. Compound **41** is a selective JNK3 inhibitor over other isoforms of JNK1 and closely related MAP kinases. The crystal structure of compound **41** in complex with JNK3 (PDB code: 2B1P) reveals that the compound binds at the ATP active site [24]. The indazole moiety of compound **41** forms two hydrogen bonding interactions with amide of Met149 and the backbone carbonyl of Glu147. The carboxyl group forms a hydrogen bonding interaction with the amide of Asn152. AstraZeneca reported a similar type of compound such as compound **42** as a selective JNK3 inhibitor [25].

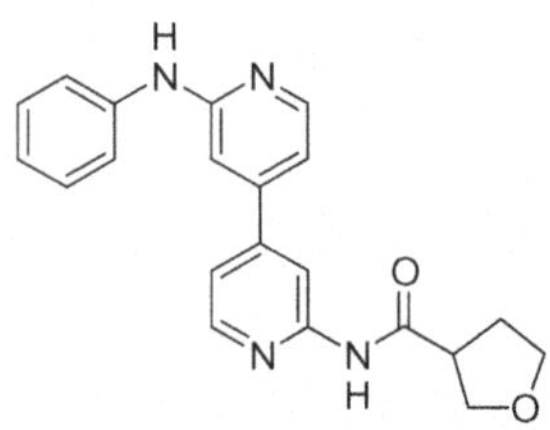

38

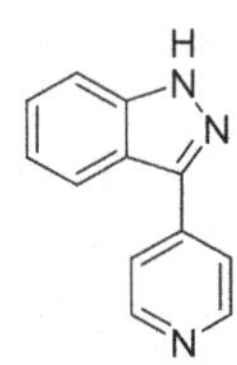

39

pIC_{50} (JNK1) = 5.8

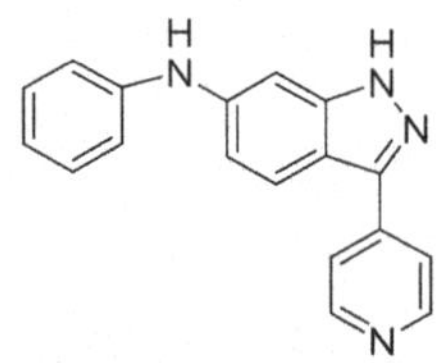

40

IC_{50} (JNK1) > 10 μM

IC_{50} (JNK3) = 4 nM

IC_{50} (p38α) = 30 nM

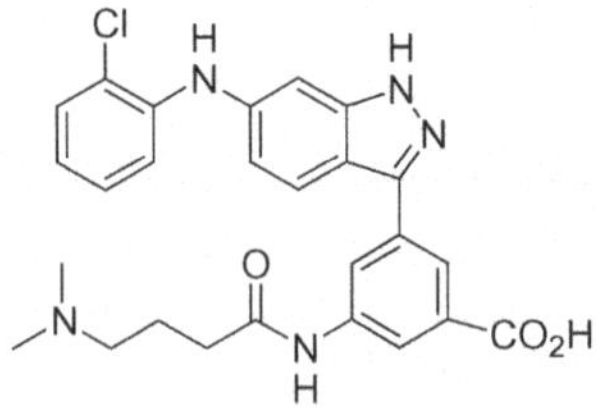

41

IC_{50} (JNK1) = 101 nM

IC_{50} (JNK3) = 3 nM

IC_{50} (p38α) 903 nM

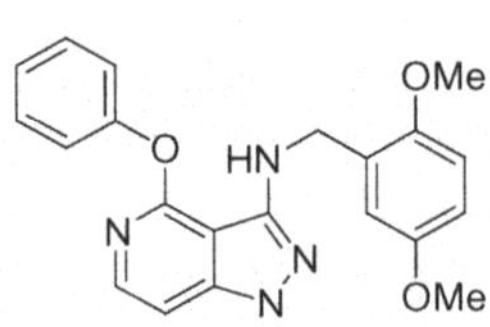

42

Fig. 3.12 2′-Anilino-4,4′-bipyridines and indazoles

3.10 Inhibitors from Vertex

Investigators at Vertex disclosed a series of *N*-benzylated isatin oximes as inhibitors of JNK3. From the HTS campaign, Vertex identified a hit compound **43** with modest JNK3 potency but selective over p38 and ERK MAP kinases [26]. Structure-based guides designed and synthesized several analogs of this series such as **44**, **45**, and **46** (Fig. 3.13). Among them, **45** is the most potent JNK3 inhibitor with K_i 90 nM and selectivity over closely related MAP kinases. The crystal structure of a compound in this series in complex with JNK3 (PDB code: 3G90) reveals that the isatin carbonyl oxygen and the oxime hydroxyl group form two hydrogen bonding interactions with the amide of Met149 and the carbonyl of Glu147 in the ATP

Fig. 3.13 Isatin oxime derivatives

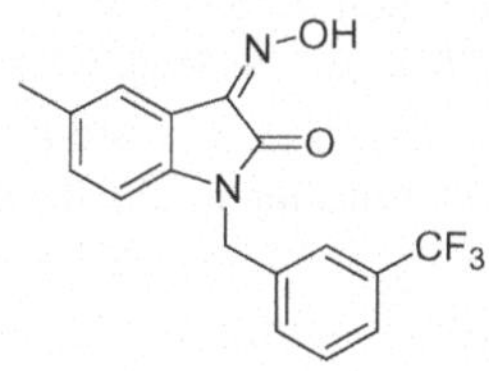

43

K_i (JNK1) = 2.7 μM

K_i (p38α) > 100 μM

K_i (ERK2) > 30 μM

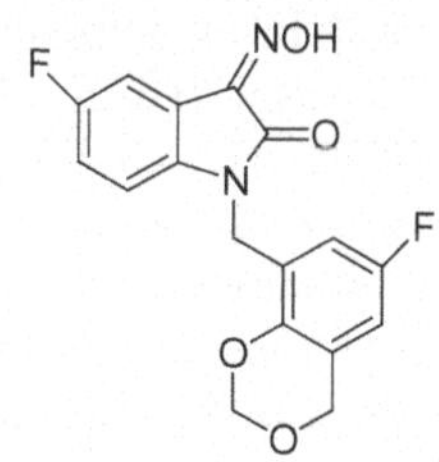

44

K_i (JNK3) = 0.44 μM

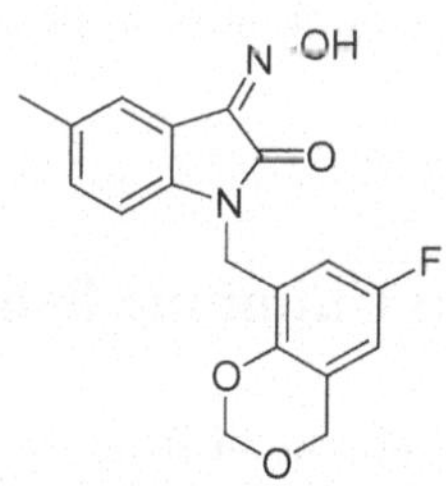

45

K_i (JNK1) = 0.45 μM

K_i (JNK2) = 0.07 μM

K_i (JNK3) = 0.09 μM

K_i (p38α) > 20 μM

K_i (ERK2) > 50 μM

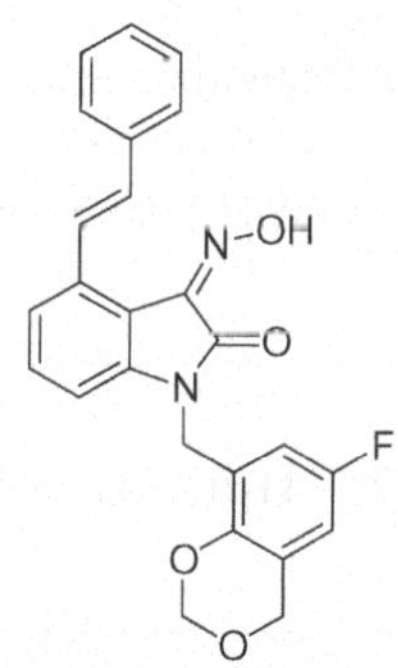

46

K_i (JNK3) = 0.74 μM

K_i (p38α) = 0.23 μM

binding site. The gatekeeper residue Met149 controls the selectivity for this series. Introducing a bulky group such as styrenyl group on isatin core moiety (compound **46**), JNK3 potency and selectivity over p38 reduces.

3.11 Inhibitors from GSK

Investigators at GlaxoSmithKline reported 1-aryl-3,4-dihydroisoquinolines as JNK3 inhibitors. Compound **47** was identified from the HTS campaign with weak potency pIC_{50} (JNK3) = 5.8 [27]. The optimization of this series resulted in potent compounds such as **48–50** (Fig. 3.14) as JNK3 inhibitors. This series of compounds is JNK isoform-selective inhibitor exhibiting 10 folds over JNK2 and 1000 folds over JNK1. Compound also shows selectivity against wider MAPK family (ERK, p-38α) as well as few kinases tested including b-Raf, c-FMS, CKD-2, EGFR, ErB2, GSK3β, IIK, MLK3, PLK1, ROCK1, and SGK1. GSK also provides a co-crystal structure of compound **48** with JNK3 at 2.4 A° resolution (PDB code: 2WAJ). The crystal structure reveals a different binding mode from the docking experiments. The H-bond acceptor interaction with the hinge region occurs by the 7-chloro group of **48** with Met149, not by the dihydroisoquinoline nitrogen and a halogen bond with the carbonyl of Glu147.

GSK identified a series of *N*-(3-cyano-4,5,6,7-tetrahydro-1-benzothien-2-yl) amides inhibitors of JNK2 and JNK3 kinases [28]. The hit compound **51** is optimized resulting in compounds **52** and **53** (Fig. 3.14). Compounds are selective against JNK1, p38α, and EK2. The X-ray crystallography of compound **53** in complex with JNK3 (PDB code: 2O2U) reveals a unique binding mode. The 3-cyano group forms an H bond acceptor with the amide of Met149 in the hinge region of the ATP binding site. Although compounds are very selective within JNK isoform or against other kinases, there are no reports about cellular activities. Nevertheless, the design of JNK isoform-selective inhibitors is possible and will have greater benefit in JNK-related therapeutics.

3.12 Inhibitors from Union Chimique Belge (UCB)

Investigators at UCB identified compound **54** from the HTS of collection of compounds as a selective JNK inhibitor [29]. This compound exhibits potency against CDK2 (Fig. 3.15). SAR studies resulted in early lead compound **55** which shows strong potency against JNK1,2, 3. Compound **55** is selective over p38 and CDK2 and exhibits strong cellular activity. The crystal structure of compound **55** in complex with JNK3 (PDB code: 2P33) reveals that the aminopyrimidine moiety forms two hydrogen bonding interactions with the backbone of Met149 in the hinge region. Indole stays in the ribose binding site. The chloro group sterically interacts with the sulfur of the gatekeeper Met 146. Compound **56** also shows potency in vitro and cell-based assays.

Fig. 3.14 Isoform-selective JNK inhibitors

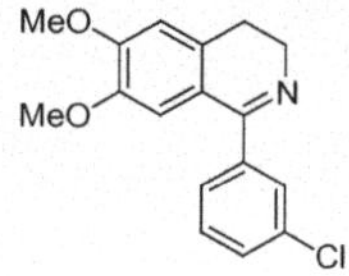

47

pIC_{50} (JNK3) = 5.0

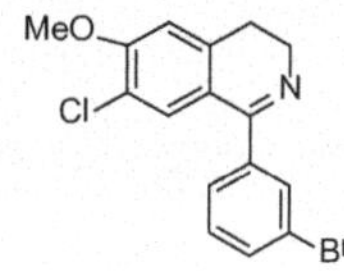

48

pIC_{50} (JNK3) = 6.2

pIC_{50} (p38α) < 4.8

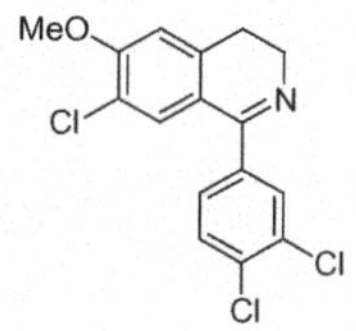

49

pIC_{50} (JNK1) = 4

pIC_{50} (JNK2) = 6.1

pIC_{50} (JNK3) = 7.3

pIC_{50} (p38α) = 5.1

pIC_{50} (ERK2) < 4

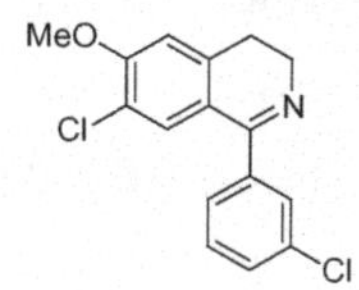

50

pIC_{50} (JNK1) < 4

pIC_{50} (JNK2) = 5.9

pIC_{50} (JNK3) = 6.9

pIC_{50} (p38α) = 5.1

pIC_{50} (ERK2) < 4

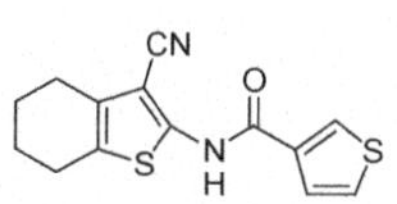

51

pIC_{50} (JNK1) < 5

pIC_{50} (JNK2) = 5.9

pIC_{50} (JNK3) = 6.4

pIC_{50} (p38α) < 4.8

pIC_{50} (ERK2) <4

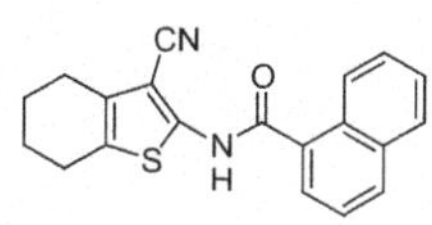

52

pIC_{50} (JNK1) < 5

pIC_{50} (JNK2) = 6.5

pIC_{50} (JNK3) = 6.7

pIC_{50} (p38α) <5

pIC_{50} (ERK2) <4

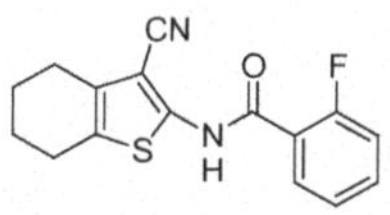

53

pIC_{50} (JNK1) <5

pIC_{50} (p38α) < 4.8

pIC_{50} (JNK3) = 6.4

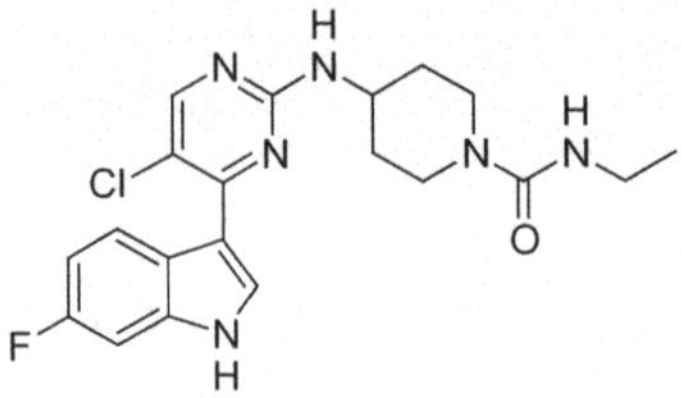

54

IC_{50} (JNK1) = 92 nM

IC_{50} (JNK2) = 67 nM

IC_{50} (JNK3) = 412 nM

IC_{50} (p38) = 4.86 μM

IC_{50} (CDK2) = 0.41 μM

IC_{50} (c-Jun) = 3.70 μM

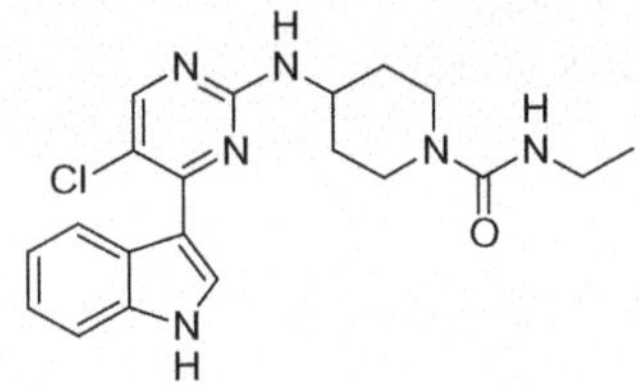

55

IC_{50} (JNK1) = 13 nM

IC_{50} (JNK2) = 25 nM

IC_{50} (JNK3) = 57 nM

IC_{50} (CDK2) = 1.57 μM

IC_{50} (p38) = 5.42 μM

IC_{50} (c-Jun) = 0.704 μM

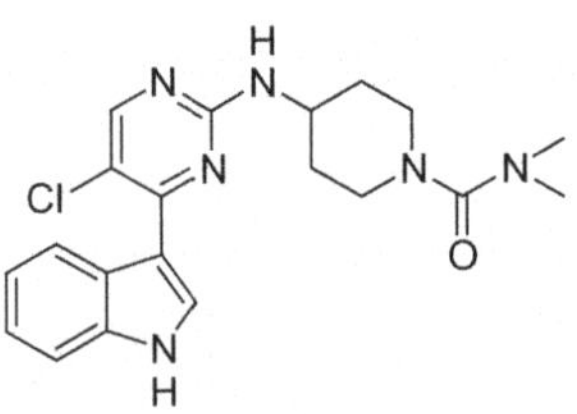

56

IC_{50} (JNK1) = 15 nM

IC_{50} (JNK2) = 37 nM

IC_{50} (JNK3) = not reported

IC_{50} (p38) = not reported

IC_{50} (CDK2) = 4.35 μM

IC_{50} (c-Jun) = 0.741 μM

Fig. 3.15 2-Aminopyrimidines as JNK inhibitors

3.13 Inhibitors from Universität Tübingen

Scientists at Universität Tübingen reported a series of imidazole derivatives as selective JNK3 inhibitors [30]. They started from known potent dual p38α and JNK3 inhibitors such as compound **57a** with the IC_{50} values 17 nM and 24 nM respectively (Fig. 3.16). Modifying pyridinylimidazole core resulted in compound **58** as a potent JNK3 and inactive over p38α. The X-ray crystal structure of

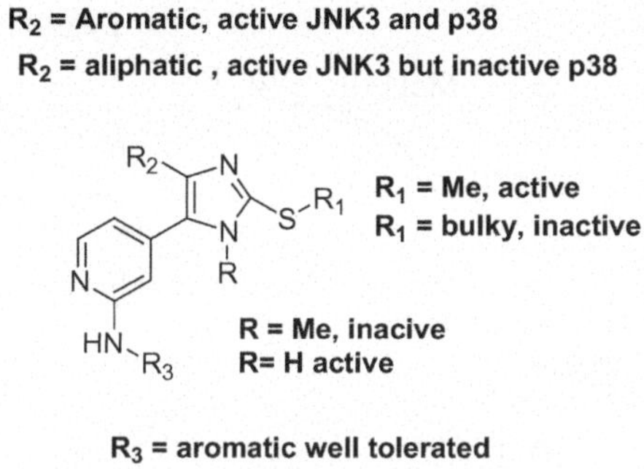

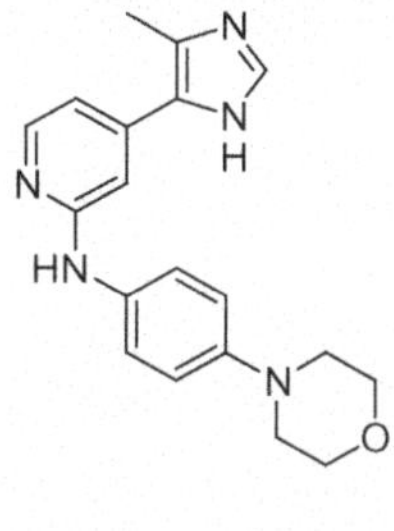

57

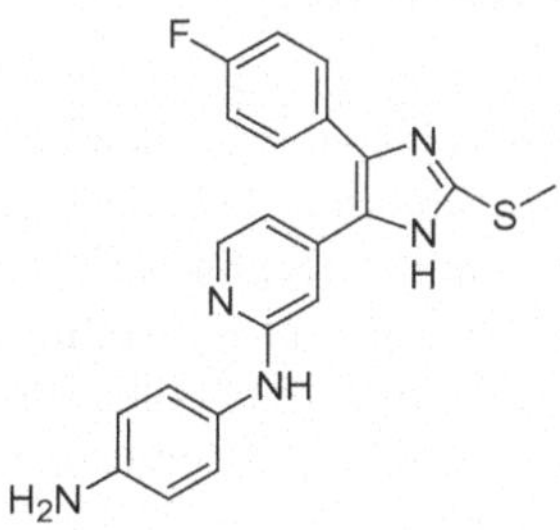

57a

IC_{50} (p38 α) = 17 nM

IC_{50} (JNK3) = 24 nM

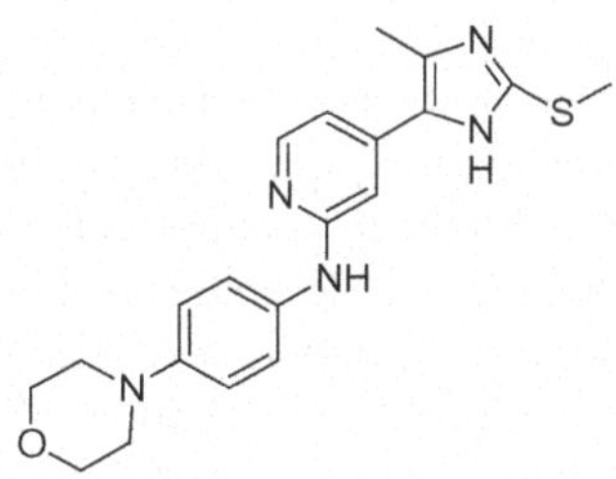

58

IC_{50} (p38α) > 10 μM

IC_{50} (JNK3) = 146 nM

IC_{50} (JNK2) = 468 nM

IC_{50} (JNK1) = 119 nM

Fig. 3.16 Imidazole derivatives

compound **58** in complex with JNK3 (PDB code: 6EKD) reveals that the compound **58** forms two hydrogen bonding interactions with carbonyl and backbone amine groups of Met149 in the ATP binding site (Fig. 3.17). Compound **58** forms a hydrophobic interaction with Ile70, Val78, Met146, Val196, and Leu206. Imidazole moiety forms a water-mediated hydrogen bond network. The N-H of imidazole makes a water-mediated hydrogen bond with Asn152 and that is why N-Me of imidazole is inactive.

These interactions are favorable for JNK3 selectivity as these cannot be formed in the larger binding pocket of p38α MAPK.

Compound **58** inhibits the JNK3 in the low triple-digit nanomolar range and slight selectivity against JNK2. It shows minimal interference with human-ether-à-go-go-related gene (hERG) IC_{50} > 10 μM, potassium channels, and liver cytochromes P450 (CYP450) resulting in it not having potential liabilities.

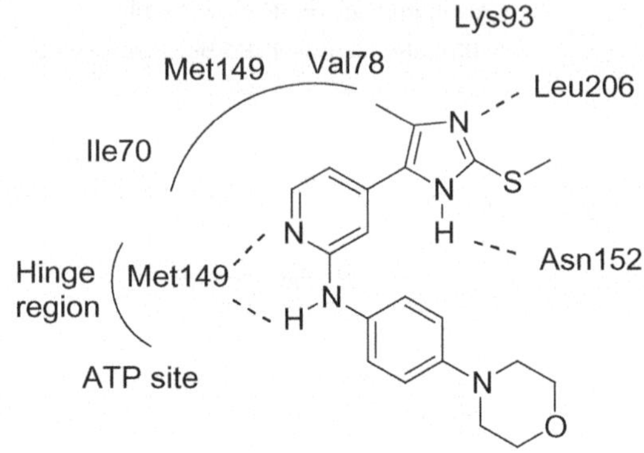

Fig. 3.17 Hydrogen bonding network of JNK3 in complex with compound **58**

3.14 Inhibitors from Scripps Research Institute

Researchers at Scripps-Florida identified from the HTS campaign of their collection of compounds the piperazine amides as potent JNK inhibitors [31] such as compound **59** with the IC_{50} values of 0.49 μM and 1 μM against JNK1 and JNK3 respectively. Extensive SAR studies resulted in compound **60** as potent JNK inhibitor and selectivity over closely related p38 MAPK. The crystal structure of compound **60** in complex with JNK3 (PDB code: 3FV8) reveals that the inhibitor binds at the ATP site and there is a critical hydrogen bonding interaction between the NH of Met149 and the carbonyl of the inhibitor in the hinge region. The bromo furan ring occupied deep within the adenosine binding region. The furan pointed against the gatekeeper Met 146, interacting via van der Waals contacts only. The piperazine forms van der Waals contacts with Ile70 and Val78 and probably with Val196 and Leu206 to a lesser extent. These factors help the stabilization of the p-loop. These features may help in designing more potent and selective JNK3 inhibitors. Some analogs of this series exhibit good cellular activities.

Scripps Floride also reported aminopyrimidines [32] as potent and selective JNK inhibitors such as compound **61** (Fig. 3.18). Compound **61** showed an IC_{50} of 0.8 nM for inhibition of reactive oxygen species (ROS) and had good PK properties in rats with a favorable brain/plasma ratio of 0.75. Compound **61** demonstrates good oral bioavailability (45%), good oral exposure (AUC = 1.4 mM 3 h at 2 mg/kg dose), and reasonable half-life in rat model. The crystal structure of compound **61** in complex with JNK3 (PDB code: 3KVX) reveals that the inhibitor binds at the ATP pocket. Salient feature about this structure is that compound **61** does not form any hydrogen bond interactions with any residue of JNK3 due to lack of potential donor and acceptor groups on the inhibitor. Highly planar nature of **61** nicely occupies into a groove of hydrophobic residues with arrangement of the aromatic rings. The aminopyrimidine portion of compound **61** stays in the adenosine binding region and morpholine on the phenyl ring occupies in the hydrophilic sugar region of the ATP binding site. These compounds are highly selective against a panel of kinases tested.

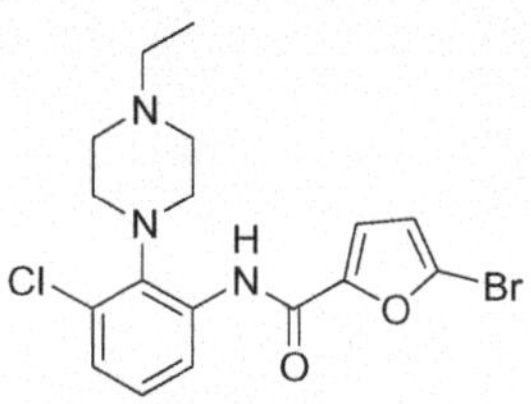

59

IC_{50} (JNK1) = 0.49 µM

IC50 (JNK3) = 1 µM

IC_{50} (p38) > 20 µM

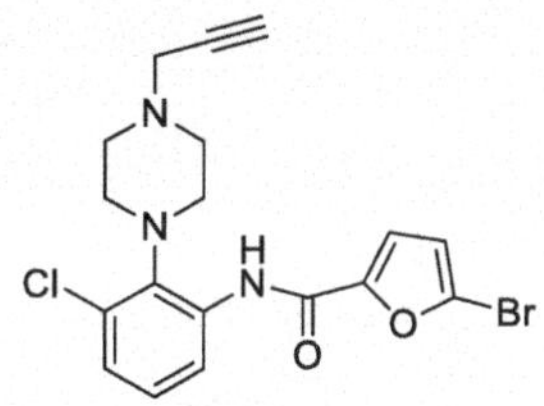

60

IC_{50} (JNK1) = 140 nM

IC_{50} (JNK3) = 160 nM

C_{50} (p38) > 20 µM

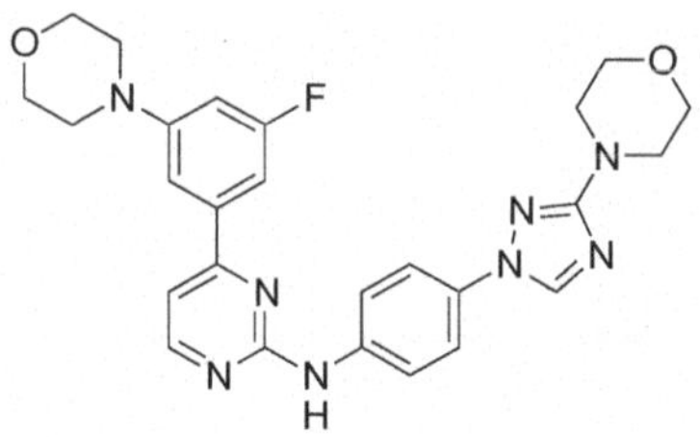

61

IC_{50} (JNK1) = 99 nM

IC_{50} (JNK3) = 148 nM

IC_{50} (p38α) > 20 µM

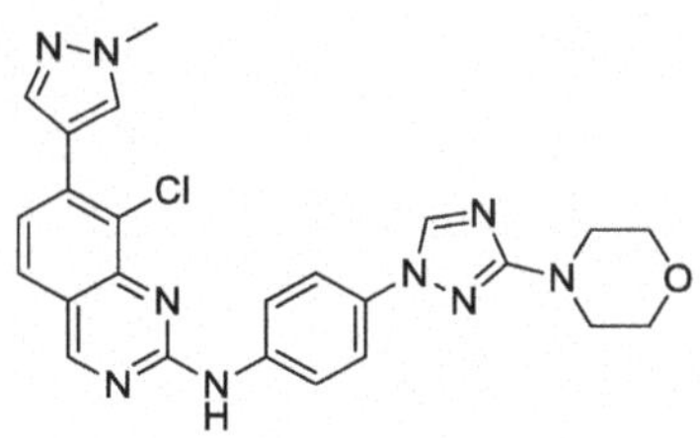

62

IC_{50} (JNK3) = 40 nM

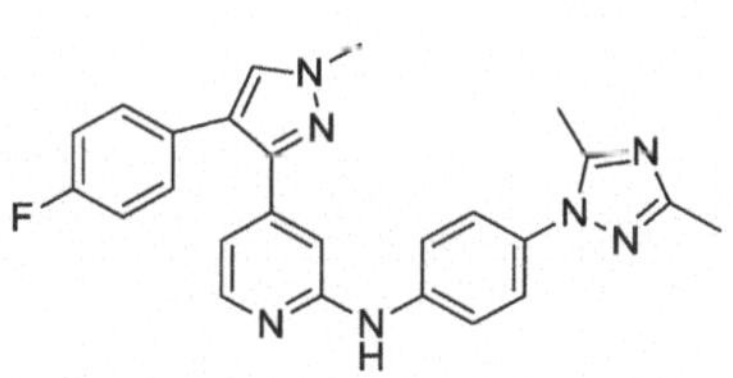

63

IC_{50} (JNK3) = 40 nM

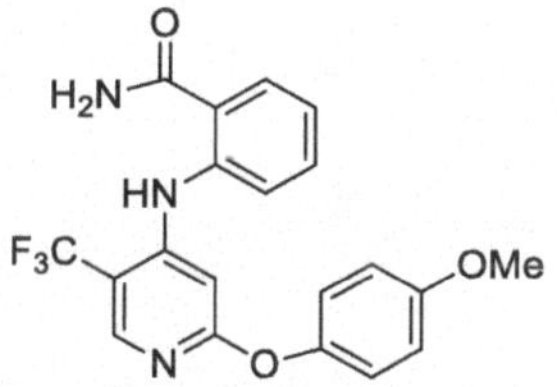

64

IC_{50} (JNK3) = 37 nM

IC_{50} (JNK1) = 27 nM

Fig. 3.18 JNK inhibitors from Scripps Florida

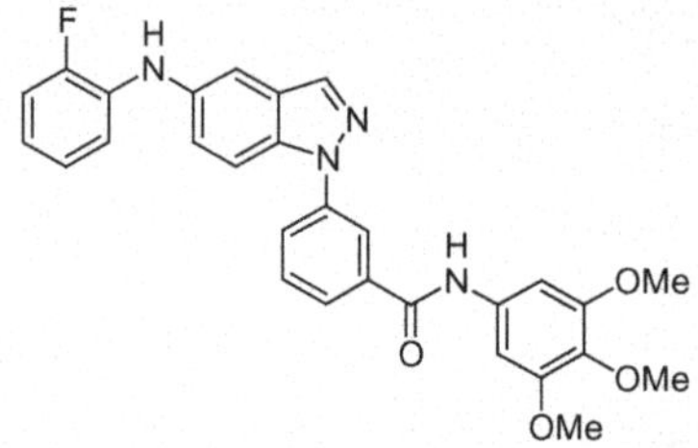

65

IC_{50} (JNK1) = 73 nM

IC_{50} (JNK3) = 12 nM

IC_{50} (p38α) = 3 nM

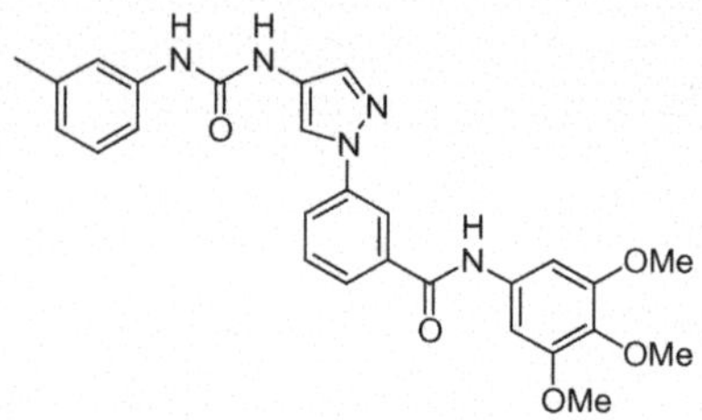

66

IC_{50} (JNK1) = 170 nM

IC_{50} (JNK3) = 7 nM

IC_{50} (p38α) > 20 μM

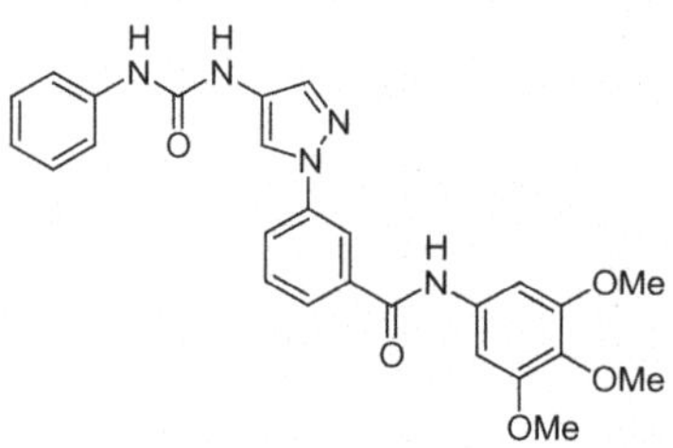

67

IC_{50} (JNK1) = 400 nM

IC_{50} (JNK3) = 25 nM

IC50 (p38α) = 3.7 μM

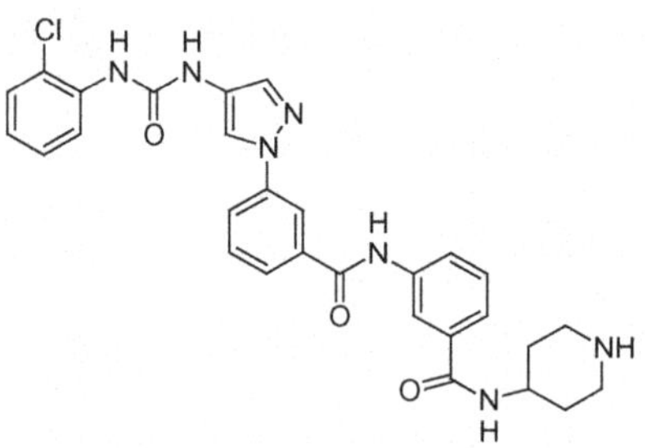

68

IC_{50} (JNK1) >10 μM

IC_{50} (JNK3) = 0.20 μM

IC_{50} (p38α) > 10 μM

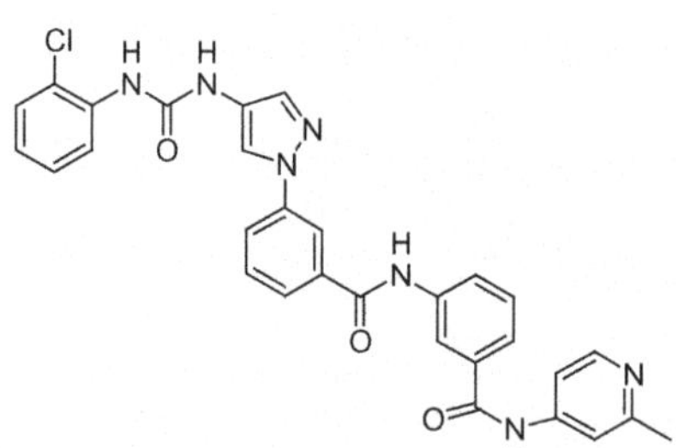

69

IC_{50} (JNK1) = 2.74 μM

IC_{50} (JNK3) = 0.115 μM

IC_{50} (p38α) >10 μM

Fig. 3.18 (continued)

Along the same line, Scripps Florida reported quinazolines [33] as potent and brain-penetrant JNK inhibitors such as compound **62** where pyrimidine was replaced with quinazoline moiety in the central part and morpholine of the left side also replaced with pyrazole but right side morpholine part remained as the same of the previous compound **61**. Nevertheless, compound **62** showed good PK properties with half-life of 3 h and bioavailability was 19% and low clearance. Compound **62** had >500 folds selectivity over p-38α.

Investigators at Scripps-Florida discovered 4-pyraz-3-yl-pyridines as novel JNK inhibitors [34] where the central core was replaced from pyrimidine to pyridine. Proper optimization of this series afforded compound **63** and its analogs (Fig. 3.18). Several compounds from this scaffold showed good PK parameters such as low clearance, long half-line, and favorable bioavailability (51–75%) in mice and rats.

Investigators at Scripps-Florida disclosed 2-phenoxypyridines [35] as JNK inhibitors such as compound **64**. This series of compounds showed good PK properties such as half-life was 1 h, low clearance, and favorable bioavailability (79%) and brain penetration. Some compounds also were tested in the 6-hydroxydopamine model of Parkinson's disease [36–38] and data suggested that these orally bioavailable JNK inhibitors may serve as a tool for further development of neurodegenerative diseases based on JNK biology. These compounds from Scripps have a good chance to move on to further advanced-level studies (phase 1) in Parkinson's disease. As other small molecules reported in the literature from other institutes/industries did not target CNS therapeutic indication in animal models at that time, it is difficult to compare this class of compounds with others. Nevertheless, these disclosed an opportunity for further studies in the field of neurodegenerative disorders particularly, Parkinson's disease based on the JNK pathway with a drug-like small molecule.

Kamenecka et al. [38] reported aminopyrazoles as JNK3 inhibitors. Starting from indazole derivative (compound **65**), they modified the structure to aminopyrazole derivatives. Some analogs of this series such as compounds **66–68** exhibit very promising JNK3 inhibition with >2800 times selectivity over p-38 MAPK. The crystal structure of one compound from this series with JNK3 and p-38 showed different binding modes. Highly planar of the pyrazole and *N*-phenyl group which better occupy the smaller active site of JNK3 than the larger active site of p-38; the gatekeeper residue in JNK3 is different than p-38, these are possible explanations for selectivity of JNK3 over p-38 [38].

Zhen et al. reported pyridopyrimidinone derivatives [39] as potent JNK inhibitors. Through structure-activity relationship studies, they found several lead compounds with low nanomolar IC_{50} as shown in Table 3.1. The crystal structure of one of the key compounds with human JNK3 at 2.0 A° showed Type-1 kinase binding mode with main hydrogen bonding interactions at Met149 and Gln155 (PDB code: 4Y5H) at the ATP binding site.

Selected key compounds had very favorable PK profiles such as good oral bioavailability (43–100%). These compounds were also studied with 6-OHDA-induced mitochondrial membrane depolarization in SHSY5Y cells, which represents the Parkinson's disease model.

Table 3.1 SAR studies for the *N*-substitutions

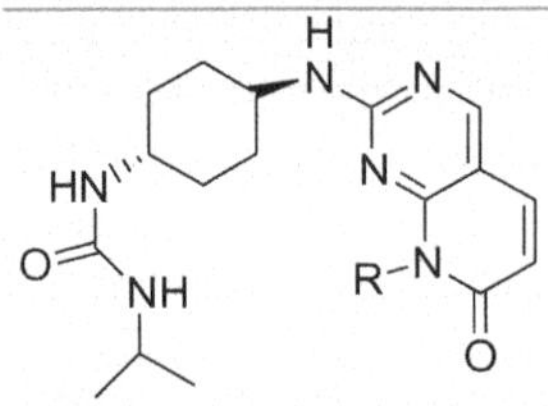

R	JNK3 IC_{50} (nM)	JNK2 IC_{50} (nM)	JNK1 IC_{50} (nM)
H	2166	3450	1988
Me	238	463	303
Isopropyl	15	66	21
Cyclopentyl	11	12	7
Tetrahydropyran	204	434	438

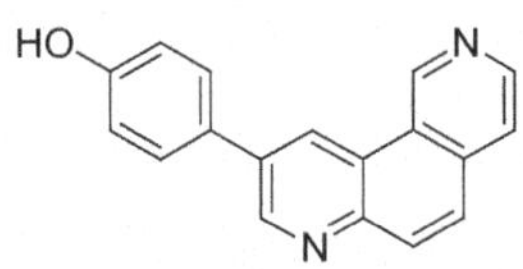

70

IC_{50} (JNK3) = 0.59 μM

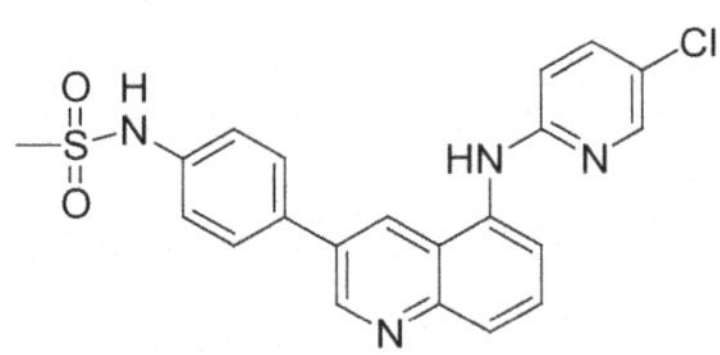

71

IC_{50} (JNK3) = 120 nM

Fig. 3.19 3,5-Disubstituted quinolines

Investigators at Scripps Florida reported a new series of 3,5-disubstituted quinolines as JNK3 inhibitors [40]. They identified phenanthroline derivatives such as compound **70** from a HTS campaign with modest potency against JNK3. Extensive SAR studies resulted in compound **71** with an IC_{50} value of 120 nM for JNK3 (Fig. 3.19). The crystal structure of compound **71** in complex with JNK3 (PDB code: 2R9S) reveals an unusual binding mode. Two nitrogen atoms of the aminopyridinyl moiety form two hydrogen bonding interactions with Met149 at the hinge region of the ATP binding site.

3.15 Inhibitors from Hanyang University

Jang et al [40] reported 1-pyrimidinyl-2-aryl-4, 6-dihydropyrrolo [3,4-*d*] imidazole-5(1*H*)-carboxamide derivatives as selective inhibitors of c-Jun-N-terminal Kinase 3 (JNK3), targeting for the treatment of Alzheimer's diseases (AD). The key compound (**72**) showed an IC_{50} value 2.69 nM against JNK3, it was isoform selective over JNK1, JNK2, and other kinases including RIPK3 and GSK3β

Fig. 3.20 1-Pyrimidinyl-2-aryl-4, 6-dihydropyrrolo [3,4-*d*] imidazole-5(1*H*)-carboxamide

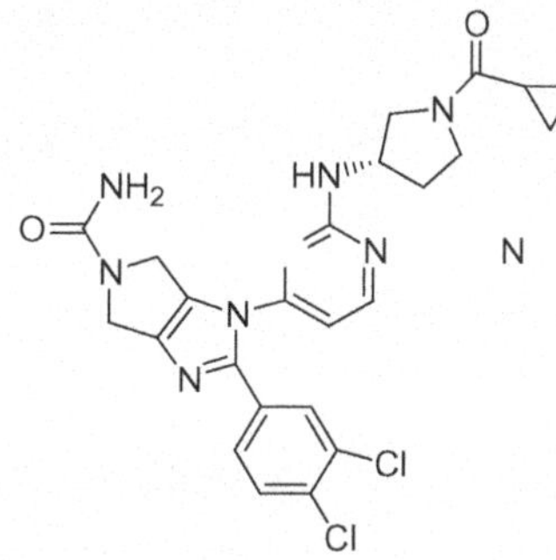

72

IC_{50} (JNK2) = 225 nM

IC_{50} (JNK3) = 2.69 nM

IC_{50} (GSK3β) = 5530 nM

IC_{50} (RIPK3) = 269 nM

(Fig. 3.20). They performed a docking study to explain the binding mode on this compound with JNK3 crystal structure (PDB code: 3OY1). It shows that amino pyridine moiety forms two hydrogen interactions with the Met149 of JNK3 in the hinge region. Furthermore, the oxygen of cyclopropyl carboxamide moiety could form hydrogen bond interaction with Gln155, and the oxygen of 2,5-dihydropyrrole carboxamide forms two hydrogen bonds with Asn152 and Ser72 respectively.

3.16 Inhibitors from Celgene (Now BMS)

Investigators at Celgene reported [41] aminopurine-based JNK inhibitors for the prevention of ischemia-reperfusion injury. Initial hit compound **73** (Fig. 3.21) was optimized by introducing fluoro group to give highly potent JNK inhibitors such as **74** (**CC-359**) with improved solubility at pH 5 buffer. The X-ray crystal structure of **73** in complex with JNK3 reveals that the compound binds at the ATP binding site (PDB code: 3TTJ). Selected residues of JNK3 ATP site are participated with H-bonding interactions such as Met149 in the hinge region, Met146, Lys93, Leu144 in the lipophilic back pocket, Ser 193, Asn152 in the ribose-binding pocket, and Gln155 in the solvent-exposed region. These important insights helped to discover **74**. Compound 74 has 10 folds selective against p-38α and high selectivity maintained against other kinases tested. Compound **74** inhibits c-Jun phosphorylation in Jurkat cells with an IC_{50} value of 0.40 μM. Compound showed excellent in vivo efficacy in a mouse concanavalinA (ConA)-induced liver injury model, compound **74** administered 1 h prior to ConA at 30 and 60 mg/Kg ip, inhibited the elevation of ALT enzymes by 41% and 98% respectively.

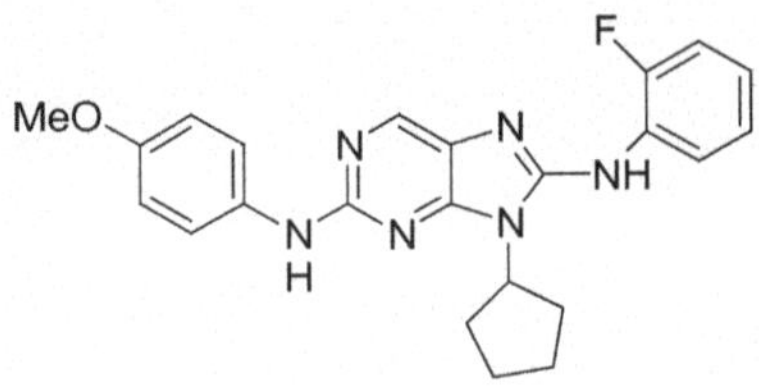

73

IC_{50} (JNK1) = 1100 nM

IC_{50} (JNK2) = 120 nM

IC_{50} (JNK3) = 210 nM

IC_{50} (p38α) >30 μM

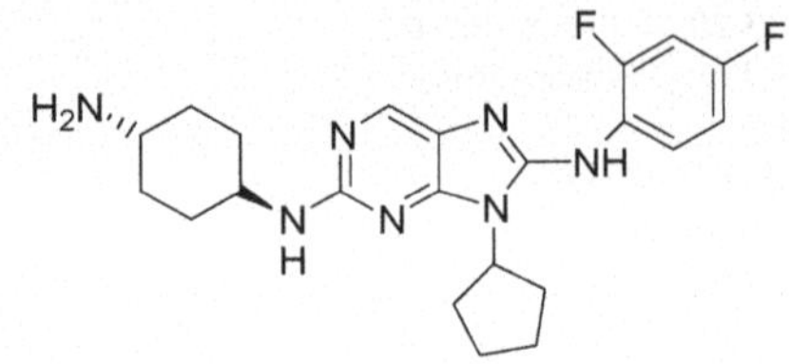

74 (CC-359)

IC_{50} (JNK1) = 270 nM

IC_{50} (JNK3) = 25 nM

IC_{50} (JNK3) = 24 nM

IC_{50} (p38α) = 220 nM

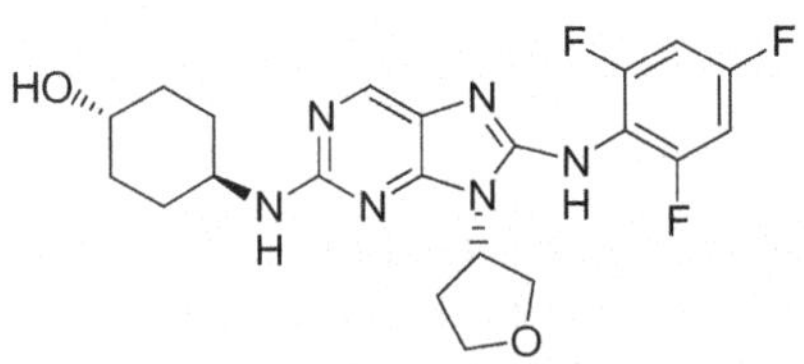

75 (CC-930)

IC_{50} (JNK1) = 61 nM

IC_{50} (JNK2) = 7 nM

IC_{50} (JNK3) = 6 nM

IC_{50} (p38α) = 3.4 μM

Fig. 3.21 JNK inhibitor from Celgene

Compound **74** also demonstrated in vivo efficacy in a rat ischemia-reperfusion model at doses 3 and 10 mg/Kg iv bolus in citrate formulation, single dose, 15 min prior to ischemia, a 60% at 3 mg/Kg and 80% at 10 mg/Kg survival rate as compared to 50% survival for vehicle treated were observed. The PK profile of **74** was determined in rat, dog, and monkey and showed favorable clearance in monkeys (5 mL/min/ Kg at 1 mg/Kg dose) compared to others. Based on these favorable properties, compound **74** entered into further clinical studies.

Along the same line, Celgene reported [42] the discovery of compound **75** (**CC-930**) for the first time in public literature, starting from the same hit compound **73**. It is also noteworthy that *R* isomer of **75** is less active than *S* isomer. This was rationalized by X-ray crystal structure of **75** in complex with JNK3 (PDB code: 3TTI). The *S*-isomer makes favorable electrostatic interaction between side chain NH_2 of Asn152 and oxygen of tetrahydrofuran (THF) whereas in *R* isomer oxygen of THF stays away from Asn152, cannot make favorable interaction. PK parameters for **75** were good with 30% bioavailability in rats and 77% in monkeys and favorable clearance in monkeys and dogs. Based on these results, **75** (**CC-930**) entered into a phase II clinical trial in patients with idiopathic pulmonary fibrosis.

3.17 Inhibitors from CHDI Foundation

In 2015, the investigators from CHDI foundation and others reported 4-(1*H*-pyrazol-4-yl)pyrimidine as a class of pan-JNK inhibitor [43]. Compound **76** (Fig. 3.22) was optimized based on potency and ADME-PK profile resulting in compound **77**. Traditional medicinal chemistry approaches used for the lead optimization process with the help of X-ray crystallography and computational modeling, they discovered potent JNK inhibitor **77** which showed sub-micromolar activity in cell-based assay. Compound **77** showed good brain penetrant in oral dose but it was high plasma clearance. Nevertheless, this compound was studied on Huntington's disease which is another target based on JNK inhibition.

3.18 Inhibitors from Peking University

Alzheimer's disease (AD) is a neurodegenerative disease that begins slowly and progressively worsens. AD is approximately 60–70% cases of dementia. The symptoms include problems with language, memory loss, mood swings, loss of motivation, self-neglect, and behavioral issues. Dou et al [44] reported 3-substituted indolin-2-one derivatives as the isoform-selective JNK3 inhibitors from multistage screening. The key compound (**78**) showed an IC_{50} value of 40 nM against JNK3, which was 2500 times isoform selective over JNK1, JNK2 and it was also selective over other kinases tested (Fig. 3.23). Compound **78** showed neuroprotective activity in vitro to alleviate the spatial memory impairment. It inhibits the phosphorylation of JNKs, Aβ precursor protein, and Tau protein. Compound **78** is docked into the ATP binding site of JNK3 (PDB code: 4WHZ). The indolin-2-one group forms two hydrogen bonding interactions with NH and oxygen atom of the carbonyl to Met149. Thiazol-4-one forms a hydrophobic interaction with Val7, Ala91, Met149, Val196,

Fig. 3.22 JNK inhibitor from CHDI foundation

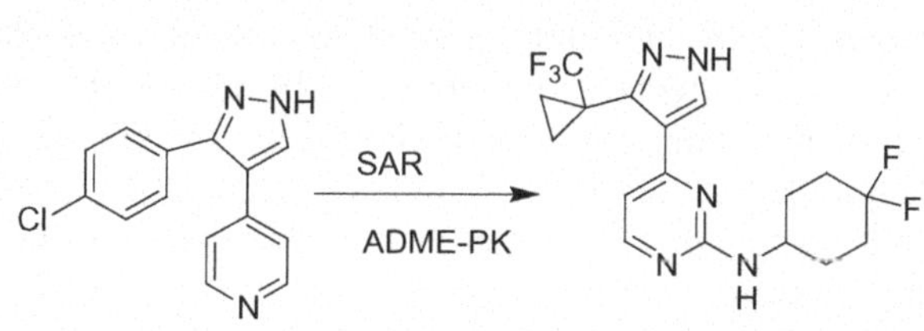

76 **77**

Fig. 3.23 Indolinone derivative as a selective JNK3 inhibitor

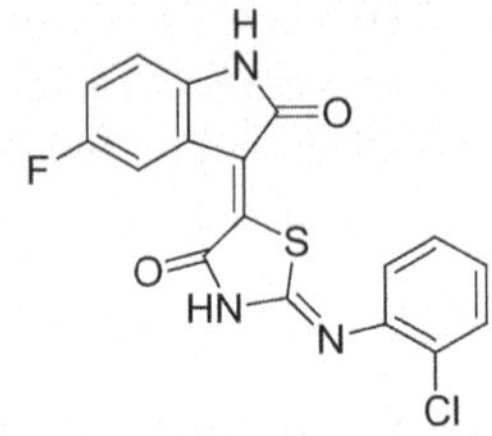

78

IC_{50} (JNK1) > 100 μM

IC_{50} (JNK2) > 100 μM

IC_{50} (JNK3) = 40 nM

and Leu206. The 2-chlorophenyl group makes an interaction with Ala91, Ile92, Lys93, Ile124, Leu126, Leu144, Val145, Met146, and Leu206.

Compound **78** showed neuroprotective activity with spatial memory impairment in in vivo model.

3.19 Inhibitors from Montana State University

Schepetkin et al recently reported a new tryptanthrin series of compounds as selective JNK3 inhibitors [45]. Compounds **79** and **80** are selective JNK3 inhibitors (Fig. 3.24). Compound **81** inhibits all isoforms of JNKs (JNK1, JNK2, JNK3) as shown in Table 3.2. These compounds inhibited lipopolysaccharide (LPS)-induced nuclear factor-κB/activating protein 1 (NF-κB/AP-1) transcription activity in THP-1Blue cells and interleukin-6 (IL-6) production in MonoMac-6 monocytic cells, respectively.

3.20 Inhibitors from Nanjing Tech University

Inhibition of JNK activation in mitochondria may provide an alternative pathway for the treatment of Parkinson's disease (PD). There is no specific mitochondrial-targeted JNK inhibitor reported. Investigators at Nanjing Tech University have designed a mitochondrial-targeted JNK inhibitor by linking a mitochondrial-specific cell-penetrating peptide with SP600125 by a linker as shown in Fig. 3.25. This compound particularly inhibits mitochondrial JNK phosphorylation instead of nuclear JNK signaling. This compound exhibits rescued PD phenotypes both in vitro and in vivo [46].

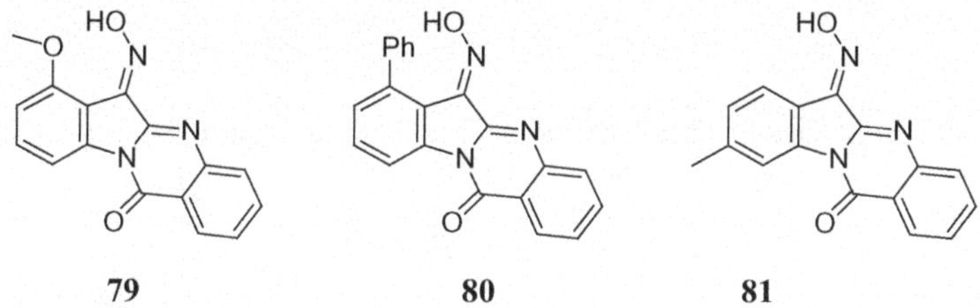

Fig. 3.24 Tryptanthrin derivatives as JNK3 inhibitors

Table 3.2 Key compounds' K_d values and cell-based assay data

Compound	JNK1 K_d (μM)	JNK2 K_d (μM)	JNK3 K_d (μM)	AP production in THP–1Blue IC_{50} (μM)	IL–6 production in MonoMac–6 IC_{50} (μM)
79	>30	>30	0.34	0.8	0.7
80	>30	>30	0.33	1.6	0.7
81	0.67	0.57	0.23	0.9	0.3

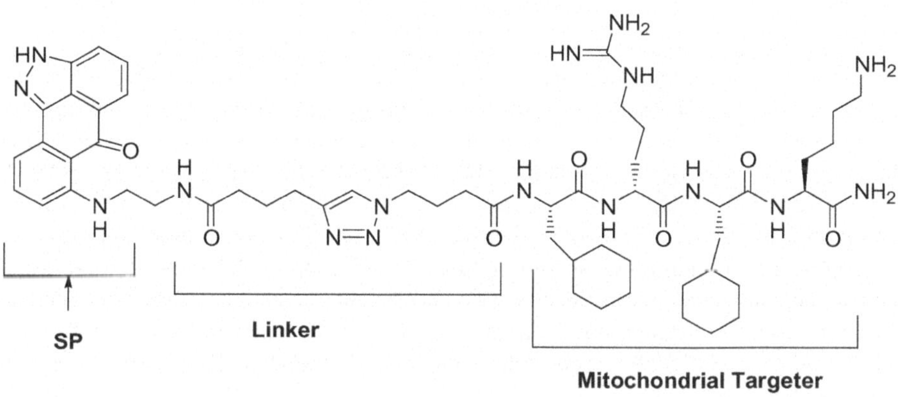

Fig. 3.25 SP600125 attached with mitochondrial-targeted peptide with a linker

3.21 Inhibitors from Sichuan University

Neurodegenerative disorders are CNS diseases characterized by synaptic loss, neuronal atrophy, and death as common pathological hallmarks. Parkinson's disease (PD) is one of the most common neurological disorders, affecting over 6.3 million people globally, and 1.5 million people in North America. Parkinson's disease (PD) belongs to a group of conditions called motor system disorders, which are the result of the loss of dopamine-producing neurons in the substantia nigra. The abnormal aggregation of misfolded α-synuclein in the Lewy body is the primary pathological condition of Parkinson's disease. JNK3 is mainly expressed in the central nervous system and is involved in the process of neurodegeneration [47–55].

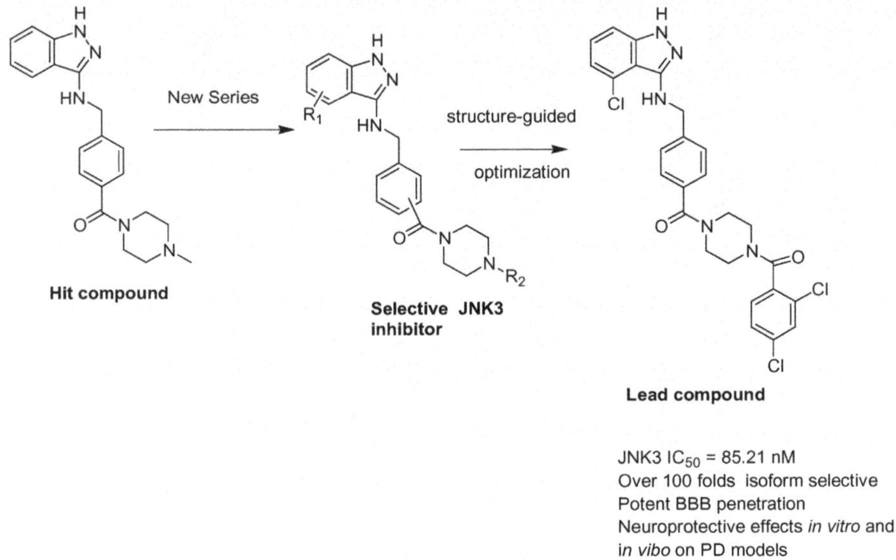

Fig. 3.26 Hit to lead optimization of indazole series

Most recently, Shuai et al reported indazole analogs as selective JNK3 inhibitors for the treatment of Parkinson's disease [47]. From docking-based high-throughput virtual screening provided a hit compound with an indazole moiety (Fig. 3.26). The hit compound is an isoform-selective JNK3 inhibitor with an IC_{50} value of 0.89 μM (JNK1 IC_{50} > 10 μM). From structure-activity relationship study results in lead compound **82**, which is isoform-selective JNK3 inhibitor. Shuai et al. synthesized several advanced analogs as shown in Table 3.3. Compound **82** showed good brain penetrating and lower cardiotoxicity. This compound also showed selectivity against 376 kinases tested. Compound **82** exerted significant neuroprotective effects on MPP^+-induced SH-SY5Y cell lines and MPTP-induced Parkinson's disease mice models.

3.22 Development of New Clinical Candidate CC-90001 as a JNK Inhibitor

BMS previously reported the discovery and development of Tanzisertib (CC-930) with strong JNK2 potency for the treatment of idiopathic pulmonary fibrosis (IPF). However, this clinical candidate has been withdrawn due to severe toxicity compared with benefit. Most recently, BMS started another campaign on IPF which focused on more JNK1 potency. The ongoing evidence suggested that JNK1 is the major crucial isoform for fibrotic disease in both hepatic and lung fibrosis in animal models. From the library of compounds' high-throughput screening, a strong hit

Table 3.3 Inhibitory activities of indazole series

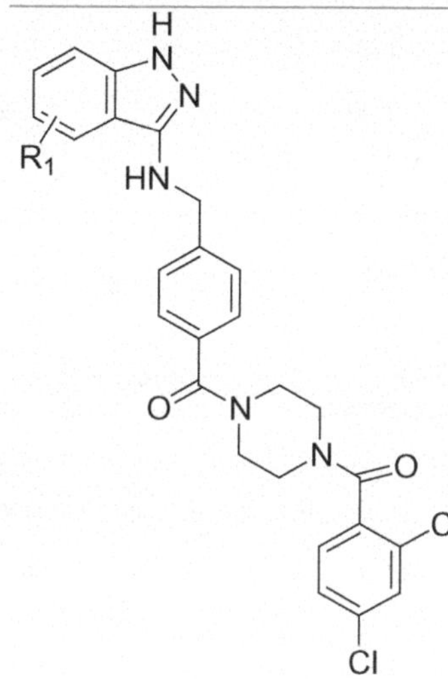

R_1	JNK1 IC_{50} (nM)	JNK3 IC_{50} (nM)
1-me	>10000	>10000
7-me	>10000	>10000
5-NO_2-7-cl	>10000	>10000
6,7-di-F	8237	136
5-Br	9172	288
5-cl	8534	202
6-Br	8578	348
6-F	8873	141
6-cl	7247	238
4-F	7798	188
4-I	9357	129
4-Cl (**82**)	>10000	85

compound was identified as shown in Fig. 3.27. This hit compound exhibited almost similar biochemical potency against JNK1 and JNK2, and an approximately equal potency in cell-based assay with CC-930, as determined by phosphorylation of the direct JNK substrate, c-Jun [56]. The hit compound displays very poor kinase selectivity over tested 70 kinases.

Investigators from BMS studied an overlay of the structural binding mode of the hit compound with tanzisertib in a JNK3 co-crystal structure (PDB code: 3TTI). They found that structure-activity relationship could be applied on the hit compound by replacing C2 and C4 aryl groups to aliphatic moieties. Introducing C2-position with the trans-4-hydroxycyclohexylamino group resulted in compound **83** with similar potency but poor kinase selectivity. The methoxy derivative of **83** with C4 aliphatic group furnished compound **84** showed a better kinase selectivity (Fig. 3.28). A number of aliphatic groups on C4 position with aiming in hydrogen bonding interaction in the JNK1 ribose pocket was explored, resulting in compound

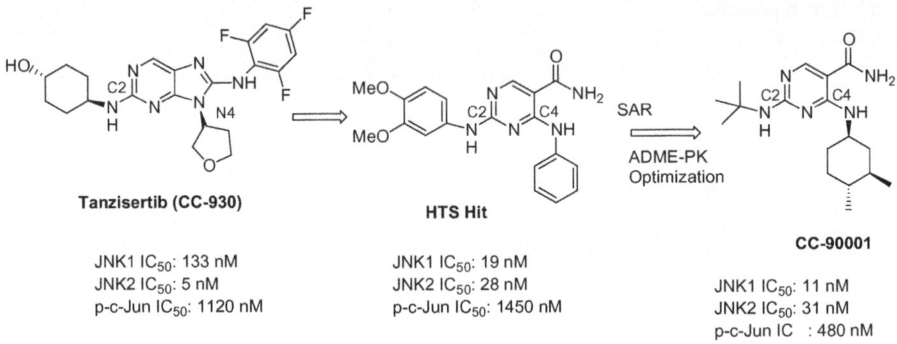

Fig. 3.27 Discovery of new clinical candidate CC-90001

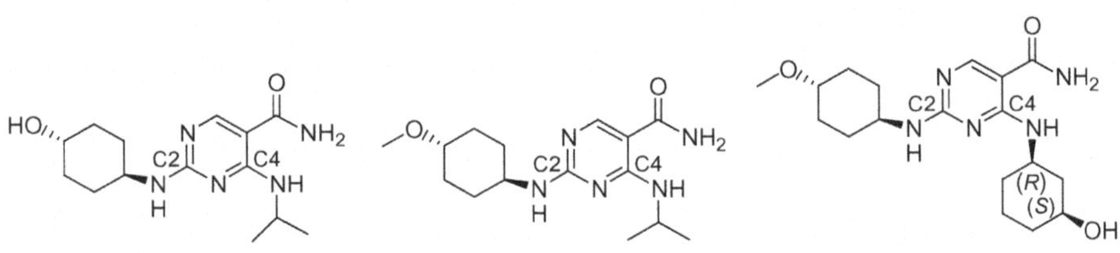

83 **84** **85**

Fig. 3.28 Discovery of selective inhibitors

85. Compound **85** with proper stereochemistry (*R*, *S*) exhibited good biochemical and cellular potency with improved kinase selectivity and favorable rat clearance. Further optimization both on C2 and C4 positions resulted in compounds **86–90** (Fig. 3.29). Reducing overall polar surface area (PSA) resulted in clinical candidate **90** (**CC-90001**) with improved oral exposure in rats (Table 3.4). **CC-90001** exhibited an excellent kinase selectivity, acceptable safety profile, and efficacy in a disease-relevant model of fibrosis [56–61].

Metabolism of CC-90001 occurs through glucuronidation mediated by uridine diphosphate glucuronosyltransferase (UGT) 1A9, UGT1A1, and UGT1A4. CC-90001 also undergoes oxidative metabolism catalyzed by CYP1A1, CYP3A1, CYP2C9, CYP2B6, and CYP2D6 [61]. From the Phase 1b trial, CC-90001 oral dose of 200 mg daily for 12 weeks in patients with pulmonary fibrosis was generally safe [60].

3.23 ATP-Competitive ERK1/2 Inhibitors

Investigators at Vertex Pharmaceutical identified a hit compound with a Ki value of 2.3 μM [62]. The X-ray crystal structure of the hit compound in complex with ERK2 (PDB code: 2OJG) reveals that the pyrazole moiety forms two hydrogen

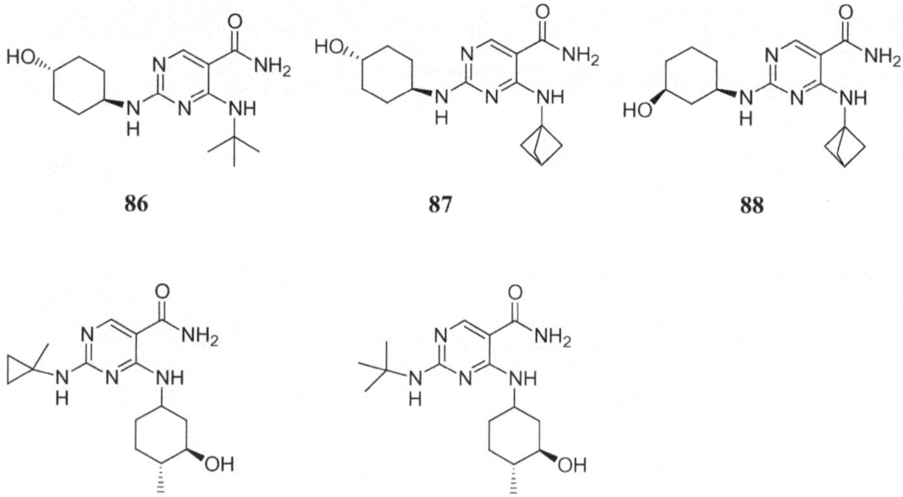

86 **87** **88**

89 **90** (CC-90001)

Fig. 3.29 Selection of clinical candidate

Table 3.4 Advanced analogs with improved PK properties

Compound number	JNK1 IC_{50} (μM)	JNK2 IC_{50} (μM)	p-c-Jun IC_{50} (μM)	Rat CI (clearance) mL/min/kg	Rat PO AUC_{inf} (μM.h) (Dose 10 mg/Kg)	PSA/ALogP
83	0.050	0.079	1.77	26.5	45.8	NR
84	0.035	0.113	2.39	NR	NR	NR
85	0.003	0.005	0.40	42.9	2.3	122/1.26
86	0.013	0.018	0.35	11	69.8	113/1.28
87	0.011	0.013	0.17	7	41.5	113/0.99
88	0.010	0.12	0.21	30	35.2	113/0.87
89	0.011	0.021	0.55	22	42.6	113/1.24
90 **CC-90001**	0.011	0.031	0.48	11	35.5	113/1.48

NR Not reported

bonds in the hinge region. The amide carbonyl forms a hydrogen bond with the Lys52 and the pyrazole-linked phenyl makes a hydrophobic interaction with Val37. The NH on the pyrrole forms a hydrogen bond with the gatekeeper residue Gln103. The optimization of the hit compound results in the first-generation lead compound with a K_i value of 2 nM as shown in Fig. 3.30. This lead compound shows the selectivity over other kinases tested. The crystal structure of the first-generation compound with EKK2 (PDB code: 2OJJ) reveals that the hydroxyl group forms two hydrogen bonding interactions with the carboxylate oxygen of Asp165 and the

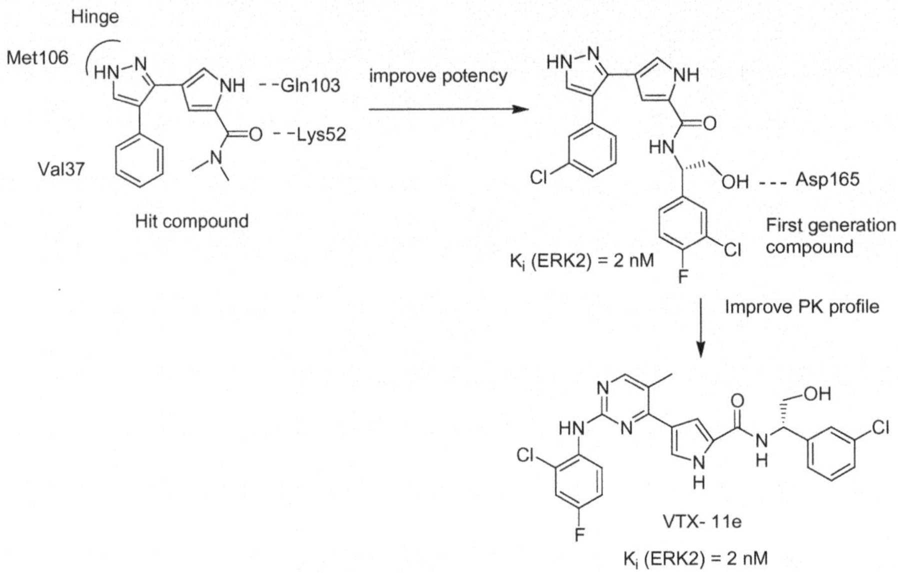

Fig. 3.30 Pyrazole and aminopyrimidine derivates as ATP-competitive ERK2 inhibitors

carboxamide oxygen of Asn152. Four hydrogen bonds previously observed in the hit compound also exist. Six hydrogen bonding interactions are responsible for the potency and ERK selectivity (JNK3 $K_i = 4$ μM; CDK2 $K_i > 4$ μM; p38α $K_i > 4$ μM; MEK1 $K_i > 4$ μM; AKT3 $K_i = 1.8$ μM; AURA $K_i > 4$ μM; FLT3 $K_i = 0.75$ μM). This compound is active in the Colo205 cell proliferation assay ($IC_{50} = 0.54$ μM). Due to insufficient cellular activity, they replaced pyrazole to aminopyrimidine-based alternative hinge binding moiety, resulting in compound **VTX-11e.** The lead compound (**VTX-11e**) K_i value is less than 2 nM. It is over 200-time selective against GSK3, CDK2, and AuroraA and 500-fold selective against other kinases tested [62]. It shows a significant anti-tumor activity ($IC_{50} = 48$ nM) in the HT29 cell proliferation assay. It is orally bioavailable in both rats (F = 65%; $T_{1/2} = 3$ h) and mice (F = 67%; $T_{1/2} = 4.4$ h).

Investigators at Fujisawa Pharmaceutical identified compound **FR180204** (Fig. 3.31) as selective ERK1 and ERK2 inhibitors with K_i values of 0.31 μM and 0.14 μM respectively [63]. This compound inhibits TGFβ-induced luciferase expression in mink lung epithelial Mv1Lu cells. The X-ray crystal structure of this compound in complex with human ERK2 shows that the 3-amino group of the pyrazolopyridazine ring forms two hydrogen bonding interactions with the carbonyl groups of Gln105 and Asp106 in the hinge region of the ATP binding site. The pyrazolopyridine ring forms a hydrogen bond with Lys54. Met108, Leu156, and Cys166 also play important roles in the compound/protein interaction.

Liu et al. developed a new ERK2 inhibitor such as compound **BL-EI001** (Fig. 3.32). Docking study on this compound reveals that aromatic rings form

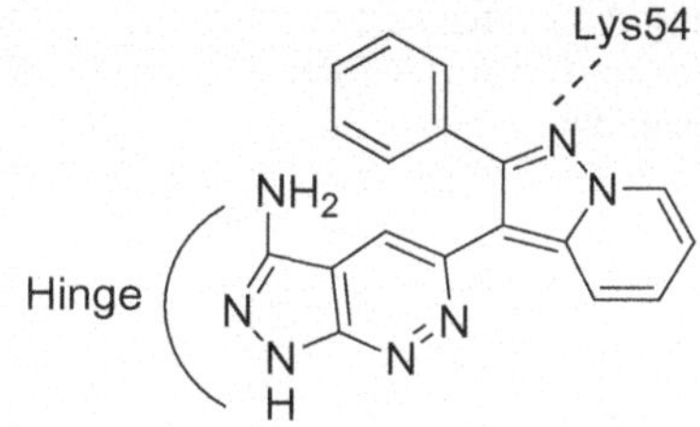

Fig. 3.31 Chemical structure of FR180204

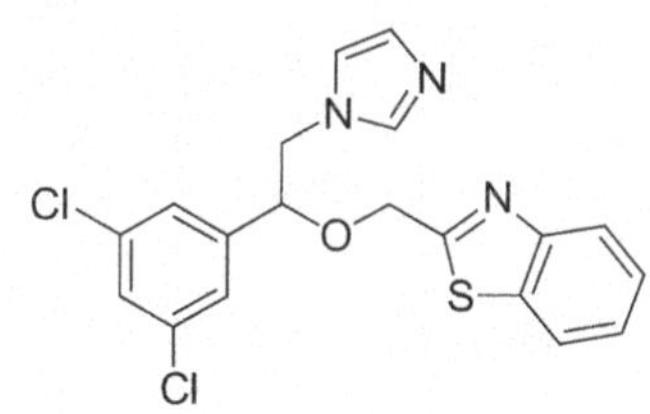

Fig. 3.32 Chemical structure of BL-EI001

hydrophobic interactions with Ile48, Val156, Ala69, and Met125 [64]. In addition, the compound forms two hydrogen bonding interactions with the nitrogen of Lys71 and two π-π interactions with Tyr53. **BL-EI001** shows moderate anti-proliferative activity with an IC_{50} value of 5 µM in MCF-7 human breast carcinoma cells.

Investigators at Astex Pharmaceuticals developed a potent ERK1/2 inhibitor through a fragment-based drug design strategy (Fig. 3.33). This compound shows excellent antiproliferative activity with an IC_{50} of 4.9 nM in BRAF V600E mutant A375 human melanoma cells and an IC_{50} of 7.5 nM in Colo205 human colorectal cancer cells [65]. The compound inhibits the phosphorylation of ERK with an IC_{50} value of 2.3 nM and it shows tumor regression in BRAF mutant xenograft models.

Ji et al. developed a novel ERK inhibitor using a clinical candidate BVD-523 as a starting point [66]. Replacing the isopropyl group with pyran and pyrrole to isoindolin-1-one results in a lead compound as shown in Fig. 3.34. The lead compound inhibits ERK1 and ERK2 with IC_{50} values of 1.5 nM and 0.7 nM respectively. The lead compound has an acceptable pharmacokinetic profile and exhibits in vivo antitumor efficacy in the HCT-116 xenograft model.

Investigators at Novartis discovered a very potent ERK2 inhibitor with an IC_{50} value of 0.026 nM using a structure-based drug design strategy [67]. Replacing the isopropyl group with the cyclopentanol and seven-member ring results in improving potency. Introducing the fluoro group on the cyclopentanol moiety improves microsomal stability (Fig. 3.35).

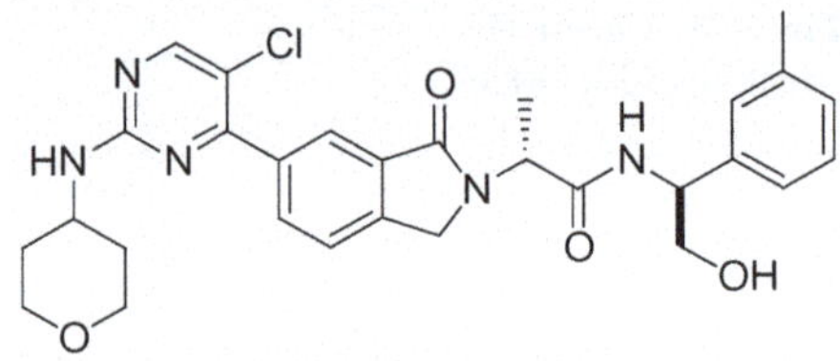

ASTX-ERK inhibitor

IC_{50} (ERK1/2) = 3 nM

IC_{50} (pERK) = 2.3 nM

IC_{50} (A375 cells) = 4.9 nM

Fig. 3.33 Chemical structure of ASTX-ERK inhibitor

BVD-523

IC_{50} (ERK1) = 1.5 nM
IC_{50} (ERK2) = 0.7 nM

Fig. 3.34 New ERK inhibitors

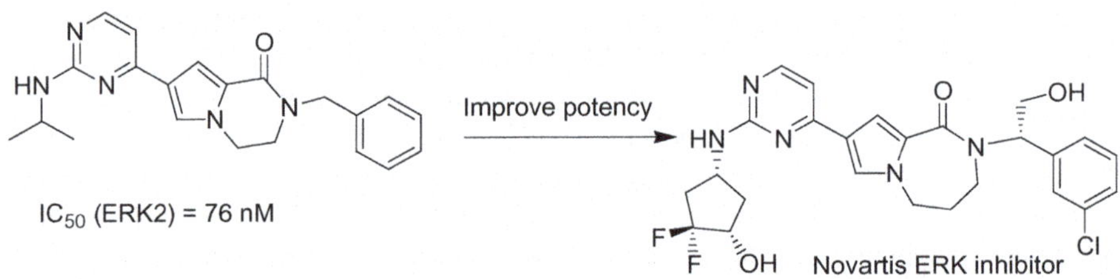

IC_{50} (ERK2) = 76 nM

IC_{50} (ERK2) = 0.026 nM

Fig. 3.35 Tetrahydropyrrole-diazepam derivative as ERK2 inhibitor

Scientists from Array BioPharma developed a novel series of pyrimidine derivatives using a structure-based design approach [68]. The lead compound is a potent, selective, and orally bioavailable ERK inhibitor (Fig. 3.36). It shows an inhibition of tumor growth in mouse xenograft models.

Aly et al. reported the design and synthesis of novel series of fused naphthofuro[3,2-*c*] quinoline-6,7,12-triones and pyrano[3,2-*c*]quinoline-6,7,8,13-tetraones as potential ERK inhibitors [69]. The lead compound inhibits ERK1 and ERK2 with the IC_{50} values of 0.50 μM and 0.60 μM respectively (Fig. 3.37). It inhibits the proliferation of the BRAF mutant A375 melanoma cells with an IC_{50}

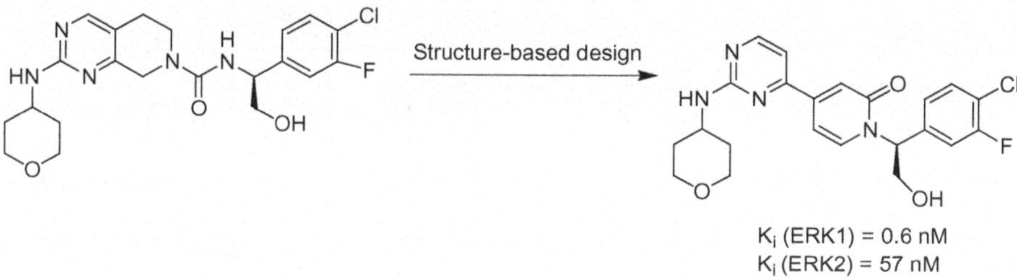

Fig. 3.36 Pyrimidine derivatives as ERK1/2 inhibitors

Fig. 3.37 Naphthofuro [3,2-*c*] quinoline-6,7,12-triones as ERK inhibitors

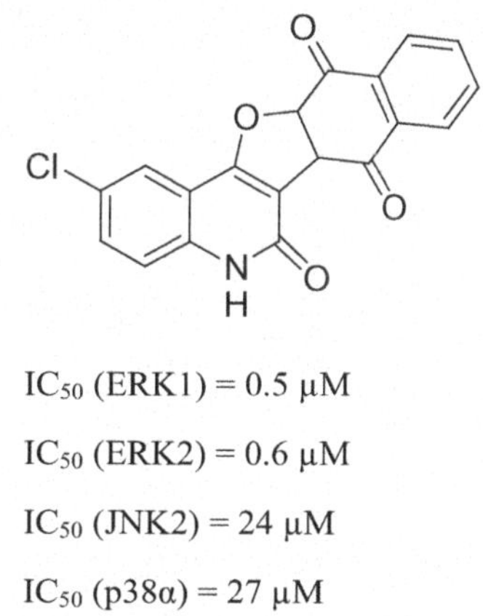

value of 3.7 μM and inhibits the phosphorylation of ERK substrates p-90RSK and ELK-1. The molecular docking study reveals that the compound binds within the ATP catalytic site.

Investigators at Genentech applied a fragment-based lead discovery strategy to find a new ERK2 inhibitor using a surface plasmon resonance (SPR) assay [70]. The crystal structure of fragment *N*-benzyl-9*H*-purin-6-amine (Fig. 3.38) in complex with ERK2 reveals that it binds at the hinge region of the ATP site. The optimization of this fragment leads to a potent ERK2 inhibitor with a K_i of 37 nM, 6600 times improvement in potency. The crystal structure of the lead compound in complex with ERK2 (PDB code: 4QPA) reveals that the 4-position *N* atom forms a hydrogen bond with Leu107. The pyridine forms a hydrogen bond with Lys114 and pyrazole forms a hydrogen bond with Lys54.

Investigators from Green Valley Pharmaceuticals reported the discovery of imidazole amides as ERK inhibitors (Fig. 3.39). The optimized lead compound inhibits ERK1 and ERK2 with IC_{50} values of 19.4 nM and 9.5 nM respectively. It inhibits the growth of BRAF mutant HT29 cells [71].

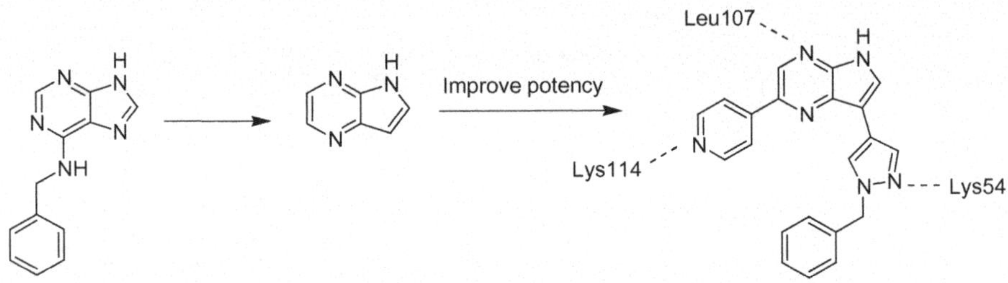

K_d (ERK2) = 247 μM K_d (ERK2) = 347 μM K_i (ERK2) = 37 nM

LE = 0.29 LE = 0.53 LE = 0.378

Fig. 3.38 5*H*-pyrrolo[3,2-*b*]pyrazine derivatives as ERK2 inhibitors

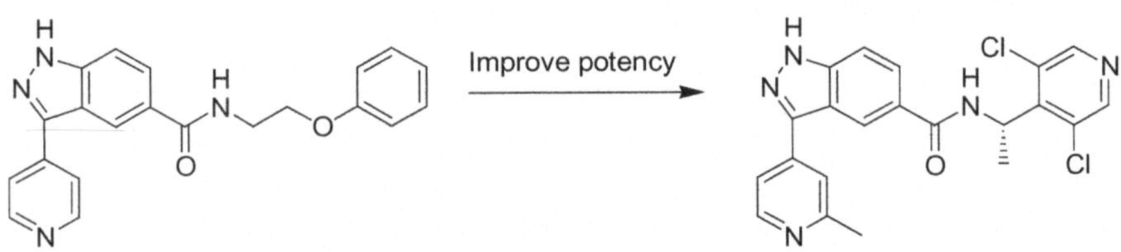

IC_{50} (ERK2) = 63 nM IC_{50} (ERK2) = 9 nM

IC_{50} (HT29 cells) = 9.10 μM IC_{50} (HT29 cells) = 0.48 μM

Fig. 3.39 Indazole derivatives as ERK2 inhibitors

3.24 Development of Clinical Candidates as ERK1/2 Inhibitors

Several small molecule ERK1/2 inhibitors are undergoing clinical trials for the treatment of patients with cancer and other diseases [72, 73]. Investigators at BioMed Valley Discoveries developed a highly potent and selective ATP-competitive ERK1/2 inhibitor BVD-523 (ulixertinib) based on high-throughput screening and optimization of the hit compound (Fig. 3.40). BVD-523 binds in the ATP site of ERK1/2 with K_i values of 0.3 nM and 0.04 nM respectively [74]. BVD-523 inhibits tumor growth in BRAF V600E mutated A375 malignant melanoma cell line and Colo205 colorectal cancer cell line xenograft models. It has low toxicity and it demonstrates excellent pharmacokinetic and pharmacodynamic properties. The clinical trial on BVD-523 has completed or it is undergoing in patients with acute myeloid leukemia or myelodysplastic syndrome, melanoma, and metastatic pancreatic cancers.

Fig. 3.40 Pyrrole derivative as ERK1/2 inhibitor

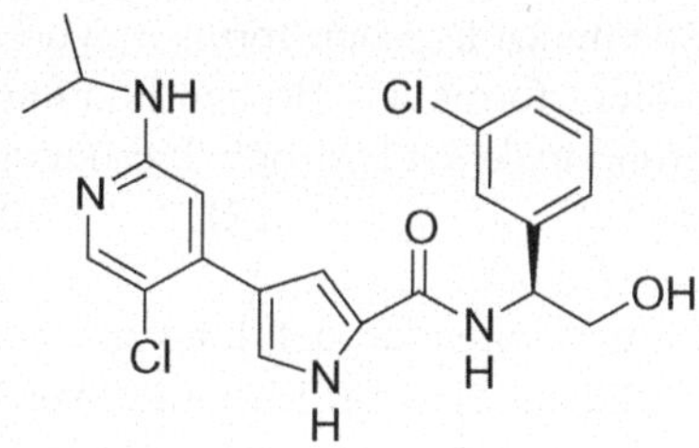

BVD-523

K_i (ERK1) = 0.3 nM

K_i (ERK2) = 0.04 nM

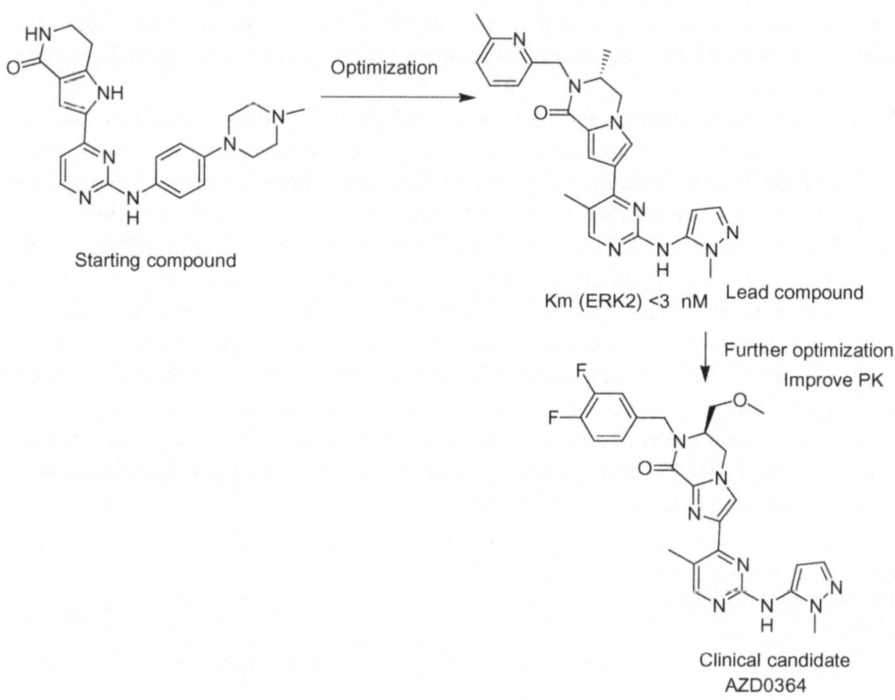

IC_{50} (pRSK) = 5.6 nM

IC_{50} (pERK) = 1.7 nM

Fig. 3.41 Lead optimization to clinical candidate

Investigators at AstraZeneca discovered AZD0364 through a structure-based design technique as a potent and selective inhibitor of ERK1 and ERK2 as shown in Fig. 3.41. It exhibits high cellular potency with IC_{50} values of 5.6 nM (pRSK) and 1.7 nM (pERK). It demonstrates an acceptable pharmacokinetic profile (ADME).

The pyrazole group forms a hydrogen bond with Lys114 of ERK2 and pyrimidine moiety forms a hydrogen bonding interaction with Met108. The methoxymethyl group makes a hydrophobic interaction with Val39 and Ile 31. AZD0364 binds in the ATP site of the ERK2. AZD0364 shows excellent antiproliferative effects in *BRAF V600E* mutation A375 cell line (GI_{50} = 0.06 μM) and *KRAS Q61K* mutation Calu-6 non-small-cell lung cancer (NSCLC) cell line [75, 76].

Scientists at Eli Lilly discovered LY3214996 as a potent and selective ERK1/2 inhibitor based on lead optimization of hit strategy (Fig. 3.42). LY3214996 is a reversible ATP-competitive inhibitor with an IC_{50} value of 5 nM for both ERK1 and ERK2. It shows strong binding with K_i values of 0.65 nM for ERK1 and 0.06 nM for ERK2, and > 40-fold selectivity against 512 kinases and > 1000-fold selectivity against 486 kinases [77, 78]. It induces cell cycle arrest in the G1 phase and apoptosis in a dose- and time-dependent manner in *KRAS*-mutant HCT116 cells. It exhibits antitumor activity in *KRAS*-mutant HCT116 colorectal cancer xenograft model as well as *BRAF, NRAS,* and *MEK*-mutant models as a single agent. Currently, it is undergoing alone or in combination with other drugs in a phase II clinical trial for the treatment of patients with advanced and metastatic cancer (NCT04534283).

Scientists at Astex Pharmaceuticals reported the discovery of the clinical candidate **ASTX029** using structure-guided optimization strategy of previously reported isoindolinone lead compound (Fig. 3.43). The previous lead compound was metabolically not stable. Replacing several CYP3A4-mediated metabolism groups resulted in a clinical candidate with favorable physicochemical properties [79, 80]. ASTX029 inhibits ERK2 with an IC_{50} value of 2.7 nM. It demonstrates excellent antiproliferative activity with IC_{50} values of 2.70 nM in A375 melanoma cell line and 3.60 nM in Colo205 human CRC cell line respectively. It shows antitumor activity in MAPK-activated tumor xenograft in vivo models using oral doses. It shows an acceptable PK profile with half-life of 2.9 h and 42% oral bioavailability. Currently, ASTX029 is undergoing a phase I-II clinical trial for the treatment of patients with advanced solid tumors (NCT03520075).

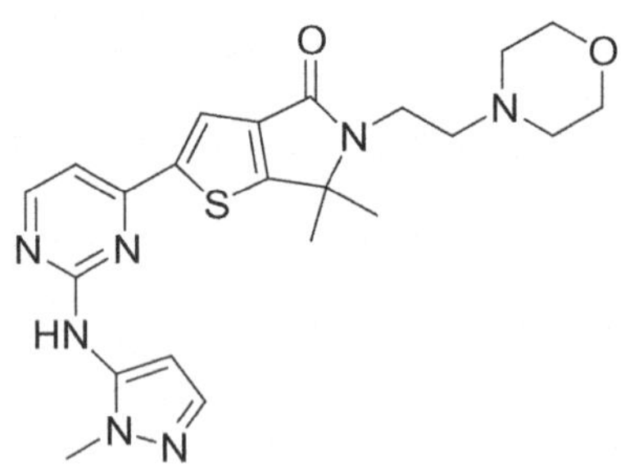

Fig. 3.42 2-Aminopyrimidine derivative as ERK1/2 inhibitor

LY3214996

K_i (ERK1) = 0.65 nM

K_i (ERK2) = 0.06 nM

Fig. 3.43 2-Aminopyrimidine derivative

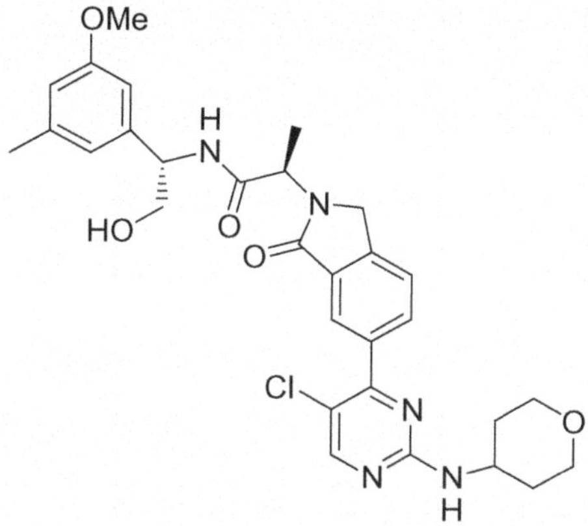

ASTX029

IC_{50} (ERK1/2) = 2.7 nM

Fig. 3.44 Imidazole derivative as ERK1/2 inhibitor

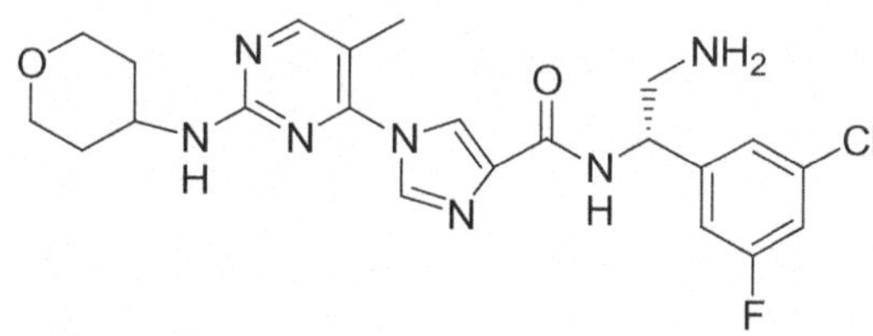

ASN007

IC_{50} (ERK1/2) = 2 nM

ASN007 is an orally bioavailable potent and selective inhibitor with IC_{50} values of 2 nM for both ERK1 and ERK2 (Fig. 3.44). It shows excellent antiproliferative activity in tumors in $BRAF^{V600E}$ mutant lymphoma cell line U-2973, $KRAS^{A146T}$ mutant cell line NU-DHL-1, $NRAS^{G13D}$ mutant cell line MINO with IC_{50} values of 100, 200, and 300 nM respectively [81, 82]. It induces cell cycle arrest in the Go/G1 phase and apoptosis in a dose- and time-dependent manner. ASN007 exerts antiproliferative activities in vitro and in vivo $BRAF^{V600E}$ mutant melanoma tumor models that are resistant to BRAF and MEK inhibitors. Currently, it is undergoing phase I/II clinical trials in combination with PI3K inhibitor or EGFR inhibitor.

Burrows et al. developed **KO-947** a potent and selective ERK1/2 inhibitor with an IC_{50} value of 10 nM (Fig. 3.45). It shows favorable pharmacokinetics properties and strong antiproliferative activity in vivo models [83, 84]. It is now in phase I/II clinical trials in patients with advanced tumors (NCT03051035).

ONC201 (Fig. 3.46) is a potent and selective dual inhibitor of AKT and ERK [85]. It shows favorable pharmacokinetics properties with $T_{1/2}$ = 8.4 h; C_{max} = 2.3 μg/mL, CL = 27.5 L/h, AUC = 16.4 hμg/mL. Its safety profile is acceptable. Currently, it is undergoing a phase I clinical trial as a single agent in pediatric patients with diffuse midline gliomas following radiotherapy (NCT02420795).

Fig. 3.45 Indazole derivative as ERK1/2 inhibitor

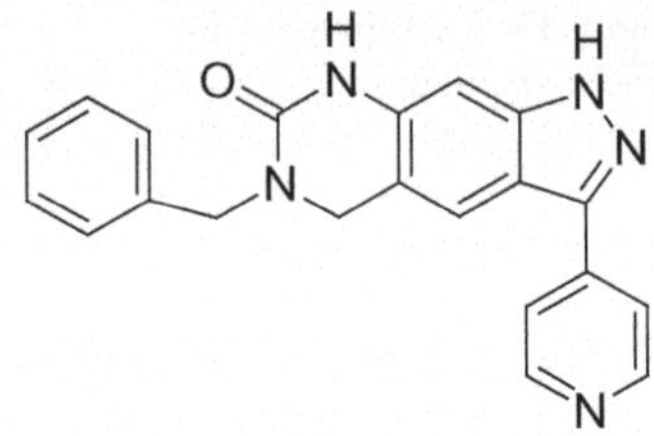

KO-947
IC_{50} (ERK1/2) = 10 nM

Fig. 3.46 Chemical structure of **ONC201**

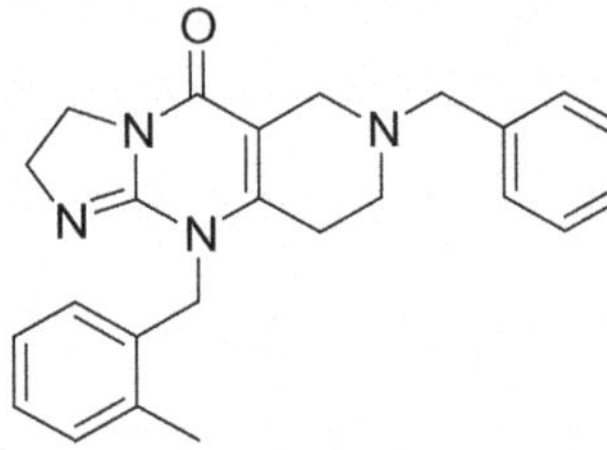

ONC201

Fig. 3.47 Chemical structure of **LTT462**

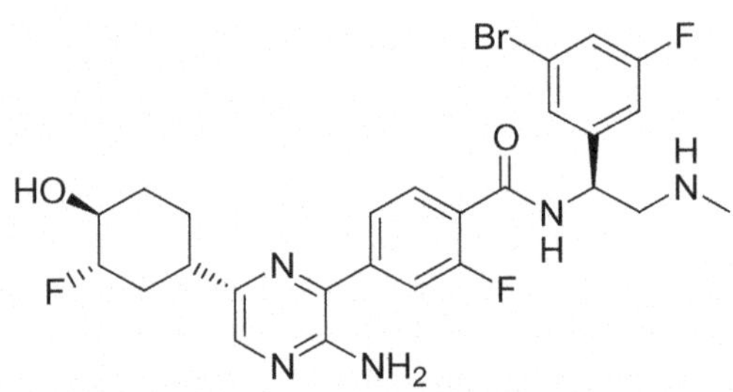

Janku et al. developed **LTT462** (Fig. 3.47) as an ATP-competitive ERK inhibitor that demonstrated strong preclinical activity in multiple MAPK-activated cancer cells and xenograft models [86]. Currently, LTT462 alone or in combination with other drugs are undergoing clinical trials for the treatment of colorectal cancer with *BRAF-V600E* mutation (NCT02711345).

Investigators from Array BioPharma identified a hit compound from their high-throughput screening campaign with an IC_{50} value of 106 nM for ERK2 (Fig. 3.48). The optimization strategy resulted in first-generation lead compounds with strong affinity in both enzymatic and cell-based assays (ERK2 $IC_{50} = 0.6$ nM, $IC_{50} = 45$ nM in HepG2 cells). The lead compound exhibited high clearance and low oral bioavailability [87]. They performed metabolic stability of lead compound with human, mouse, dog, rat, and cynomolgus monkey hepatocytes for 3 h and it showed that the

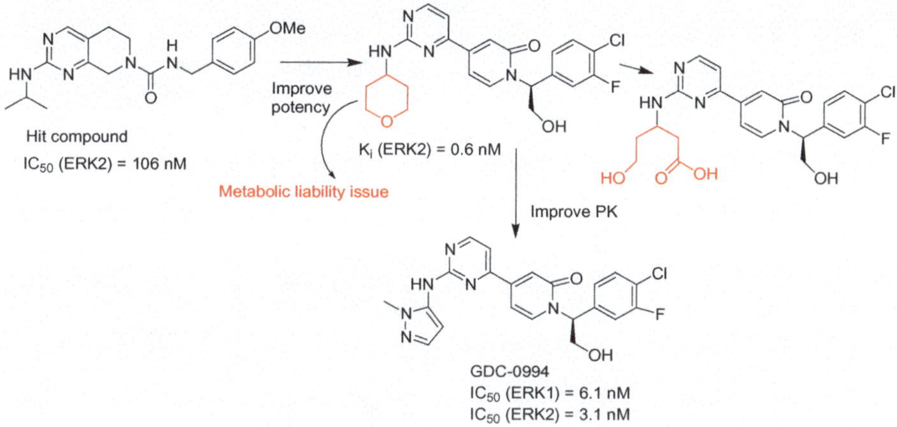

Fig. 3.48 Hit to lead to clinical candidate

metabolism occurring on the tetrahydropyran (THP) ring to form a hydroxy acid. The replacing metabolically unstable THP ring with an *N*-methylpyrazole resulted in the clinical candidate **GDC-0994** with IC_{50} values of 6.1 nM and 3.1 nM against ERK1 and ERK2 respectively [68, 88]. The X-ray crystal structure of **GDC-0994** in complex with ERK2 (PDB code: 5K4I) reveals that the 2-aminopyrimidine forms two hydrogen bonding interactions with Met108 and Leu107 of ERK2. The *N*-methylpyrazole forms a hydrogen bonding interaction with Lys114 and 3-fluoro-4-chloro-phenyl ring makes a hydrophobic interaction in the glycine-rich loop. GDC-0994 shows excellent antitumor efficacy and low toxicity. Currently, it is undergoing clinical trials alone (NCT01875705) and in combination with cobimetinib (a MEK inhibitor) in patients with locally advanced or metastatic solid tumors (NCT02457793).

Boga et al. identified a hit compound (Fig. 3.49) from their library screening. After extensive SAR studies, they found a lead compound **SCH772984** with IC_{50} values of 4 nM and 1 nM against ERK1 and ERK2 respectively. This lead compound is very potent, selective, and exhibits good anti-tumor cell proliferation effects in several xenograft mouse models with KRAS or BRAF mutant cell lines. However, the lead compound has very poor bioavailability due to proteolytic hydrolysis of pyrrolidine amide. Introducing a steric hindrance group on the 3-position of the pyrrolidine such as a thiomethyl group alleviates amide metabolism and improves PK profile [89, 90]. The clinical candidate **MK-8353** shows excellent activity against ERK1 and ERK2 with IC_{50} values of 41 nM and 7 nM. This clinical candidate exhibits significant antiproliferative activity with IC_{50} values of 118 nM and 45 nM in HT-29 and Colo205 human CRC cells respectively. The X-crystal study of **MK-8353** in complex with ERK (PDB code: 6DCG) reveals that it binds in the ATP site. The indazole moiety forms hydrogen bonding interactions with Met106 in the hinge region. The nitrogen atom on pyrrolidine makes a hydrogen bonding interaction with Lys52. **MK-8353** is well tolerated in patients with BRAF

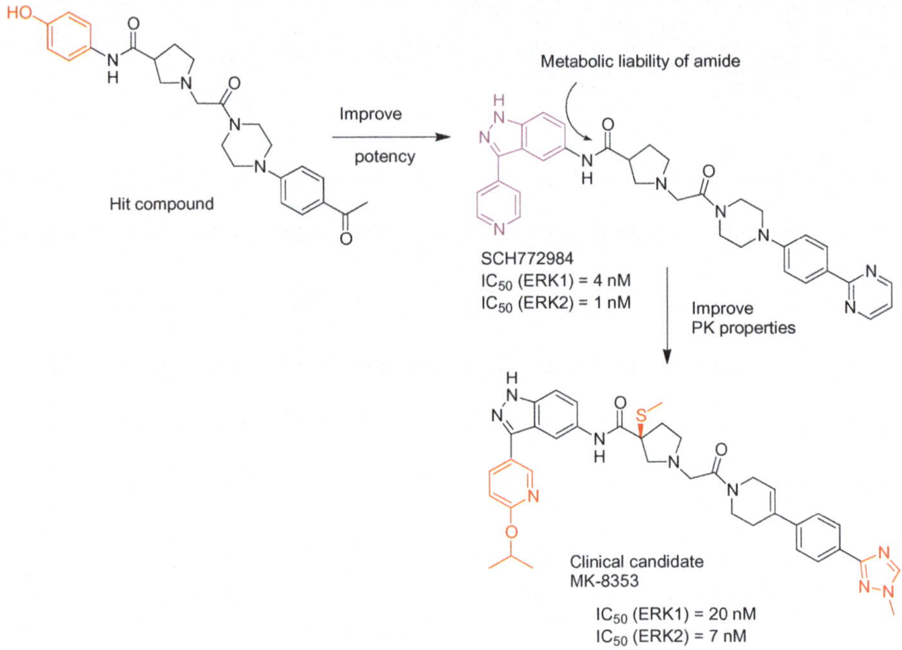

Fig. 3.49 Lead optimization with improved PK profile resulting in clinical candidate

V600E mutant melanoma and shows antitumor efficacy at an oral dose of 400 mg twice a day. Currently, alone or in combination with selumetinib is undergoing clinical trials (NCT03745989) in patients with advanced malignant tumors and colorectal cancer.

3.25 Development of Clinical Candidates as p38 MAPK Inhibitors

Roy et al. identified a pyridazine fragment (Fig. 3.50). Introducing the piperazine-pyrimidine group resulted in a hit compound with a K_i value of 0.62 μM for p38α. Further optimization of the hit compound provided a lead compound with improved potency, K_i of 180 nM against p38α. However, this lead compound exhibited a poor PK profile. After improving PK profile they delivered a clinical candidate MW01-18-150SRM (**MW150**). This clinical candidate is a selective and brain penetrant p38α inhibitor and it shows excellent efficaciousness in various animal models for the treatment of Alzheimer's disease and other neurologic and neuropsychiatric disorders [91–93]. Currently, it is undergoing phase I clinical trials. The same group developed **AM151** which is a more stable, more water soluble, and brain-penetrant p38α inhibitor [94].

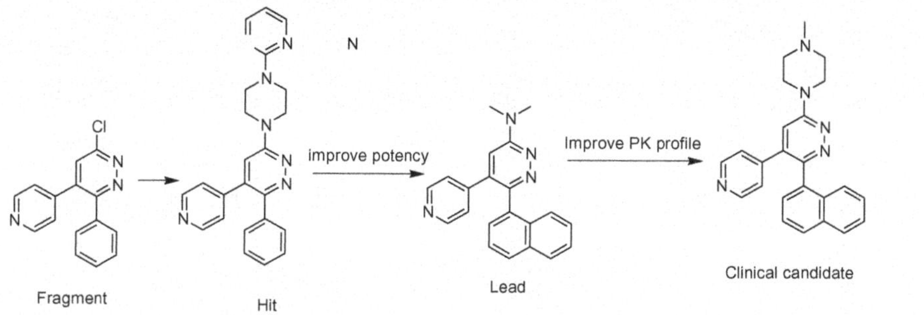

MW-150

K_i (p38α) = 0.62 μM Ki (p38α) = 0.18 μM Ki (p38α) = 0.10 μM

K_i (p38β) > 10 μM K_i (p38β) = 0.32 μM K_i (p38β) = 1.85

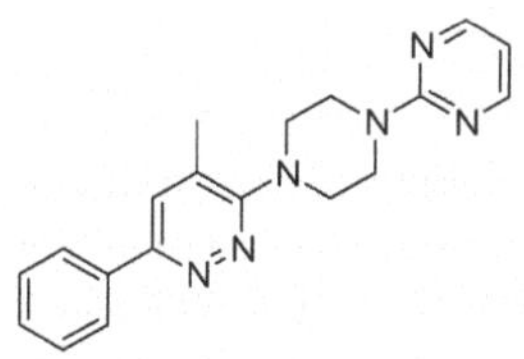

MW-151

Fig. 3.50 Fragment to lead to clinical candidate of p38 MAPK inhibitor

Investigators at Roche reported imidazole derivative **RO-3201195** as a selective p38α inhibitor with an IC_{50} value of 700 nM (Fig. 3.51). The benzoyl oxygen atom forms a hydrogen bonding interaction with Met109 at the hinge region of the ATP binding site [95]. The 5 position amino group on the pyrazole forms a hydrogen bonding interaction with Thr106 of p38α which provides selectivity. This compound inhibits the production of cytokine IL-6 and TNF-α induced by LPS stimulation. Introducing 2,3-dihydroxypropoxy group improves solubility, drug-like properties, and oral bioavailability. Roche also developed azaindole derivative **RO-3206145** as a selective and very potent p38α inhibitor with an IC_{50} value of 1.29 nM [96]. This compound also inhibits cytokines such as IL-1 and TNF-α. This compound exhibits a favorable PK profile. **Pamapimod** (RO-4402257) was developed by Roche. It is a selective p38α/β inhibitor with the IC_{50} values of 14 nM and 480 nM respectively [97, 98]. It is inactive against p38γ and p38δ. It is also selective for other 350 kinases tested. It binds in the ATP site. It inhibits the production of cytokines such as IL-1β and TNF-α in patients with rheumatoid arthritis.

Investigators from GSK developed Adezmapimod (SB-203580) as a potent, cell-permeable, reversible, and ATP-competitive p38α and p38β inhibitor with IC_{50} values of 50 nM and 500 nM respectively (Fig. 3.52). It is inactive against p38γ and

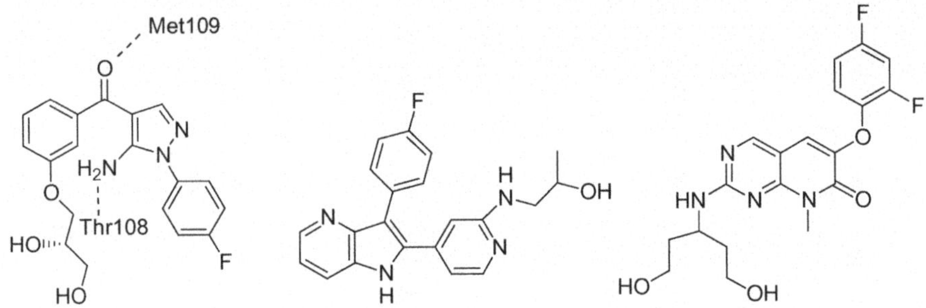

RO-3201195 — IC_{50} (p38α) = 700 nM

RO-3206145 — IC_{50} (p38α) = 1.29 nM

Pamapimod (RO-4402257) — IC_{50} (p38α) = 14 nM; IC_{50} (p38β) = 480 nM

Fig. 3.51 Clinical candidates from Roche

p38δ and other kinases tested [99–101]. It inhibits IL-1 and TNF-α production from LPS-stimulated human monocyte cell line THP-1 with IC_{50} value of 50–100 nM. It is also effective in vivo in both mice and rats for various disease models. The nitrogen atom of the pyridine ring forms a hydrogen bond with Met109 at the ATP binding site. The 3-position nitrogen atom of the imidazole ring forms a hydrogen bonding interaction with Lys53 [101]. The fluorophenyl ring makes a hydrophobic interaction with Thr106 and Leu104.

GSK developed another imidazole derivative **SB-220025** as a potent p38α inhibitor with an IC_{50} value of 60 nM [102]. It is a reversible, ATP-competitive, and cell-permeable inhibitor. It shows 50-fold selectivity over PKC, 500-fold over PKA, and 1000-fold over EGFR, 2000-fold over ERK. It exhibits in vivo as a potent inhibitor of angiogenesis and inhibits the production of LPS-stimulated TNF-α with an ED_{50} value of 7.5 mg/kg. **SB 220025** at 30 mg/kg orally twice daily was able to prevent the progression of established arthritis. From docking studies, the 2-aminopyrimidine forms a hydrogen bonding interaction with Met109 at the hinge region of the ATP site. The piperidine *N* atom forms a hydrogen bonding interaction with Tyr35 and the fluoro group on the phenyl ring forms a hydrogen bonding interaction with Val105.

SB-202190 was developed by GSK [103, 104]. It is similar to previous imidazole derivatives as mentioned above. It is a potent and selective p38α and p38β inhibitor with the IC_{50} values of 50 nM and 100 nM respectively. It is inactive for JNK, ERK, and other kinases tested. **SB 202190** binds to the ATP site of the p38 kinase with a K_d value of 38 nM. It inhibits the p38 phosphorylation of activating transcription factor 2 (ATF-2) with an IC_{50} value of 280 nM. It induces cell death via nucleus condensation and intranucleosomal DNA fragmentation.

GSK disclosed the clinical candidate **Dilmapimod** (SB-681323) as a potent and selective p38α inhibitor. It binds to the ATP site of the p38α. It inhibits the

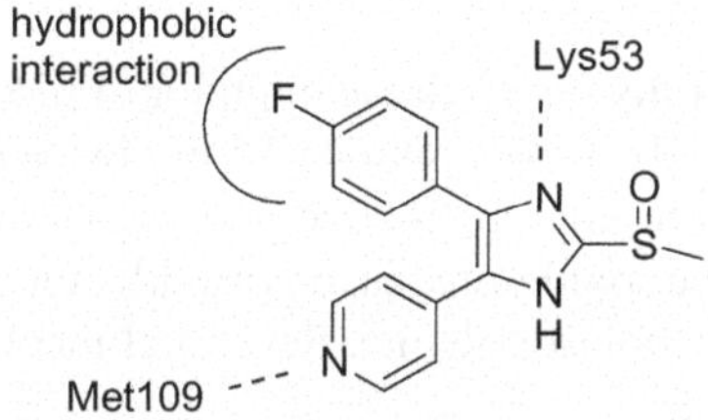

Adezmapimod (SB-203580)

IC_{50} (p38α) = 50 nM

IC_{50} (p38β) = 500 nM

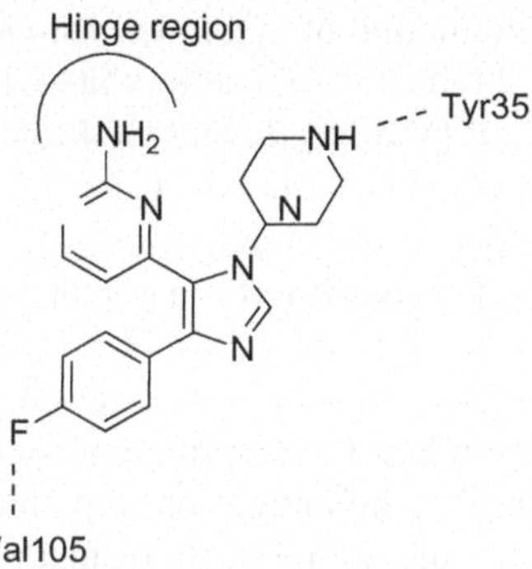

SB-220025

IC_{50} (p38α) = 60 nM

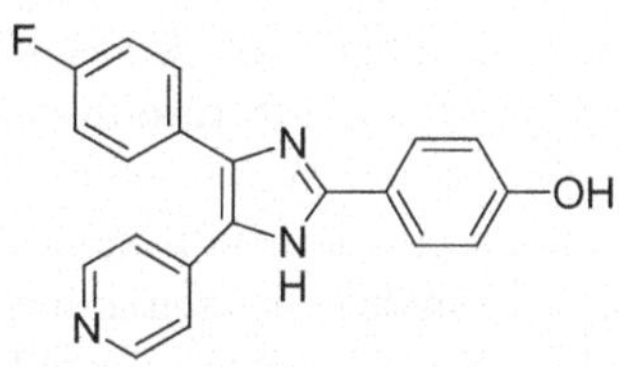

SB-202190

IC_{50} (p38α) = 50 nM

IC_{50} (p38β) = 100 nM

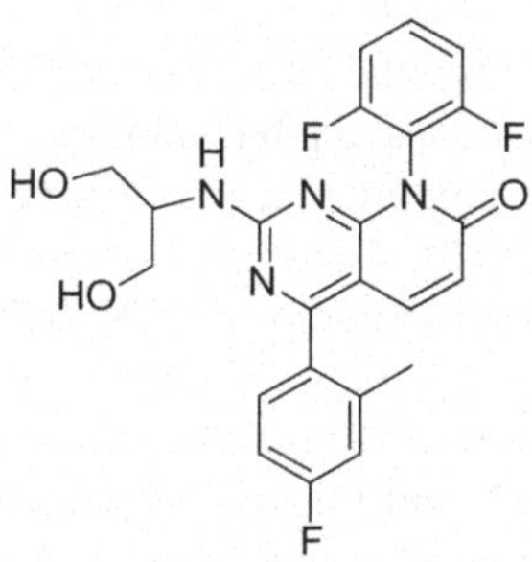

Dilmapimod (SB-681323)

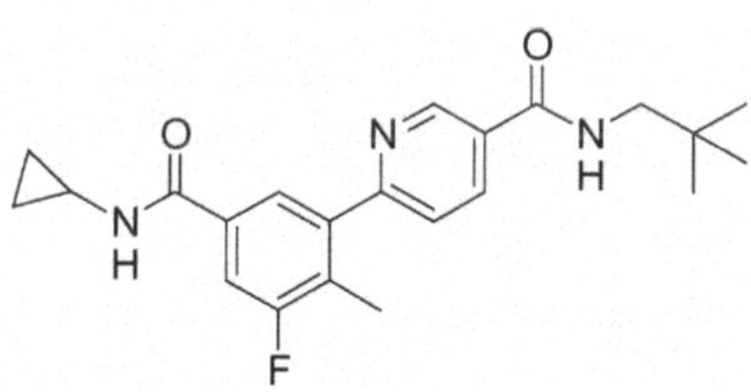

Losmapimod (SB-856553)

pK_i (p38α) = 8.1

pK_i (p38β) = 7.6

Fig. 3.52 p38 clinical candidates from GSK

production of cytokines such as IL-1β, IL-6, IL-8, TNF-α, and others. Dilmapimod inhibits the phosphorylated heat shock protein 27 (pHSP-27) in patients with COPD. It is safe and well tolerated and an effective drug for neuropathic pain and rheumatoid arthritis. Currently, several clinical trials are undergoing with this drug alone or in combination with other drugs [105–107].

Losmapimod is a potent, selective, and orally bioavailable inhibitor of p38α and p38β with pK_i values of 8.1 and 7.6 respectively. Losmapimod (7.5 mg twice daily) reduces plasma fibrinogen in patients with chronic obstructive pulmonary disease (COPD). Losmapimod also improves nitric oxide-mediated vasodilatation and inhibits inflammation in patients with hypercholesterolemia. Several clinical trials are undergoing using losmapimod [108–112].

Investigators from Vertex Pharmaceuticals developed **VX-702** as a potent and selective p38α inhibitor with IC_{50} values in the range 4–20 nM. This compound inhibits the production of LPS-stimulated cytokines such as IL-1β, IL-6, and TNF-α with IC_{50} values of 122 nM, 59 nM, and 99 nM respectively (Fig. 3.53). It has entered a clinical trial for the patients of RA but failed to improve significant disease conditions [113–115].

Vertex pharmaceuticals developed their second clinical candidate **Neflamapimod** (**VX-745**) as a selective p38α inhibitor with an IC_{50} value of 9 nM. Extensive SAR studies and introducing proper substituents on two phenyl rings resulted in the clinical candidate [116]. There are no significant inhibitions with other MAP kinases. The X-ray crystal structure of **VX-745** in complex with p38α at 2.4 Å resolution reveals that the 2,6-dichloro phenyl ring makes favorable van der Waals interactions. Each chlorine atom stays in a hydrophobic pocket and contacts with Val30, Leu108, Ala157, and Leu167 of p38α. The phenyl ring also makes an interaction with backbone residues of Gly110, Ala111, and Asp112. The 2,4-difluorophenyl ring stays in a hydrophobic pocket and makes a favorable van der Waals interaction with the gatekeeper residue Thr106. This compound shows a favorable PK profile with good oral bioavailability (56% in rats, 78% in mice, 69% in dogs). Several

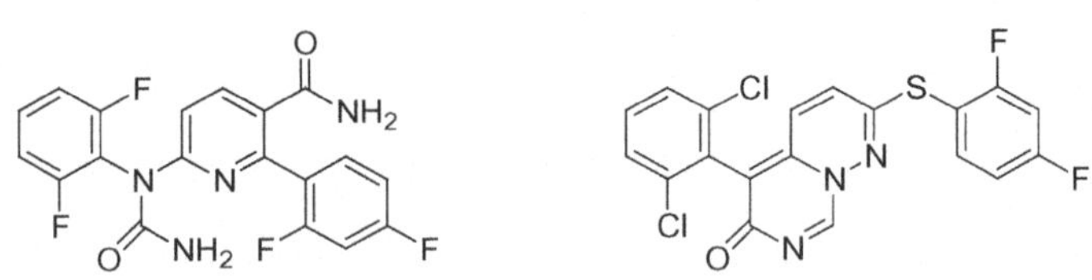

VX-702

IC_{50} (p38α) = 4-20 nM

IC_{50} (IL-1β) = 122 nM

IC_{50} (TNF-α) = 99 nM

Neflamapimod (VX-745)

IC_{50} (p38α) = 9 nM

IC_{50} (IL-1β) = 45 nM

IC_{50} (TNF-α) = 51 nM

Fig. 3.53 Clinical candidates from Vertex

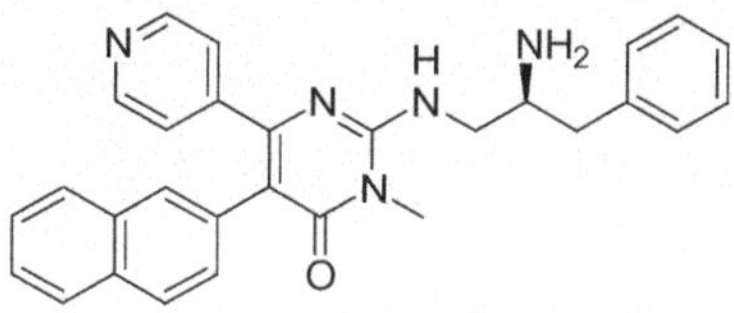

AMG-548

K_i (p38α) = 0.5 nM

K_i (p38β) = 36 nM

Fig. 3.54 Clinical candidate of p38 MAPK inhibitor from Amgen

clinical trials are undergoing with this compound for the treatment of Alzheimer's disease and others [117].

Investigators at Amgen developed **AMG-548** as a potent, orally bioavailable, and selective inhibitor of p38α with a K_i value of 0.5 nM (Fig. 3.54) [115, 118, 119]. It is slightly selective over p38β (K_i = 36 nM) and > 1000 fold selective against p38γ (K_i = 2600 nM) and p38δ (K_i = 4100 nM). It is modestly selective over JNK2 (K_i = 39 nM) and JNK3 (K_i = 61 nM). It shows very strong inhibition of whole blood LPS stimulated cytokines such as TNF-α (IC_{50} = 3 nM), IL-1β (IC_{50} = 7 nM), IL-6 (IC_{50} = 1.3 nM), and IL-8 (IC_{50} = 0.7 nM). The half-life of **AMG-548** is 4.6 h in rats and 7.3 h in dogs. It has good oral bioavailability with 62% in rats and 47% in dogs. It has entered clinical trials.

Investigators at Pfizer identified a hit compound as a p38α inhibitor with an IC_{50} value of 680 nM (Fig. 3.55). However, it showed poor metabolic stability due to *N*-benzyl group [120, 121]. The replacing *N*-benzyl group with *N*-phenyl resulted in the first lead compound with improved metabolic stability and potency with an IC_{50} value of 12 nM (p38α). The 2,4-difluoro substitution on the benzyloxy group prevents *O*-debenzylation. Further optimization on lead compound furnished in clinical candidate **PH-797804**, as a racemate (±) with an IC_{50} value of 2.5 nM against p38α [122–126]]. The (-) **PH-79804** is more potent with 2.3 nM IC_{50} value over p38α than (+) isomer IC_{50} value of 247 nM over p38α. This compound also inhibits the production of LPS-stimulated TNF-α with an IC_{50} value of 3.4 nM. Several clinical trials have been completed using this compound for the treatment of patients with COPD, rheumatic arthritis, and osteoarthritis [126].

Talmapimod, also known as **SCIO-469**, was developed by SCIOS as a potent and selective p38α inhibitor with an IC_{50} value of 14 nM (Fig. 3.56). It is 34 times selective against p38β and 2000 times selective over other kinases tested including MAP kinases. It is an orally bioavailable and ATP-competitive inhibitor [114, 115, 127–129]. The inhibition of the phosphorylation of p38 MAPK results in the induction of tumor cell apoptosis, the inhibition of tumor cell proliferation, and the inhibition of tumor angiogenesis. It inhibits the production of LPS-stimulated TNF-α with an IC_{50} value of 300 nM. Several clinical trials have been completed using this compound for the treatment of patients with rheumatic arthritis [130], multiple myeloma, and others.

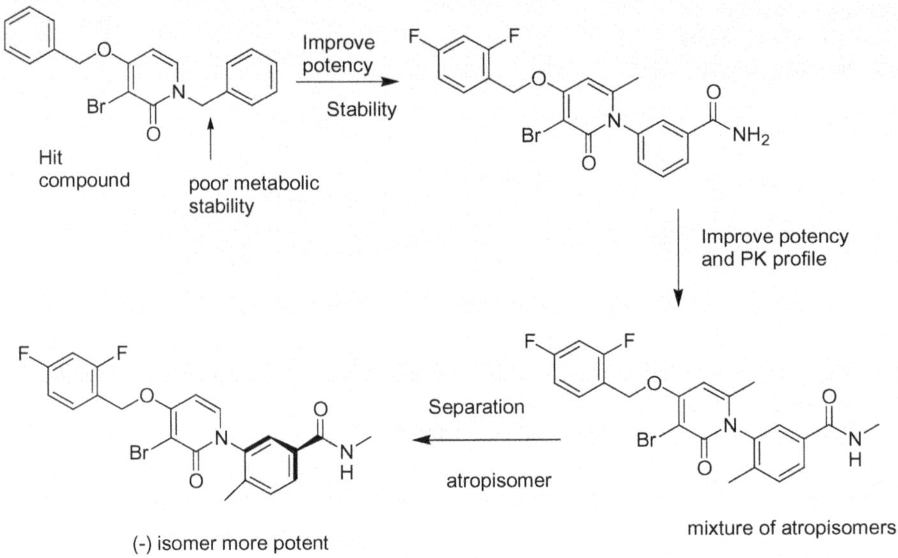

PH-797804

IC_{50} (p38α) = 2.3 nM

PH-797804

IC_{50} (p38α) = 2.5 nM

Fig. 3.55 Hit to lead to clinical candidate as *N*-substituted pyridinone derivative

Fig. 3.56 Indole derivative as p38 MAPK inhibitor

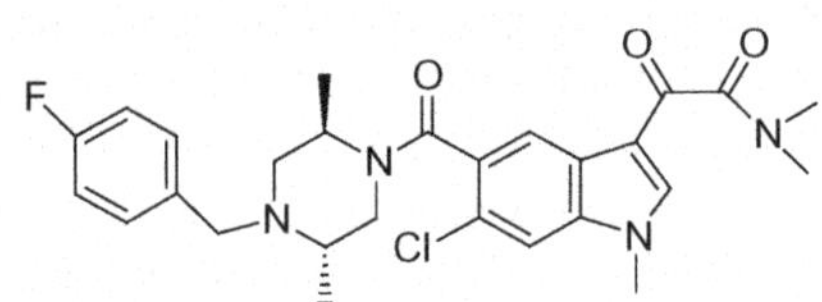

Talmapimod (SCIO-469)

IC_{50} (p38α) = 14 nM

IC_{50} (p38β) = 480 nM

Investigators at BMS developed **BMS-582949** as a potent p38α inhibitor with an IC_{50} value of 13 nM (Fig. 3.57). It shows 2000 times selectivity over p38γ, p38δ, nonreceptor tyrosine kinases, receptor tyrosine kinases, and kinases tested [131, 132]. It is 450 times selective over JNK2 which is associated with inflammation. It inhibits LPS-induced TNF-α production with an IC_{50} value of 50 nM. It exhibits favorable oral bioavailability 90% in mice and 60% in rats. It is effective in both the acute murine model of inflammation and pseudoestablished rats AA model. The X-ray crystal structure of BMS-582949 in complex with p38α (PDB code: 3MVM) reveals that it binds in the hinge region of the ATP site. The oxygen atom of the

Fig. 3.57 Clinical candidate from BMS

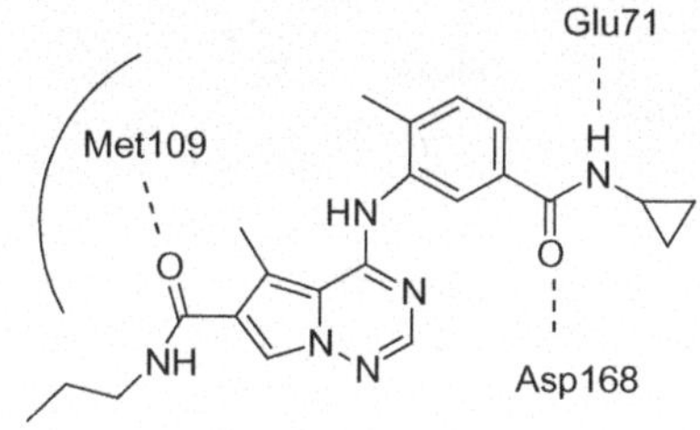

BMS-582949

IC_{50} (p38α) = 13 nM

IC_{50} (TNF-α) = 2.9 nM

Fig. 3.58 Clinical candidate from Takeda

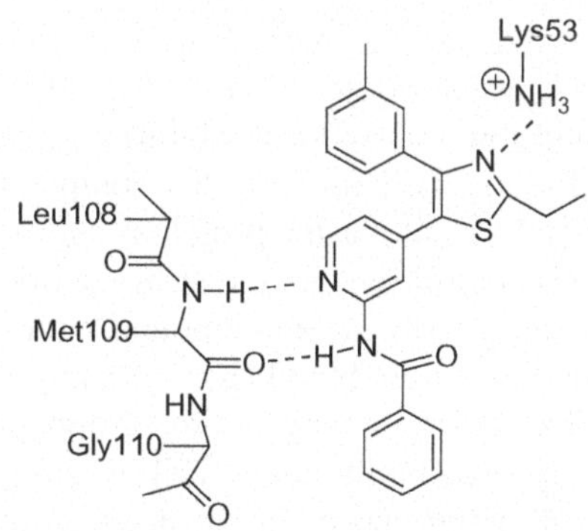

benzamide moiety forms a hydrogen bonding interaction with Asp168 and the NH forms a hydrogen bonding interaction with Glu71. The oxygen atom of the 6-carboxamide forms a hydrogen bonding interaction with Met109 in the hinge region. It is currently in phase II clinical trials for the treatment of rheumatoid arthritis [133].

Investigators from Takeda Pharmaceuticals developed **TAK-715** as a potent and selective p38α inhibitor with an IC50 value of 7.1 nM (Fig. 3.58). **TAK-715** inhibits the production of LPS-induced cytokine TNF-α with an IC_{50} value of 48 nM in human monocytic THF-1 cells [134–136]. It shows selectivity over p38γ, p38δ, JNK, and other kinases tested. This compound is not a CYP inhibitor. The X-ray crystal structure of **TAK-715** in complex with p-38α (PDB code: 3ZSG) reveals that the nitrogen atom on the pyridine ring forms a hydrogen bond with NH of Met109. The amide hydrogen atom of **TAK-715** forms a hydrogen bond with the carbonyl amide of Met109. The nitrogen atom on thiazole forms a hydrogen bond with Lys53. It exhibits good oral bioavailability and efficacy in a rat adjuvant-induced arthritis model. It is undergoing a clinical trial for the treatment of rheumatoid arthritis.

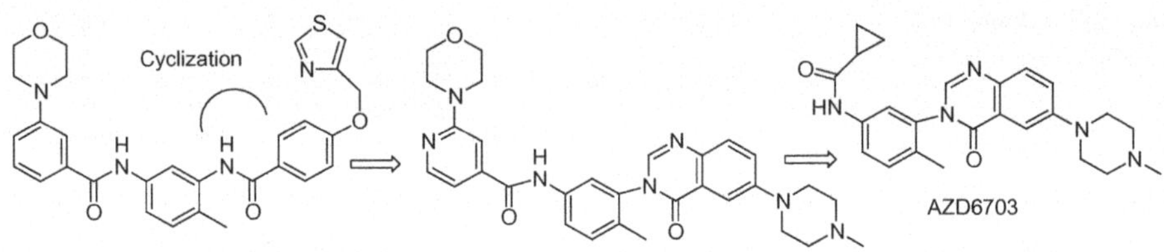

IC_{50} (p38α) = 18 nM IC_{50} (p38α) = 35 nM IC_{50} (p38α) = 16 nM

IC_{50} (HWB) = 1.5 μM IC_{50} (HWB) = 0.42 μM IC_{50} (HWB) = 0.027 μM

Fig. 3.59 Development of clinical candidate AZD6703 from bisamide

Scientists at AstraZeneca identified a novel bisamide inhibitor of p38α with an IC_{50} value of 18 nM [137, 138]. This compound shows a potent inhibition of p38α and low micromolar inhibition of LPS-stimulated TNF-α in human whole blood (Fig. 3.59). However, this compound's water solubility is very poor (0.54 μM at pH 7.4) and plasma protein binding is very high (only 2% free). On the other hand, it is possible enzymatic cleavage of one amide bond. The cyclization strategy results in an intermediate compound with improved inhibition of LPS-induced TNF-α in human whole blood (IC_{50} = 0.42 μM). Further optimization including reducing its size and reversing amide group produces a clinical candidate **AZD6703** with improved inhibition of TNF-α in human whole blood (IC_{50} = 27 nM). It is active in both p38α and p38β but inactive against p38γ and p38δ. It also shows 100 times selective over other kinases tested. **AZD6703** exhibits favorable oral bioavailability 68% in dogs and 49% in rats. This compound has entered in a phase I clinical trial for the treatment of inflammatory diseases.

Investigators at AstraZeneca developed **AZD7624** as a potent and selective inhibitor of p38α inhibitor with an IC_{50} value of 0.1 nM (Fig. 3.60). It is inactive against p38γ, p38δ, and other kinases tested. It inhibits the production of LPS-stimulated TNF-α with an IC_{50} value of 3.5 nM. Due to its strong potency and safety profile, it has entered a phase II clinical trial for the treatment of patients with COPD [139–142].

Investigators at Eli Lilly developed **Ralimetinib** (LY2228820) as a dual MEK and p38 MAPK inhibitor (Fig. 3.61). It is an ATP-competitive and selective inhibitor of p38α and p38β with the IC_{50} values of 5.3 nM and 3.2 nM respectively [143–145]. It inhibits the production of LPS-stimulated cytokine TNF-α in A549 cells and IL-6 production from bone marrow stromal cells and bone marrow mononuclear cells (BMMNC) in patients with multiple myeloma. Several clinical trials are undergoing using this compound alone or in combination with other drugs for the treatment of patients with glioblastoma and other advanced or metastatic cancers [146–149].

Fig. 3.60 Clinical candidate from AstaZeneca

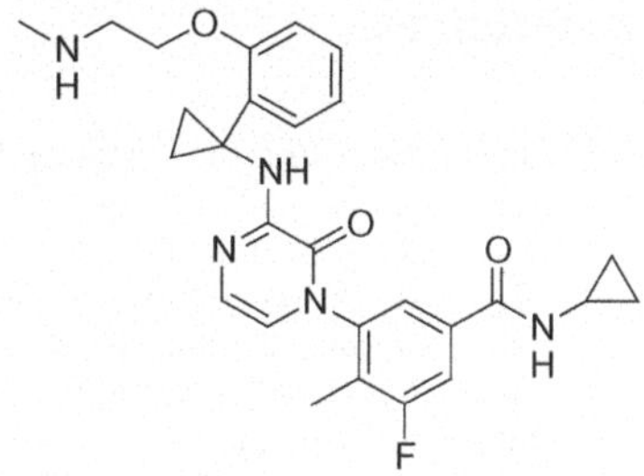

AZD7624

IC_{50} (p38α) = 0.1 nM

IC_{50} (TNF-α) = 3.5 nM

Fig. 3.61 Clinical candidate from Eli Lilly

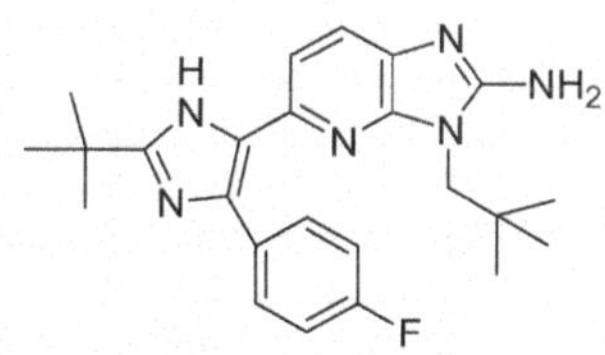

Ralimetinib (LY-2228820)

IC_{50} (p38α) = 5.3 nM

IC_{50} (p38β) = 3.2 nM

In summary, ATP-competitive MAPK inhibitors show a promising class of compounds for modulating the activity of MAP kinases, which are key regulators of various cellular processes such as proliferation, differentiation, apoptosis, and inflammation. By targeting the ATP-binding site of MAPKs, these inhibitors effectively block kinase activity and have potential therapeutic applications in a wide range of diseases where dysregulated MAPK signaling is implicated. However, challenges such as selectivity, pharmacokinetics, and off-target effects remain important considerations in the development of ATP-competitive MAPK inhibitors for clinical use. Further research and optimization efforts are needed to enhance the therapeutic potential and safety profile of these compounds. Overall, ATP-competitive MAPK inhibitors grasp promise as valuable tools for studying MAPK signaling pathways and have the potential to lead to the development of novel therapeutic interventions for various diseases. Several clinical trials are undergoing using several novel small molecules.

References

1. Bennett BL, Sasaki DT, Murray BW, O'Leary EC, Sakata ST, Xu W, Leisten JC, Motiwala A, Pierce S, Satoh Y, Bhagwat SS, Manning AM, Anderson DW. SP600125, an anthrapyrazolone inhibitor of JNK. Proc Natl Acad Sci USA. 2001;98:13681–6.
2. Duong MTH, Lee JH, Ahn HC. C-Jun N-terminal kinase inhibitors: structural insight into kinase-inhibitor complexes. Comput Struct Biotechnol J. 2020;18:1440–57. https://doi.org/10.1016/j.csbj.2020.06.013. PMID: 32637042; PMCID: PMC7327381
3. Ruckle T, Biamonte M, Grippi-Vallotton T, Arkinstall S, Cambet Y, Camps M, Chabert C, Church DJ, Halazy S, Jiang X, Martinou I, Nichols A, Sauer W, Gotteland JP. Design and synthesis and biological activity of novel, potent, and selective, (benzoylaminomethyl)thiophene sulfonamide inhibitors of JNK. J Med Chem. 2004;47:6921–34.
4. Gaillard P, Jeanclaude-Etter I, Ardissone V, Arkinstall S, Cambet Y, Camps M, Chabert C, Church D, Cirillo R, Gretener D, Halazy S, Nichols A, Szyndralewiez C, Vitte PA, Gotteland JP. Design and synthesis of the first generation of novel potent, selective, and in vivo active benzothiazol-2-yl-acetonitrile inhibitors of JNK. J Med Chem. 2005;48:4596–607.
5. Graczyk PP, Khan A, Bhatia GS, Palmer V, Medland D, Numata H, Oinuma H, Catchick J, Dunne A, Ellis M, Smales C, Whitfield J, Neame SJ, Shah B, Wilton D, Morgan L, Patel T, Chung R, Desmond H, Staddon JM, Sato N, Inoue A. The neuroprotective action of JNK3 inhibitors based on the 6,7-dihydro-5H-pyrrolo[1,2-a]imidazole scaffold. Bioorg Med Chem Lett. 2005;15(21):4666–70. https://doi.org/10.1016/j.bmcl.2005.07.076.
6. Humphries PS, Lafontaine JA, Agree CS, Alexander D, Chen P, Do QQ, Li LY, Lunney EA, Rajapakse RJ, Siegel K, Timofeevski SL, Wang T, Wilhite DM. Synthesis and SAR of 4-substituted-2-aminopyrimidines as novel c-Jun N-terminal kinase (JNK) inhibitors. Bioorg Med Chem Lett. 2009;19(8):2099–102. https://doi.org/10.1016/j.bmcl.2009.03.023. Epub 2009 Mar 13
7. Liu M, Xin Z, Clampit JE, Wang S, Gum RJ, Haasch DL, Trevillyan JM, Abad-Zapatero C, Fry EH, Sham HL, Liu G. Synthesis and SAR of 1,9-dihydro-9-hydroxypyrazolo[3,4-b]quinolin-4-ones as novel, selective c-Jun N-terminal kinase inhibitors. Bioorg Med Chem Lett. 2006;16(10):2590–4. https://doi.org/10.1016/j.bmcl.2006.02.046. Epub 2006 Mar 9
8. Liu M, Wang S, Clampit JE, Gum RJ, Haasch DL, Rondinone CM, Trevillyan JM, Abad-Zapatero C, Fry EH, Sham HL, Liu G. Discovery of a new class of 4-anilinopyrimidines as potent c-Jun N-terminal kinase inhibitors: synthesis and SAR studies. Bioorg Med Chem Lett. 2007;17(3):668–72. https://doi.org/10.1016/j.bmcl.2006.10.093. Epub 2006 Nov 2
9. Szczepankiewicz BG, Kosogof C, Nelson LT, Liu G, Liu B, Zhao H, Serby MD, Xin Z, Liu M, Gum RJ, Haasch DL, Wang S, Clampit JE, Johnson EF, Lubben TH, Stashko MA, Olejniczak ET, Sun C, Dorwin SA, Haskins K, Abad-Zapatero C, Fry EH, Hutchins CW, Sham HL, Rondinone CM, Trevillyan JM. Aminopyridine-based c-Jun N-terminal kinase inhibitors with cellular activity and minimal cross-kinase activity. J Med Chem. 2006;49(12):3563–80. https://doi.org/10.1021/jm060199b.
10. Zhao H, Serby MD, Xin Z, Szczepankiewicz BG, Liu M, Kosogof C, Liu B, Nelson LT, Johnson EF, Wang S, Pederson T, Gum RJ, Clampit JE, Haasch DL, Abad-Zapatero C, Fry EH, Rondinone C, Trevillyan JM, Sham HL, Liu G. Discovery of potent, highly selective, and orally bioavailable pyridine carboxamide c-Jun NH2-terminal kinase inhibitors. J Med Chem. 2006;49(15):4455–8. https://doi.org/10.1021/jm060465l.
11. Hom RK, Bowers S, Sealy JM, Truong AP, Probst GD, Neitzel ML, Neitz RJ, Fang L, Brogley L, Wu J, Konradi AW, Sham HL, Toth G, Pan H, Yao N, Artis DR, Quinn K, Sauer JM, Powell K, Ren Z, Bard F, Yednock TA, Griswold-Prenner I. Design and synthesis of disubstituted thiophene and thiazole based inhibitors of JNK. Bioorg Med Chem Lett. 2010;20:7303–7.
12. Bowers S, Truong AP, Neitz RJ, Netzel M, Probst GD, Hom RK, Peterson B, Galemmo RA, Koradi AW, Sham HL, Toth G, Pan H, Yao N, Artis DR, Brigham EF, Quinn KP, Sauer JM, Powell K, Ruslim L, Ren Z, Bard F, Yednock TA, Griswold-Prenner I. Design and synthe-

sis of a novel, orally active, brain penetrant, tri-substituted thiophene based JNK inhibitor. Bioorg Med Chem Lett. 2011;21:1838–43.

13. Bowers S, Truong AP, Jeffrey Neitz R, Hom RK, Sealy JM, Probst GD, Quincy D, Peterson B, Chan W, Galemmo RA Jr, Konradi AW, Sham HL, Tóth G, Pan H, Lin M, Yao N, Artis DR, Zhang H, Chen L, Dryer M, Samant B, Zmolek W, Wong K, Lorentzen C, Goldbach E, Tonn G, Quinn KP, Sauer JM, Wright S, Powell K, Ruslim L, Ren Z, Bard F, Yednock TA, Griswold-Prenner I. Design and synthesis of brain penetrant selective JNK inhibitors with improved pharmacokinetic properties for the prevention of neurodegeneration. Bioorg Med Chem Lett. 2011;21(18):5521–7. https://doi.org/10.1016/j.bmcl.2011.06.100. Epub 2011 Jul 7
14. Probst GD, Bowers S, Sealy JM, Truong AP, Hom RK, Galemmo RA Jr, Konradi AW, Sham HL, Quincy DA, Pan H, Yao N, Lin M, Tóth G, Artis DR, Zmolek W, Wong K, Qin A, Lorentzen C, Nakamura DF, Quinn KP, Sauer JM, Powell K, Ruslim L, Wright S, Chereau D, Ren Z, Anderson JP, Bard F, Yednock TA, Griswold-Prenner I. Highly selective c-Jun N-terminal kinase (JNK) 2 and 3 inhibitors with in vitro CNS-like pharmacokinetic properties prevent neurodegeneration. Bioorg Med Chem Lett. 2011;21(1):315–9. https://doi.org/10.1016/j.bmcl.2010.11.010. Epub 2010 Nov 5
15. Neitz RJ, Konradi AW, Sham HL, Zmolek W, Wong K, Qin A, Lorentzen C, Nakmura D, Quinn KP, Sauer JM, Powell K, Ruslim L, Chereau D, Ren Z, Anderson J, Bard F, Yednock TA, Griswold,-Prenner, I. Highly selective c-Jun N-terminal kinase 3 inhibitors with in vitro CNS-like pharmacokinetic properties II. Central core replacement. Bioog Med Chem Lett. 2011;21:3726–9.
16. Kuglstatter A, Ghate M, Tsing S, Villasenor AG, Shaw D, Barnett JW, Browner MF. X-ray crystal structure of JNk2 complexed with the p-38 inhibitor BIRB796: insights into the rational design of DFG-out binding MAP kinase inhibitors. Bioorg Med Chem Lett. 2010;20:5217–20.
17. Palmer WS, Alam M, Arzeno HB, Chang KC, Dunn JP, Goldstein DM, Gong L, Goyal B, Hermann JC, Hogg JH, Hsieh G, Jahangir A, Janson C, Jin S, Ursula Kammlott R, Kuglstatter A, Lukacs C, Michoud C, Niu L, Reuter DC, Shao A, Silva T, Trejo-Martin TA, Stein K, Tan YC, Tivitmahaisoon P, Tran P, Wagner P, Weller P, Wu SY. Development of amino-pyrimidine inhibitors of c-Jun N-terminal kinase (JNK): kinase profiling guided optimization of a 1,2,3-benzotriazole lead. Bioorg Med Chem Lett. 2013;23(5):1486–92.
18. Gong L, Han X, Silva T, Tan YC, Goyal B, Tivitmahaisoon P, Trejo A, Palmer W, Hogg H, Jahagir A, Alam M, Wagner P, Stein K, Filonova L, Loe B, Makra F, Rotstein D, Rapatova L, Dunn J, Zuo F, Dal Porto J, Wong B, Jin S, Chang A, Tran P, Hsieh G, Niu L, Shao A, Reuter D, Hermann J, Kuglstatter A, Goldstein D. Development of indole/indazole-aminopyrimidines as inhibitors of c-Jun N-terminal kinase (JNK): optimization for JNK potency and physicochemical properties. Bioorg Med Chem Lett. 2013;23(12):3565–9.
19. Gong L, Tan YC, Boice G, Abbot S, McCaleb K, Iyer P, Zuo F, Dal Porto J, Wong B, Jin S, Chang A, Tran P, Hsieh G, Niu L, Shao A, Reuter D, Lukacs CM, Ursula Kammlott R, Kuglstatter A, Goldstein D. Discovery of a novel series of 4-quinolone JNK inhibitors. Bioorg Med Chem Lett. 2012;22(24):7381–7. https://doi.org/10.1016/j.bmcl.2012.10.066. Epub 2012 Oct 24
20. Haynes NE, Scott NR, Chen LC, Janson CA, Li JK, Lukacs CM, Railkar A, Tozzo E, Whittard T, Brown NF, Cheung AW. Identification of an Adamantyl Azaquinolone JNK selective inhibitor. ACS Med Chem Lett. 2012;3(9):764–8. https://doi.org/10.1021/ml300175c. PMID: 24900545; PMCID: PMC4025727
21. Asano Y, Kitamura S, Ohra T, Aso K, Igata H, Tamura T, Kawamoto T, Tanaka T, Sogabe S, Matsumoto S, Yamaguchi M, Kimura H, Itoh F. Discovery, synthesis and biological evaluation of isoquinolones as novel and highly selective JNK inhibitors (1). Bioorg Med Chem. 2008;16(8):4715–32. https://doi.org/10.1016/j.bmc.2008.02.027. Epub 2008 Feb 13

22. Asano Y, Kitamura S, Ohra T, Itoh F, Kajino M, Tamura T, Kaneko M, Ikeda S, Igata H, Kawamoto T, Sogabe S, Matsumoto S, Tanaka T, Yamaguchi M, Kimura H, Fukumoto S. Discovery, synthesis and biological evaluation of isoquinolones as novel and highly selective JNK inhibitors (2). Bioorg Med Chem. 2008;16(8):4699–714. https://doi.org/10.1016/j.bmc.2008.02.028. Epub 2008 Feb 13
23. Swahn BM, Xue Y, Arzel E, Kallin E, Magnus A, Plobeck N, Viklund J. Design and synthesis of 2'-anilino-4,4-bipyridines as selective inhibitors of JNK-3. Bioorg Med Chem Lett. 2006;16:1397–401.
24. Swahn BM, Huerta F, Kallin E, Malmstrom J, Weigelt T, Viklund J, Womack P, Xue Y, Ohberg L. Design and synthesis of 6-anilinoindazoles as selective inhibitors of JNK. Bioorg Med Chem Lett. 2005;15:5095–9.
25. Stocks MJ, Barber S, Ford R, Leroux F, St-Gallay S, Teague S, Xue Y. Structure-driven HtL: design and synthesis of novel aminoindazole inhibitors of c-Jun N-terminal kinase activity. Bioorg Med Chem Lett. 2005;15(14):3459–62. https://doi.org/10.1016/j.bmcl.2005.05.008.
26. Cao J, Gao H, Bemis G, Salituro F, Ledeboer M, Harrington E, Wilke S, Taslimi P, Pazhanisamy S, Xie X, Jacobs M, Green J. Structure-based design and parallel synthesis of N-benzyl isatin oximes as JNK3 MAP kinase inhibitors. Bioorg Med Chem Lett. 2009;19(10):2891–5. https://doi.org/10.1016/j.bmcl.2009.03.043. Epub 2009 Mar 17
27. Christopher JA, Atkinson FL, Bax BD, Brown MJ, Champigny AC, Chuang TT, Jones EJ, Mosley JE, Musgrave JR. 1-Aryl-3,4-dihydroisoquinoline inhibitors of JNK3. Bioorg Med Chem Lett. 2009;19(8):2230–4. https://doi.org/10.1016/j.bmcl.2009.02.098. Epub 2009 Feb 28
28. Angell RM, Atkinson FL, Brown MJ, Chuang TT, Christopher JA, Cichy-Knight M, Dunn AK, Hightower KE, Malkakorpi S, Musgrave JR, Neu M, Rowland P, Shea RL, Smith JL, Somers DO, Thomas SA, Thompson G, Wang R. N-(3-Cyano-4,5,6,7-tetrahydro-1-benzothien-2-yl)amides as potent, selective, inhibitors of JNK2 and JNK3. Bioorg Med Chem Lett. 2007;17(5):1296–301. https://doi.org/10.1016/j.bmcl.2006.12.003. Epub 2006 Dec 15
29. Alam M, Beevers RE, Ceska T, Davenport RJ, Dickson KM, Fortunato M, Gowers L, Haughan AF, James LA, Jones MW, Kinsella N, Lowe C, Meissner JW, Nicolas AL, Perry BG, Phillips DJ, Pitt WR, Platt A, Ratcliffe AJ, Sharpe A, Tait LJ. Synthesis and SAR of aminopyrimidines as novel c-Jun N-terminal kinase (JNK) inhibitors. Bioorg Med Chem Lett. 2007;17(12):3463–7. https://doi.org/10.1016/j.bmcl.2007.03.078. Epub 2007 Mar 30
30. Ansideri F, Macedo JT, Eitel M, El-Gokha A, Zinad DS, Scarpellini C, Kudolo M, Schollmeyer D, Boeckler FM, Blaum BS, Laufer SA, Koch P. Structural optimization of a Pyridinylimidazole scaffold: shifting the selectivity from p38α mitogen-activated protein kinase to c-Jun N-terminal kinase 3. ACS Omega. 2018;3(7):7809–31. https://doi.org/10.1021/acsomega.8b00668. Epub 2018 Jul 12. PMID: 30087925; PMCID: PMC6072243
31. Shin Y, Chen W, Habel J, Duckett D, Ling YY, Koenig M, He Y, Vojkovsky T, LoGrasso P, Kamenecka TM. Synthesis and SAR of piperazine amides as novel c-Jun N-terminal kinase (JNK) inhibitors. Bioorg Med Chem Lett. 2009;19(12):3344–7. https://doi.org/10.1016/j.bmcl.2009.03.086. Epub 2009 Mar 26. PMID: 19433357; PMCID: PMC2737472
32. Kamenecka T, Jiang R, Song X, Duckett D, Chen W, Ling YY, Habel J, Laughlin JD, Chambers J, Figuera-Losada M, Cameron MD, Lin L, Ruiz CH, LoGrasso PV. Synthesis, biological evaluation, X-ray structure, and pharmacokinetics of aminopyrimidine c-Jun-N-terminal kinase (JNK) inhibitors. J Med Chem. 2010;53(1):419–31. https://doi.org/10.1021/jm901351f. Erratum in: J Med Chem. 2010 Feb 25;53(4):1882. PMID: 19947601; PMCID: PMC2804074
33. He Y, Kamenecka TM, Shin Y, Song X, Jiang R, Noel R, Duckett D, Chen W, Ling YY, Cameron MD, Lin L, Khan S, Koenig M, LoGrasso PV. Synthesis and SAR of novel quinazolines as potent and brain-penetrant c-Jun N-terminal kinase (JNK) inhibitors. Bioorg Med Chem Lett. 2011;21(6):1719–23. https://doi.org/10.1016/j.bmcl.2011.01.079. Epub 2011 Jan 22. PMID: 21316221; PMCID: PMC3052630

34. Noël R, Shin Y, Song X, He Y, Koenig M, Chen W, Ling YY, Lin L, Ruiz CH, LoGrasso P, Cameron MD, Duckett DR, Kamenecka TM. Synthesis and SAR of 4-(pyrazol-3-yl)-pyridines as novel c-Jun N-terminal kinase inhibitors. Bioorg Med Chem Lett. 2011;21(9):2732–5. https://doi.org/10.1016/j.bmcl.2010.11.104. Epub 2010 Dec 1. PMID: 21185177; PMCID: PMC3081976
35. Song X, Chen W, Lin L, Ruiz CH, Cameron MD, Duckett DR, Kamenecka TM. Synthesis and SAR of 2-phenoxypyridines as novel c-Jun N-terminal kinase inhibitors. Bioorg Med Chem Lett. 2011;21(23):7072–5. https://doi.org/10.1016/j.bmcl.2011.09.090. Epub 2011 Sep 29
36. Chambers JW, Pachori A, Howard S, Ganno M, Hansen D Jr, Kamenecka T, Song X, Duckett D, Chen W, Ling YY, Cherry L, Cameron MD, Lin L, Ruiz CH, Lograsso P. Small molecule c-Jun-N-terminal kinase (JNK) inhibitors protect dopaminergic neurons in a model of Parkinson's disease. ACS Chem Neurosci. 2011;2(4):198–206. https://doi.org/10.1021/cn100109k. PMID: 21666839; PMCID: PMC3110074
37. Crocker CE, Khan S, Cameron MD, Robertson HA, Robertson GS, Lograsso P. JNK inhibition protects dopamine neurons and provides Behavioral improvement in a rat 6-hydroxydopamine model of Parkinson's disease. ACS Chem Neurosci. 2011;2(4):207–12. https://doi.org/10.1021/cn1001107. PMID: 21666838; PMCID: PMC3110072
38. Kamenecka T, Habel J, Duckett D, Chen W, Ling YY, Frackowiak B, Jiang R, Shin Y, Song X, LoGrasso P. Structure-activity relationships and X-ray structures describing the selectivity of aminopyrazole inhibitors for c-Jun N-terminal kinase 3 (JNK3) over p38. J Biol Chem. 2009;284(19):12853–61. https://doi.org/10.1074/jbc.M809430200. Epub 2009 Mar 4. PMID: 19261605; PMCID: PMC2676016
39. Zheng K, Iqbal S, Hernandez P, Park H, LoGrasso PV, Feng Y. Design and synthesis of highly potent and isoform selective JNK3 inhibitors: SAR studies on aminopyrazole derivatives. J Med Chem. 2014;57(23):10013–30. https://doi.org/10.1021/jm501256y. Epub 2014 Nov 21. Erratum in: J Med Chem. 2016 Oct 13;59(19):9276. PMID: 25393557; PMCID: PMC4266361
40. Jiang R, Duckett D, Chen W, Habel J, Ling YY, LoGrasso P, Kamenecka TM. 3,5-Disubstituted quinolines as novel c-Jun N-terminal kinase inhibitors. Bioorg Med Chem Lett. 2007;17(22):6378–82. https://doi.org/10.1016/j.bmcl.2007.08.054. Epub 2007 Aug 26
41. Krenitsky VP, Delgado M, Nadolny L, Sahasrabudhe K, Ayala L, Clareen SS, Hilgraf R, Albers R, Kois A, Hughes K, Wright J, Nowakowski J, Sudbeck E, Ghosh S, Bahmanyar S, Chamberlain P, Muir J, Cathers BE, Giegel D, Xu L, Celeridad M, Moghaddam M, Khatsenko O, Omholt P, Katz J, Pai S, Fan R, Tang Y, Shirley MA, Benish B, Blease K, Raymon H, Bhagwat S, Henderson I, Cole AG, Bennett B, Satoh Y. Aminopurine based JNK inhibitors for the prevention of ischemia reperfusion injury. Bioorg Med Chem Lett. 2012;22(3):1427–32. https://doi.org/10.1016/j.bmcl.2011.12.028. Epub 2011 Dec 11
42. Krenitsky VP, Nadolny L, Delgado M, Ayala L, Clareen SS, Hilgraf R, Albers R, Hrgde S, D'Sidocky N, Sapienza J, Wright J, McCarrick M, Bahmanyar S, Chamberlin P, Delker SL, Muir J, Giegel D, Xu L, Celeridad M, Lachowitzer J, Bennet B, Moghaddam M, Khatsenko O, Katz J, Fan R, Bai A, Tang Y, Shirley MA, Benish B, Bodine T, Blease K, Raymon H, Cathers BE, Satoh Y. Discovery of CC-930-, an orally active anti-fibrotic JNK inhibitor. Bioorg Med Chem Lett. 2012;22:1433.
43. Wityak J, McGee KF, Conlon MP, Song RH, Duffy BC, Clayton B, Lynch M, Wang G, Freeman E, Haber J, Kitchen DB, Manning DD, Ismail J, Khmelnitsky Y, Michels P, Webster J, Irogoyen M, Luche M, Hultman M, Bai M, Kuok ID, Newell R, Lamers M, Leonard P, Yates D, Matthews K, Ongeri L, Clifton S, Mead T, Deupree S, Wheelan P, Lyons K, Wilson C, Kiselyov A, Toledo-Sherman L, Beconi M, Munoz-Sanjuan I, Bard J, Dominguez C. Lead optimization toward proof-of-concept tools for Huntington's disease within a 4-(1H-pyrazol-4-yl)pyrimidine class of pan-JNK inhibitors. J Med Chem. 2015;58:2967.

44. Dou X, Huang H, Li Y, Jiang L, Wang Y, Jin H, Jiao N, Zhang L, Zhang L, Liu Z. Multistage screening reveals 3-substituted Indolin-2-one derivatives as novel and isoform-selective c-Jun N-terminal kinase 3 (JNK3) inhibitors: implications to drug discovery for potential treatment of neurodegenerative diseases. J Med Chem. 2019;62(14):6645–64. https://doi.org/10.1021/acs.jmedchem.9b00537. Epub 2019 Jul 15
45. Schepetkin IA, Karpenko OS, Kovrizhina AR, Kirpotina LN, Khlebnikov AI, Chekal SI, Radudik AV, Shybinska MO, Quinn MT. Novel Tryptanthrin derivatives with selectivity as c-Jun N-terminal kinase (JNK) 3 inhibitors. Molecules. 2023;28(12):4806. https://doi.org/10.3390/molecules28124806. PMID: 37375361; PMCID: PMC10304880
46. Zhou Z, Zhang D, Du F, An W, Ge J, Yu C, Yang N, Zhang C, Lim K, Li L. Design and synthesis of a mitochondrial-targeted JNK inhibitor and its protective effect on Parkinson's disease phenotypes. Chembiochem. 2023;24(13):e202200748. https://doi.org/10.1002/cbic.202200748. Epub 2023 Jun 2
47. Shuai W, Bu F, Zhu Y, Wu Y, Xiao H, Pan X, Zhang J, Sun Q, Wang G, Ouyang L. Discovery of novel Indazole Chemotypes as isoform-selective JNK3 inhibitors for the treatment of Parkinson's disease. J Med Chem. 2023;66(2):1273–300. https://doi.org/10.1021/acs.jmedchem.2c01410. Epub 2023 Jan 17
48. Resnick L, Fennell M. Targeting JNK3 for the treatment of neurodegenerative disorders. Drug Discov Today. 2004;9(21):932–9. https://doi.org/10.1016/S1359-6446(04)03251-9.
49. Cuny GD. Kinase inhibitors as potential therapeutics for acute and chronic neurodegenerative conditions. Curr Pharm Des. 2009;15(34):3919–39. https://doi.org/10.2174/138161209789649330.
50. Yang XL, Gao CM. C-Jun N-terminal kinase 3 signalling serves a potential role as a biomarker for determining the pathogenesis of Parkinson's disease. Mol Med Rep. 2018;17(2):3255–9. https://doi.org/10.3892/mmr.2017.8244. Epub 2017 Dec 11
51. Christensen DP, Ejlerskov P, Rasmussen I, Vilhardt F. Reciprocal signals between microglia and neurons regulate α-synuclein secretion by exophagy through a neuronal cJUN-N-terminal kinase-signaling axis. J Neuroinflammation. 2016;13(1):59. https://doi.org/10.1186/s12974-016-0519-5. PMID: 26957005; PMCID: PMC4782388
52. Pan J, Li H, Zhang B, Xiong R, Zhang Y, Kang WY, Chen W, Zhao ZB, Chen SD. Small peptide inhibitor of JNK3 protects dopaminergic neurons from MPTP induced injury via inhibiting the ASK1-JNK3 signaling pathway. PLoS One. 2015;10(4):e0119204. https://doi.org/10.1371/journal.pone.0119204. PMID: 25856433; PMCID: PMC4391862
53. Choi WS, Klintworth HM, Xia Z. JNK3-mediated apoptotic cell death in primary dopaminergic neurons. Methods Mol Biol. 2011;758:279–92. https://doi.org/10.1007/978-1-61779-170-3_19. PMID: 21815073; PMCID: PMC3490420
54. Tönges L, Planchamp V, Koch JC, Herdegen T, Bähr M, Lingor P. JNK isoforms differentially regulate neurite growth and regeneration in dopaminergic neurons in vitro. J Mol Neurosci. 2011;45(2):284–93. https://doi.org/10.1007/s12031-011-9519-1. Epub 2011 Apr 6. PMID: 21468718; PMCID: PMC3178779
55. Hunot S, Vila M, Teismann P, Davis RJ, Hirsch EC, Przedborski S, Rakic P, Flavell RA. JNK-mediated induction of cyclooxygenase 2 is required for neurodegeneration in a mouse model of Parkinson's disease. Proc Natl Acad Sci USA. 2004;101(2):665–70. https://doi.org/10.1073/pnas.0307453101. Epub 2004 Jan 2. PMID: 14704277; PMCID: PMC327205
56. Nagy MA, Hilgraf R, Mortensen DS, Elsner J, Norris S, Tikhe J, Yoon W, Paisner D, Delgado M, Erdman P, Haelewyn J, Khambatta G, Xu L, Romanow WJ, Condroski K, Bahmanyar S, McCarrick M, Benish B, Blease K, LeBrun L, Moghaddam MF, Apuy J, Canan SS, Bennett BL, Satoh Y. Discovery of the c-Jun N-terminal kinase inhibitor **CC-90001**. J Med Chem. 2021;64(24):18193–208. https://doi.org/10.1021/acs.jmedchem.1c01716. Epub 2021 Dec 13
57. U.S. National Library of Medicine. Study Record Detail Page. https://clinicaltrials.gov/ct2/show/NCT03142191. Accessed Sept 2021

58. Ye Y, Gaudy A, Thomas M, Reyes J, Burkhardt B, Horan G, Liu L, Chen J, Ghosh A, Carayannopoulos LN, Tatosian DA, Palmisano M. Safety, pharmacokinetics, and pharmacodynamics of CC-90001 (BMS-986360), a c-Jun N-terminal kinase inhibitor, in phase 1 studies in healthy participants. Clin Pharmacol Drug Dev. 2022;11(12):1394–404. https://doi.org/10.1002/cpdd.1178. Epub 2022 Oct 18. PMID: 36256505; PMCID: PMC10092235
59. Horan G, Ye Y, Adams M, Parton A, Cedzik D, Tang S, Brown EA, Liu L, Nissel J, Carayannopoulos LN, Gaudy A, Schafer P, Palmisano M, Ramirez-Valle F. Safety, pharmacokinetics, and Antifibrotic activity of CC-90001 (BMS-986360), a c-Jun N-terminal kinase inhibitor, in pulmonary fibrosis. Clin Pharmacol Drug Dev. 2023;12(8):779–89. https://doi.org/10.1002/cpdd.1294. Epub 2023 Jun 28
60. Popmihajlov Z, Sutherland DJ, Horan GS, Ghosh A, Lynch DA, Noble PW, Richeldi L, Reiss TF, Greenberg S. CC-90001, a c-Jun N-terminal kinase (JNK) inhibitor, in patients with pulmonary fibrosis: design of a phase 2, randomised, placebo-controlled trial. BMJ Open Respir Res. 2022;9(1):e001060. https://doi.org/10.1136/bmjresp-2021-001060. PMID: 35058236; PMCID: PMC8783810
61. Tong Z, Gaudy A, Tatosian D, Ramirez-Valle F, Liu H, Chen J, Hoffmann M, Surapaneni S. Assessment of drug-drug interactions of CC-90001, a potent and selective inhibitor of c-Jun N-terminal kinase. Xenobiotica. 2021;51(12):1416–26. https://doi.org/10.1080/00498254.2022.2027553. Epub 2022 Jan 13. Erratum in: Xenobiotica. 2022 Feb;52(2):218
62. Aronov AM, Tang Q, Martinez-Botella G, Bemis GW, Cao J, Chen G, Ewing NP, Ford PJ, Germann UA, Green J, Hale MR, Jacobs M, Janetka JW, Maltais F, Markland W, Namchuk MN, Nanthakumar S, Poondru S, Straub J, ter Haar E, Xie X. Structure-guided design of potent and selective pyrimidylpyrrole inhibitors of extracellular signal-regulated kinase (ERK) using conformational control. J Med Chem. 2009;52(20):6362–8. https://doi.org/10.1021/jm900630q. PMID: 19827834
63. Ohori M, Kinoshita T, Okubo M, Sato K, Yamazaki A, Arakawa H, Nishimura S, Inamura N, Nakajima H, Neya M, Miyake H, Fujii T. Identification of a selective ERK inhibitor and structural determination of the inhibitor-ERK2 complex. Biochem Biophys Res Commun. 2005;336(1):357–63. https://doi.org/10.1016/j.bbrc.2005.08.082.
64. Liu B, Fu L, Zhang C, Zhang L, Zhang Y, Ouyang L, He G, Huang J. Computational design, chemical synthesis, and biological evaluation of a novel ERK inhibitor (BL-EI001) with apoptosis-inducing mechanisms in breast cancer. Oncotarget. 2015;6(9):6762–75. https://doi.org/10.18632/oncotarget.3105. PMID: 25742792; PMCID: PMC4466648
65. Heightman TD, Berdini V, Braithwaite H, Buck IM, Cassidy M, Castro J, Courtin A, Day JEH, East C, Fazal L, Graham B, Griffiths-Jones CM, Lyons JF, Martins V, Muench S, Munck JM, Norton D, O'Reilly M, Palmer N, Pathuri P, Reader M, Rees DC, Rich SJ, Richardson C, Saini H, Thompson NT, Wallis NG, Walton H, Wilsher NE, Woolford AJ, Cooke M, Cousin D, Onions S, Shannon J, Watts J, Murray CW. Fragment based discovery of a potent, orally bioavailable inhibitor that modulates the phosphorylation and catalytic activity of ERK1/2. J Med Chem. 2018;61(11):4978–92. https://doi.org/10.1021/acs.jmedchem.8b00421. Epub 2018 May 31
66. Ji D, Zhang L, Zhu Q, Bai Y, Wu Y, Xu Y. Discovery of potent, orally bioavailable ERK1/2 inhibitors with isoindolin-1-one structure by structure-based drug design. Eur J Med Chem. 2019;164:334–41. https://doi.org/10.1016/j.ejmech.2018.12.040. Epub 2018 Dec 21
67. Bagdanoff JT, Jain R, Han W, Zhu S, Madiera AM, Lee PS, Ma X, Poon D. Tetrahydropyrrolo-diazepenones as inhibitors of ERK2 kinase. Bioorg Med Chem Lett. 2015;25(18):3788–92. https://doi.org/10.1016/j.bmcl.2015.07.091.
68. Ren L, Grina J, Moreno D, Blake JF, Gaudino JJ, Garrey R, Metcalf AT, Burkard M, Martinson M, Rasor K, Chen H, Dean B, Gould SE, Pacheco P, Shahidi-Latham S, Yin J, West K, Wang W, Moffat JG, Schwarz JB. Discovery of highly potent, selective, and efficacious small molecule inhibitors of ERK1/2. J Med Chem. 2015;58(4):1976–91. https://doi.org/10.1021/jm501921k. Epub 2015 Feb 11. PMID: 25603482

69. Aly AA, El-Sheref EM, Bakheet MEM, Mourad MAE, Bräse S, Ibrahim MAA, Nieger M, Garvalov BK, Dalby KN, Kaoud TS. Design, synthesis and biological evaluation of fused naphthofuro[3,2-c] quinoline-6,7,12-triones and pyrano[3,2-c]quinoline-6,7,8,13-tetraones derivatives as ERK inhibitors with efficacy in BRAF-mutant melanoma. Bioorg Chem. 2019;82:290–305. https://doi.org/10.1016/j.bioorg.2018.10.044. Epub 2018 Oct 23. PMID: 30396063; PMCID: PMC6543826
70. Burdick DJ, Wang S, Heise C, Pan B, Drummond J, Yin J, Goeser L, Magnuson S, Blaney J, Moffat J, Wang W, Chen H. Fragment-based discovery of potent ERK2 pyrrolopyrazine inhibitors. Bioorg Med Chem Lett. 2015;25(21):4728–32. https://doi.org/10.1016/j.bmcl.2015.08.048. Epub 2015 Aug 22
71. Li L, Liu F, Jin N, Tang S, Chen Z, Yang X, Ding J, Geng M, Jiang L, Huang M, Cao J. Discovery and structure activity relationship study of novel indazole amide inhibitors for extracellular signal-regulated kinase1/2 (ERK1/2). Bioorg Med Chem Lett. 2016;26(11):2600–4. https://doi.org/10.1016/j.bmcl.2016.04.029. Epub 2016 Apr 12. PMID: 27106711
72. Pan X, Pei J, Wang A, Shuai W, Feng L, Bu F, Zhu Y, Zhang L, Wang G, Ouyang L. Development of small molecule extracellular signal-regulated kinases (ERKs) inhibitors for cancer therapy. Acta Pharm Sin B. 2022;12(5):2171–92. https://doi.org/10.1016/j.apsb.2021.12.022. Epub 2022 Jan 4. PMID: 35646548; PMCID: PMC9136582
73. Fu L, Chen S, He G, Chen Y, Liu B. Targeting extracellular signal-regulated protein kinase 1/2 (ERK1/2) in cancer: An update on pharmacological small-molecule inhibitors. J Med Chem. 2022;65(20):13561–73. https://doi.org/10.1021/acs.jmedchem.2c01244. Epub 2022 Oct 7
74. Germann UA, Furey BF, Markland W, Hoover RR, Aronov AM, Roix JJ, Hale M, Boucher DM, Sorrell DA, Martinez-Botella G, Fitzgibbon M, Shapiro P, Wick MJ, Samadani R, Meshaw K, Groover A, DeCrescenzo G, Namchuk M, Emery CM, Saha S, Welsch DJ. Targeting the MAPK Signaling pathway in cancer: promising preclinical activity with the novel selective ERK1/2 inhibitor BVD-523 (Ulixertinib). Mol Cancer Ther. 2017;16(11):2351–63. https://doi.org/10.1158/1535-7163.MCT-17-0456. Epub 2017 Sep 22
75. Ward RA, Bethel P, Cook C, Davies E, Debreczeni JE, Fairley G, Feron L, Flemington V, Graham MA, Greenwood R, Griffin N, Hanson L, Hopcroft P, Howard TD, Hudson J, James M, Jones CD, Jones CR, Lamont S, Lewis R, Lindsay N, Roberts K, Simpson I, St-Gallay S, Swallow S, Tang J, Tonge M, Wang Z, Zhai B. Structure-guided discovery of potent and selective inhibitors of ERK1/2 from a modestly active and promiscuous chemical start point. J Med Chem. 2017;60(8):3438–50. https://doi.org/10.1021/acs.jmedchem.7b00267. Epub 2017 Apr 14
76. Ward RA, Anderton MJ, Bethel P, Breed J, Cook C, Davies EJ, Dobson A, Dong Z, Fairley G, Farrington P, Feron L, Flemington V, Gibbons FD, Graham MA, Greenwood R, Hanson L, Hopcroft P, Howells R, Hudson J, James M, Jones CD, Jones CR, Li Y, Lamont S, Lewis R, Lindsay N, McCabe J, McGuire T, Rawlins P, Roberts K, Sandin L, Simpson I, Swallow S, Tang J, Tomkinson G, Tonge M, Wang Z, Zhai B. Discovery of a potent and selective Oral inhibitor of ERK1/2 (AZD0364) that is efficacious in both monotherapy and combination therapy in models of nonsmall cell lung cancer (NSCLC). J Med Chem. 2019;62(24):11004–18. https://doi.org/10.1021/acs.jmedchem.9b01295. Epub 2019 Nov 25
77. Bhagwat SV, McMillen WT, Cai S, Zhao B, Whitesell M, Shen W, Kindler L, Flack RS, Wu W, Anderson B, Zhai Y, Yuan XJ, Pogue M, Van Horn RD, Rao X, McCann D, Dropsey AJ, Manro J, Walgren J, Yuen E, Rodriguez MJ, Plowman GD, Tiu RV, Joseph S, Peng SB. ERK inhibitor LY3214996 targets ERK pathway-driven cancers: a therapeutic approach toward precision medicine. Mol Cancer Ther. 2020;19(2):325–36. https://doi.org/10.1158/1535-7163.MCT-19-0183. Epub 2019 Nov 19
78. Köhler J, Zhao Y, Li J, Gokhale PC, Tiv HL, Knott AR, Wilkens MK, Soroko KM, Lin M, Ambrogio C, Musteanu M, Ogino A, Choi J, Bahcall M, Bertram AA, Chambers ES, Paweletz CP, Bhagwat SV, Manro JR, Tiu RV, Jänne PA. ERK inhibitor LY3214996-based treatment strategies for RAS-driven lung cancer. Mol Cancer Ther. 2021;20(4):641–54. https://doi.org/10.1158/1535-7163.MCT-20-0531. Epub 2021 Feb 3. PMID: 33536188; PMCID: PMC8026682

79. Heightman TD, Berdini V, Bevan L, Buck IM, Carr MG, Courtin A, Coyle JE, Day JEH, East C, Fazal L, Griffiths-Jones CM, Howard S, Kucia-Tran J, Martins V, Muench S, Munck JM, Norton D, O'Reilly M, Palmer N, Pathuri P, Peakman TM, Reader M, Rees DC, Rich SJ, Shah A, Wallis NG, Walton H, Wilsher NE, Woolford AJ, Cooke M, Cousin D, Onions S, Shannon J, Watts J, Murray CW. Discovery of ASTX029, a clinical candidate which modulates the phosphorylation and catalytic activity of ERK1/2. J Med Chem. 2021;64(16):12286–303. https://doi.org/10.1021/acs.jmedchem.1c00905. Epub 2021 Aug 13
80. Munck JM, Berdini V, Bevan L, Brothwood JL, Castro J, Courtin A, East C, Ferraldeschi R, Heightman TD, Hindley CJ, Kucia-Tran J, Lyons JF, Martins V, Muench S, Murray CW, Norton D, O'Reilly M, Reader M, Rees DC, Rich SJ, Richardson CJ, Shah AD, Stanczuk L, Thompson NT, Wilsher NE, Woolford AJ, Wallis NG. ASTX029, a novel dual-mechanism ERK inhibitor, modulates both the phosphorylation and catalytic activity of ERK. Mol Cancer Ther. 2021;20(10):1757–68. https://doi.org/10.1158/1535-7163.MCT-20-0909. Epub 2021 Jul 30
81. Ku BM, Heo JY, Kim J, Sun JM, Lee SH, Ahn JS, Park K, Ahn MJ. ERK inhibitor ASN007 effectively overcomes acquired resistance to EGFR inhibitor in non-small cell lung cancer. Investig New Drugs. 2022;40(2):265–73. https://doi.org/10.1007/s10637-021-01121-6. Epub 2022 Jan 1. PMID: 34973117; PMCID: PMC8993753
82. Portelinha A, Thompson S, Smith RA, Da Silva FM, Asgari Z, Knezevic A, Seshan V, de Stanchina E, Gupta S, Denis L, Younes A, Reddy S. ASN007 is a selective ERK1/2 inhibitor with preferential activity against RAS-and RAF-mutant tumors. Cell Rep Med. 2021;2(7):100350. https://doi.org/10.1016/j.xcrm.2021.100350. PMID: 34337566; PMCID: PMC8324497
83. Burrows F, Kessler L, Chen J, Gao X, Hansen R, Li S, et al. Abstract 5168: KO-947, a potent ERK inhibitor with robust preclinical single agent activity in MAPK pathway dysregulated tumors. Cancer Res. 2017;77:5168.
84. Li L, Huang M, Li L, Chen Y, Tang S, Su Y, Dong R, Ding J, Geng M. Abstract 3760: HH2710, a highly potent and selective erk1/2 inhibitor for the treatment of mapk mutant tumors. Cancer Res. 2020;80(16_Supplement):3760.
85. Gardner SL, Tarapore RS, Allen J, McGovern SL, Zaky W, Odia Y, Daghistani D, Diaz Z, Hall MD, Khatib Z, Koschmann C, Cantor E, Kurokawa R, MacDonald TJ, Aguilera D, Vitanza NA, Mueller S, Kline C, Lu G, Allen JE, Khatua S. Phase I dose escalation and expansion trial of single agent ONC201 in pediatric diffuse midline gliomas following radiotherapy. Neurooncol Adv. 2022;4(1):vdac143. https://doi.org/10.1093/noajnl/vdac143. PMID: 36382108; PMCID: PMC9639395
86. Janku F, Elez E, Iyer G, Yamamoto N, Shao Weng Tan D, Stathis A, Stammberger UM, Evans H, Seroutou A, Mais A, Schuler MH. Meeting Abstract: 2020 ASCO Annual Meeting I
87. Blake JF, Gaudino JJ, De Meese J, Mohr P, Chicarelli M, Tian H, Garrey R, Thomas A, Siedem CS, Welch MB, Kolakowski G, Kaus R, Burkard M, Martinson M, Chen H, Dean B, Dudley DA, Gould SE, Pacheco P, Shahidi-Latham S, Wang W, West K, Yin J, Moffat J, Schwarz JB. Discovery of 5,6,7,8-tetrahydropyrido[3,4-d]pyrimidine inhibitors of Erk2. Bioorg Med Chem Lett. 2014;24(12):2635–9. https://doi.org/10.1016/j.bmcl.2014.04.068. Epub 2014 Apr 29
88. Blake JF, Burkard M, Chan J, Chen H, Chou KJ, Diaz D, Dudley DA, Gaudino JJ, Gould SE, Grina J, Hunsaker T, Liu L, Martinson M, Moreno D, Mueller L, Orr C, Pacheco P, Qin A, Rasor K, Ren L, Robarge K, Shahidi-Latham S, Stults J, Sullivan F, Wang W, Yin J, Zhou A, Belvin M, Merchant M, Moffat J, Schwarz JB. Discovery of (S)-1-(1-(4-Chloro-3-fluorophenyl)-2-hydroxyethyl)-4-(2-((1-methyl-1H-pyrazol-5-yl)amino)pyrimidin-4-yl) pyridin-2(1H)-one (GDC-0994), an extracellular signal-regulated kinase 1/2 (ERK1/2) inhibitor in early clinical development. J Med Chem. 2016;59(12):5650–60. https://doi.org/10.1021/acs.jmedchem.6b00389. Epub 2016 Jun 7

89. Boga SB, Deng Y, Zhu L, Nan Y, Cooper AB, Shipps GW Jr, Doll R, Shih NY, Zhu H, Sun R, Wang T, Paliwal S, Tsui HC, Gao X, Yao X, Desai J, Wang J, Alhassan AB, Kelly J, Patel M, Muppalla K, Gudipati S, Zhang LK, Buevich A, Hesk D, Carr D, Dayananth P, Black S, Mei H, Cox K, Sherborne B, Hruza AW, Xiao L, Jin W, Long B, Liu G, Taylor SA, Kirschmeier P, Windsor WT, Bishop R, Samatar AA. MK-8353: discovery of an orally bioavailable dual mechanism ERK inhibitor for oncology. ACS Med Chem Lett. 2018;9(7):761–7. https://doi.org/10.1021/acsmedchemlett.8b00220. PMID: 30034615; PMCID: PMC6047169
90. Moschos SJ, Sullivan RJ, Hwu WJ, Ramanathan RK, Adjei AA, Fong PC, Shapira-Frommer R, Tawbi HA, Rubino J, Rush TS 3rd, Zhang D, Miselis NR, Samatar AA, Chun P, Rubin EH, Schiller J, Long BJ, Dayananth P, Carr D, Kirschmeier P, Bishop WR, Deng Y, Cooper A, Shipps GW, Moreno BH, Robert L, Ribas A, Flaherty KT. Development of MK-8353, an orally administered ERK1/2 inhibitor, in patients with advanced solid tumors. JCI Insight. 2018;3(4):e92352. https://doi.org/10.1172/jci.insight.92352. PMID: 29467321; PMCID: PMC5916243
91. Roy SM, Minasov G, Arancio O, Chico LW, Van Eldik LJ, Anderson WF, Pelletier JC, Watterson DM. A selective and brain penetrant p38αMAPK inhibitor candidate for neurologic and neuropsychiatric disorders that attenuates Neuroinflammation and cognitive dysfunction. J Med Chem. 2019;62(11):5298–311. https://doi.org/10.1021/acs.jmedchem.9b00058. Epub 2019 Apr 23. Erratum in: J Med Chem. 2020 Aug 13;63(15):8649. doi: 10.1021/acs.jmedchem.0c01060. PMID: 30978288; PMCID: PMC6580366
92. Rutigliano G, Stazi M, Arancio O, Watterson DM, Origlia N. An isoform-selective p38α mitogen-activated protein kinase inhibitor rescues early entorhinal cortex dysfunctions in a mouse model of Alzheimer's disease. Neurobiol Aging. 2018;70:86–91. https://doi.org/10.1016/j.neurobiolaging.2018.06.006. Epub 2018 Jun 12. PMID: 30007168; PMCID: PMC6119125
93. Roy SM, Grum-Tokars VL, Schavocky JP, Saeed F, Staniszewski A, Teich AF, Arancio O, Bachstetter AD, Webster SJ, Van Eldik LJ, Minasov G, Anderson WF, Pelletier JC, Watterson DM. Targeting human central nervous system protein kinases: An isoform selective p38αMAPK inhibitor that attenuates disease progression in Alzheimer's disease mouse models. ACS Chem Neurosci. 2015;6(4):666–80. https://doi.org/10.1021/acschemneuro.5b00002. Epub 2015 Feb 23. PMID: 25676389; PMCID: PMC4404319
94. Hu W, Ralay Ranaivo H, Roy SM, Behanna HA, Wing LK, Munoz L, Guo L, Van Eldik LJ, Watterson DM. Development of a novel therapeutic suppressor of brain proinflammatory cytokine up-regulation that attenuates synaptic dysfunction and behavioral deficits. Bioorg Med Chem Lett. 2007;17(2):414–8. https://doi.org/10.1016/j.bmcl.2006.10.028. Epub 2006 Oct 17. PMID: 17079143; PMCID: PMC1868432
95. Goldstein DM, Alfredson T, Bertrand J, Browner MF, Clifford K, Dalrymple SA, Dunn J, Freire-Moar J, Harris S, Labadie SS, La Fargue J, Lapierre JM, Larrabee S, Li F, Papp E, McWeeney D, Ramesha C, Roberts R, Rotstein D, San Pablo B, Sjogren EB, So OY, Talamas FX, Tao W, Trejo A, Villasenor A, Welch M, Welch T, Weller P, Whiteley PE, Young K, Zipfel S. Discovery of S-[5-amino-1-(4-fluorophenyl)-1H-pyrazol-4-yl]-[3-(2,3-dihydroxypropoxy) phenyl]methanone (RO3201195), an orally bioavailable and highly selective inhibitor of p38 MAP kinase. J Med Chem. 2006;49(5):1562–75. https://doi.org/10.1021/jm050736c.
96. Trejo A, Arzeno H, Browner M, Chanda S, Cheng S, Comer DD, Dalrymple SA, Dunten P, Lafargue J, Lovejoy B, Freire-Moar J, Lim J, Mcintosh J, Miller J, Papp E, Reuter D, Roberts R, Sanpablo F, Saunders J, Song K, Villasenor A, Warren SD, Welch M, Weller P, Whiteley PE, Zeng L, Goldstein DM. Design and synthesis of 4-azaindoles as inhibitors of p38 MAP kinase. J Med Chem. 2003;46(22):4702–13. https://doi.org/10.1021/jm0301787.
97. Zheng Q, Li S, Wang A, Zhe M, Yang P, Wu Y, Zhao M, Zhu Y, Luo Y, Wang G, Ouyang L. MedComm Oncology. 2023;2:e53.
98. Goldstein DM, Soth M, Gabriel T, Dewdney N, Kuglstatter A, Arzeno H, Chen J, Bingenheimer W, Dalrymple SA, Dunn J, Farrell R, Frauchiger S, La Fargue J, Ghate M, Graves B, Hill RJ, Li F, Litman R, Loe B, McIntosh J, McWeeney D, Papp E, Park J, Reese HF, Roberts

RT, Rotstein D, San Pablo B, Sarma K, Stahl M, Sung ML, Suttman RT, Sjogren EB, Tan Y, Trejo A, Welch M, Weller P, Wong BR, Zecic H. Discovery of 6-(2,4-difluorophenoxy)-2-[3-hydroxy-1-(2-hydroxyethyl)propylamino]-8-methyl-8H-pyrido[2,3-d]pyrimidin-7-one (pamapimod) and 6-(2,4-difluorophenoxy)-8-methyl-2-(tetrahydro-2H-pyran-4-ylamino) pyrido[2,3-d]pyrimidin-7(8H)-one (R1487) as orally bioavailable and highly selective inhibitors of p38α mitogen-activated protein kinase. J Med Chem. 2011;54(7):2255–65. https://doi.org/10.1021/jm101423y. Epub 2011 Mar 4

99. Kumar S, Jiang MS, Adams JL, Lee JC. Pyridinylimidazole compound SB 203580 inhibits the activity but not the activation of p38 mitogen-activated protein kinase. Biochem Biophys Res Commun. 1999;263(3):825–31. https://doi.org/10.1006/bbrc.1999.1454. PMID: 10512765
100. Cuenda A, Rouse J, Doza YN, Meier R, Cohen P, Gallagher TF, Young PR, Lee JC. SB 203580 is a specific inhibitor of a MAP kinase homologue which is stimulated by cellular stresses and interleukin-1. FEBS Lett. 1995;364(2):229–33. https://doi.org/10.1016/0014-5793(95)00357-f.
101. Wang Z, Canagarajah BJ, Boehm JC, Kassisà S, Cobb MH, Young PR, Abdel-Meguid S, Adams JL, Goldsmith EJ. Structural basis of inhibitor selectivity in MAP kinases. Structure. 1998;6(9):1117–28. https://doi.org/10.1016/s0969-2126(98)00113-0.
102. Jackson JR, Bolognese B, Hillegass L, Kassis S, Adams J, Griswold DE, Winkler JD. Pharmacological effects of SB 220025, a selective inhibitor of P38 mitogen-activated protein kinase, in angiogenesis and chronic inflammatory disease models. J Pharmacol Exp Ther. 1998;284(2):687–92.
103. Nemoto S, Xiang J, Huang S, Lin A. Induction of apoptosis by SB202190 through inhibition of p38beta mitogen-activated protein kinase. J Biol Chem. 1998;273(26):16415–20. https://doi.org/10.1074/jbc.273.26.16415.
104. Davies SP, Reddy H, Caivano M, Cohen P. Specificity and mechanism of action of some commonly used protein kinase inhibitors. Biochem J. 2000;351(Pt 1):95–105. https://doi.org/10.1042/0264-6021:3510095. PMID: 10998351; PMCID: PMC1221339
105. Anand P, Shenoy R, Palmer JE, Baines AJ, Lai RY, Robertson J, Bird N, Ostenfeld T, Chizh BA. Clinical trial of the p38 MAP kinase inhibitor dilmapimod in neuropathic pain following nerve injury. Eur J Pain. 2011;15(10):1040–8. https://doi.org/10.1016/j.ejpain.2011.04.005. Epub 2011 May 14
106. Singh D, Smyth L, Borrill Z, Sweeney L, Tal-Singer R. A randomized, placebo-controlled study of the effects of the p38 MAPK inhibitor SB-681323 on blood biomarkers of inflammation in COPD patients. J Clin Pharmacol. 2010;50(1):94–100. https://doi.org/10.1177/0091270009347873. Epub 2009 Oct 30
107. Christie JD, Vaslef S, Chang PK, May AK, Gunn SR, Yang S, Hardes K, Kahl L, Powley WM, Lipson DA, Bayliffe AI, Lazaar AL. A randomized dose-escalation study of the safety and anti-inflammatory activity of the p38 mitogen-activated protein kinase inhibitor Dilmapimod in severe trauma subjects at risk for acute respiratory distress syndrome. Crit Care Med. 2015;43(9):1859–69. https://doi.org/10.1097/CCM.0000000000001132.
108. Newby LK, Marber MS, Melloni C, Sarov-Blat L, Aberle LH, Aylward PE, Cai G, de Winter RJ, Hamm CW, Heitner JF, Kim R, Lerman A, Patel MR, Tanguay JF, Lepore JJ, Al-Khalidi HR, Sprecher DL, Granger CB. SOLSTICE investigators. Losmapimod, a novel p38 mitogen-activated protein kinase inhibitor, in non-ST-segment elevation myocardial infarction: a randomised phase 2 trial. Lancet. 2014;384(9949):1187–95. https://doi.org/10.1016/S0140-6736(14)60417-7. Epub 2014 Jun 12. Erratum in: Lancet. 2014 Sep 27;384(9949):1186
109. Mellion ML, Ronco L, Berends CL, Pagan L, Brooks S, van Esdonk MJ, van Brummelen EMJ, Odueyungbo A, Thompson LA, Hage M, Badrising UA, Raines S, Tracewell WG, van Engelen B, Cadavid D, Groeneveld GJ. Phase 1 clinical trial of losmapimod in facioscapulohumeral dystrophy: safety, tolerability, pharmacokinetics, and target engagement. Br J Clin Pharmacol. 2021;87(12):4658–69. https://doi.org/10.1111/bcp.14884. Epub 2021 May 14

110. Willette RN, Eybye ME, Olzinski AR, Behm DJ, Aiyar N, Maniscalco K, Bentley RG, Coatney RW, Zhao S, Westfall TD, Doe CP. Differential effects of p38 mitogen-activated protein kinase and cyclooxygenase 2 inhibitors in a model of cardiovascular disease. J Pharmacol Exp Ther. 2009;330(3):964–70. https://doi.org/10.1124/jpet.109.154443. Epub 2009 Jun 25
111. Lomas DA, Lipson DA, Miller BE, Willits L, Keene O, Barnacle H, Barnes NC, Tal-Singer R, Losmapimod Study Investigators. An oral inhibitor of p38 MAP kinase reduces plasma fibrinogen in patients with chronic obstructive pulmonary disease. J Clin Pharmacol. 2012;52(3):416–24. https://doi.org/10.1177/0091270010397050. Epub 2011 Nov 16
112. Cheriyan J, Webb AJ, Sarov-Blat L, Elkhawad M, Wallace SM, Mäki-Petäjä KM, Collier DJ, Morgan J, Fang Z, Willette RN, Lepore JJ, Cockcroft JR, Sprecher DL, Wilkinson IB. Inhibition of p38 mitogen-activated protein kinase improves nitric oxide-mediated vasodilatation and reduces inflammation in hypercholesterolemia. Circulation. 2011;123(5):515–23. https://doi.org/10.1161/CIRCULATIONAHA.110.971986. Epub 2011 Jan 24
113. Damjanov N, Kauffman RS, Spencer-Green GT. Efficacy, pharmacodynamics, and safety of VX-702, a novel p38 MAPK inhibitor, in rheumatoid arthritis: results of two randomized, double-blind, placebo-controlled clinical studies. Arthritis Rheum. 2009;60(5):1232–41. https://doi.org/10.1002/art.24485. Erratum in: Arthritis Rheum. 2009 Oct;60(10):3071
114. Goldstein DM, Kuglstatter A, Lou Y, Soth MJ. Selective p38alpha inhibitors clinically evaluated for the treatment of chronic inflammatory disorders. J Med Chem. 2010;53(6):2345–53. https://doi.org/10.1021/jm9012906.
115. Lee MR, Dominguez C. MAP kinase p38 inhibitors: clinical results and an intimate look at their interactions with p38alpha protein. Curr Med Chem. 2005;12(25):2979–94. https://doi.org/10.2174/092986705774462914.
116. Duffy JP, Harrington EM, Salituro FG, Cochran JE, Green J, Gao H, Bemis GW, Evindar G, Galullo VP, Ford PJ, Germann UA, Wilson KP, Bellon SF, Chen G, Taslimi P, Jones P, Huang C, Pazhanisamy S, Wang YM, Murcko MA, Su MS. The discovery of VX-745: a novel and selective p38α kinase inhibitor. ACS Med Chem Lett. 2011;2(10):758–63. https://doi.org/10.1021/ml2001455. PMID: 24900264; PMCID: PMC4018046
117. Alam J, Blackburn K, Patrick D. Neflamapimod: clinical phase 2b-ready Oral small molecule inhibitor of p38α to reverse synaptic dysfunction in early Alzheimer's disease. J Prev Alzheimers Dis. 2017;4(4):273–8. https://doi.org/10.14283/jpad.2017.41.
118. Dominguez C, Powers DA, Tamayo N. p38 MAP kinase inhibitors: many are made, but few are chosen. Curr Opin Drug Discov Devel. 2005;8(4):421–30.
119. Verkaar F, van der Doelen AA, Smits JF, Blankesteijn WM, Zaman GJ. Inhibition of Wnt/β--catenin signaling by p38 MAP kinase inhibitors is explained by cross-reactivity with casein kinase Iδ/ε. Chem Biol. 2011;18(4):485–94. https://doi.org/10.1016/j.chembiol.2011.01.015.
120. Selness SR, Devraj RV, Monahan JB, Boehm TL, Walker JK, Devadas B, Durley RC, Kurumbail R, Shieh H, Xing L, Hepperle M, Rucker PV, Jerome KD, Benson AG, Marrufo LD, Madsen HM, Hitchcock J, Owen TJ, Christie L, Promo MA, Hickory BS, Alvira E, Naing W, Blevis-Bal R. Discovery of N-substituted pyridinones as potent and selective inhibitors of p38 kinase. Bioorg Med Chem Lett. 2009;19(20):5851–6. https://doi.org/10.1016/j.bmcl.2009.08.082. Epub 2009 Aug 27
121. Selness SR, Boehm TL, Walker JK, Devadas B, Durley RC, Kurumbail R, Shieh H, Xing L, Hepperle M, Rucker PV, Jerome KD, Benson AG, Marrufo LD, Madsen HM, Hitchcock J, Owen TJ, Christie L, Promo MA, Hickory BS, Alvira E, Naing W, Blevis-Bal R, Devraj RV, Messing D, Schindler JF, Hirsch J, Saabye M, Bonar S, Webb E, Anderson G, Monahan JB. Design, synthesis and activity of a potent, selective series of N-aryl pyridinone inhibitors of p38 kinase. Bioorg Med Chem Lett. 2011;21(13):4059–65. https://doi.org/10.1016/j.bmcl.2011.04.120. Epub 2011 May 13
122. Selness SR, Devraj RV, Devadas B, Walker JK, Boehm TL, Durley RC, Shieh H, Xing L, Rucker PV, Jerome KD, Benson AG, Marrufo LD, Madsen HM, Hitchcock J, Owen TJ, Christie L, Promo MA, Hickory BS, Alvira E, Naing W, Blevis-Bal R, Messing D, Yang J, Mao MK, Yalamanchili G, Vonder Embse R, Hirsch J, Saabye M, Bonar S, Webb E,

Anderson G, Monahan JB. Discovery of PH-797804, a highly selective and potent inhibitor of p38 MAP kinase. Bioorg Med Chem Lett. 2011;21(13):4066–71. https://doi.org/10.1016/j.bmcl.2011.04.121. Epub 2011 May 11

123. Hope HR, Anderson GD, Burnette BL, Compton RP, Devraj RV, Hirsch JL, Keith RH, Li X, Mbalaviele G, Messing DM, Saabye MJ, Schindler JF, Selness SR, Stillwell LI, Webb EG, Zhang J, Monahan JB. Anti-inflammatory properties of a novel N-phenyl pyridinone inhibitor of p38 mitogen-activated protein kinase: preclinical-to-clinical translation. J Pharmacol Exp Ther. 2009;331(3):882–95. https://doi.org/10.1124/jpet.109.158329. Epub 2009 Aug 31
124. Xing L, Devadas B, Devraj RV, Selness SR, Shieh H, Walker JK, Mao M, Messing D, Samas B, Yang JZ, Anderson GD, Webb EG, Monahan JB. Discovery and characterization of atropisomer PH-797804, a p38 MAP kinase inhibitor, as a clinical drug candidate. ChemMedChem. 2012;7(2):273–80. https://doi.org/10.1002/cmdc.201100439. Epub 2011 Dec 15
125. Xing L, Shieh HS, Selness SR, Devraj RV, Walker JK, Devadas B, Hope HR, Compton RP, Schindler JF, Hirsch JL, Benson AG, Kurumbail RG, Stegeman RA, Williams JM, Broadus RM, Walden Z, Monahan JB. Structural bioinformatics-based prediction of exceptional selectivity of p38 MAP kinase inhibitor PH-797804. Biochemistry. 2009;48(27):6402–11. https://doi.org/10.1021/bi900655f.
126. MacNee W, Allan RJ, Jones I, De Salvo MC, Tan LF. Efficacy and safety of the oral p38 inhibitor PH-797804 in chronic obstructive pulmonary disease: a randomised clinical trial. Thorax. 2013;68(8):738–45. https://doi.org/10.1136/thoraxjnl-2012-202744. Epub 2013 Mar 28
127. Nikas SN, Drosos AA. SCIO-469 Scios Inc. Curr Opin Investig Drugs. 2004;5(11):1205–12.
128. Medicherla S, Reddy M, Ying J, Navas TA, Li L, Nguyen AN, Kerr I, Hanjarappa N, Protter AA, Higgins LS. p38alpha-selective MAP kinase inhibitor reduces tumor growth in mouse xenograft models of multiple myeloma. Anticancer Res. 2008;28(6A):3827–33.
129. Vanderkerken K, Medicherla S, Coulton L, De Raeve H, Willems A, Lawson M, Van Camp B, Protter AA, Higgins LS, Menu E, Croucher PI. Inhibition of p38alpha mitogen-activated protein kinase prevents the development of osteolytic bone disease, reduces tumor burden, and increases survival in murine models of multiple myeloma. Cancer Res. 2007;67(10):4572–7. https://doi.org/10.1158/0008-5472.CAN-06-4361. Epub 2007 May 10
130. Genovese MC, Cohen SB, Wofsy D, Weinblatt ME, Firestein GS, Brahn E, Strand V, Baker DG, Tong SE. A 24-week, randomized, double-blind, placebo-controlled, parallel group study of the efficacy of oral SCIO-469, a p38 mitogen-activated protein kinase inhibitor, in patients with active rheumatoid arthritis. J Rheumatol. 2011;38(5):846–54. https://doi.org/10.3899/jrheum.100602. Epub 2011 Feb 1
131. Liu C, Lin J, Wrobleski ST, Lin S, Hynes J, Wu H, Dyckman AJ, Li T, Wityak J, Gillooly KM, Pitt S, Shen DR, Zhang RF, McIntyre KW, Salter-Cid L, Shuster DJ, Zhang H, Marathe PH, Doweyko AM, Sack JS, Kiefer SE, Kish KF, Newitt JA, McKinnon M, Dodd JH, Barrish JC, Schieven GL, Leftheris K. Discovery of 4-(5-(cyclopropylcarbamoyl)-2-methylphenylamino)-5-methyl-N-propylpyrrolo[1,2-f][1,2,4]triazine-6-carboxamide (BMS-582949), a clinical p38α MAP kinase inhibitor for the treatment of inflammatory diseases. J Med Chem. 2010;53(18):6629–39. https://doi.org/10.1021/jm100540x.
132. Norman P. BMS-582949: crystalline form of a p38alpha inhibitor? WO2008079857. Expert Opin Ther Pat. 2009;19(8):1165–8. https://doi.org/10.1517/13543770902816160.
133. Emami H, Vucic E, Subramanian S, Abdelbaky A, Fayad ZA, Du S, Roth E, Ballantyne CM, Mohler ER, Farkouh ME, Kim J, Farmer M, Li L, Ehlgen A, Langenickel TH, Velasquez L, Hayes W, Tawakol A. The effect of BMS-582949, a P38 mitogen-activated protein kinase (P38 MAPK) inhibitor on arterial inflammation: a multicenter FDG-PET trial. Atherosclerosis. 2015;240(2):490–6. https://doi.org/10.1016/j.atherosclerosis.2015.03.039. Epub 2015 Mar 28
134. Miwatashi S, Arikawa Y, Kotani E, Miyamoto M, Naruo K, Kimura H, Tanaka T, Asahi S, Ohkawa S. Novel inhibitor of p38 MAP kinase as an anti-TNF-alpha drug: discovery of N-[4-[2-ethyl-4-(3-methylphenyl)-1,3-thiazol-5-yl]-2-pyridyl]benzamide (TAK-715) as a potent and orally active anti-rheumatoid arthritis agent. J Med Chem. 2005;48(19):5966–79. https://doi.org/10.1021/jm050165o.

135. Azevedo R, van Zeeland M, Raaijmakers H, Kazemier B, de Vlieg J, Oubrie A. X-ray structure of p38α bound to TAK-715: comparison with three classic inhibitors. Acta Crystallogr D Biol Crystallogr. 2012;68(Pt 8):1041–50. https://doi.org/10.1107/S090744491201997X. Epub 2012 Jul 17
136. Kaieda A, Takahashi M, Takai T, Goto M, Miyazaki T, Hori Y, Unno S, Kawamoto T, Tanaka T, Itono S, Takagi T, Hamada T, Shirasaki M, Okada K, Snell G, Bragstad K, Sang BC, Uchikawa O, Miwatashi S. Structure-based design, synthesis, and biological evaluation of imidazo[1,2-b]pyridazine-based p38 MAP kinase inhibitors. Bioorg Med Chem. 2018;26(3):647–60. https://doi.org/10.1016/j.bmc.2017.12.031. Epub 2017 Dec 24
137. Brown DS, Cumming JG, Bethel P, Finlayson J, Gerhardt S, Nash I, Pauptit RA, Pike KG, Reid A, Snelson W, Swallow S, Thompson C. The discovery of N-cyclopropyl-4-methyl-3-[6-(4-methylpiperazin-1-yl)-4-oxoquinazolin-3(4H)-yl]benzamide (AZD6703), a clinical p38α MAP kinase inhibitor for the treatment of inflammatory diseases. Bioorg Med Chem Lett. 2012;22(12):3879–83. https://doi.org/10.1016/j.bmcl.2012.04.116. Epub 2012 May 2
138. Raubo P, Evans R, Willis P. The discovery and evaluation of 3-amino-2(1H)-pyrazinones as a novel series of selective p38α MAP kinase inhibitors. Bioorg Med Chem Lett. 2020;30(18):127412. https://doi.org/10.1016/j.bmcl.2020.127412. Epub 2020 Jul 15
139. Higham A, Karur P, Jackson N, Cunoosamy DM, Jansson P, Singh D. Differential anti-inflammatory effects of budesonide and a p38 MAPK inhibitor AZD7624 on COPD pulmonary cells. Int J Chron Obstruct Pulmon Dis. 2018;13:1279–88. https://doi.org/10.2147/COPD.S159936. PMID: 29719383; PMCID: PMC5914546
140. Patel NR, Cunoosamy DM, Fagerås M, Taib Z, Asimus S, Hegelund-Myrbäck T, Lundin S, Pardali K, Kurian N, Ersdal E, Kristensson C, Korsback K, Palmér R, Brown MN, Greenaway S, Siew L, Clarke GW, Rennard SI, Make BJ, Wise RA, Jansson P. The development of AZD7624 for prevention of exacerbations in COPD: a randomized controlled trial. Int J Chron Obstruct Pulmon Dis. 2018;13:1009–19. https://doi.org/10.2147/COPD.S150576. PMID: 29628759; PMCID: PMC5877500
141. Pehrson R, Hegelund-Myrbäck T, Cunoosamy D, Asimus S, Jansson P, Patel N, Borde A, Lundin S. AZD7624, an inhaled p38 inhibitor, demonstrates local lung inhibition of LPS-induced TNFα with minimal systemic exposure. J Pharmacol Exp Ther. 2018;365(3):567–72. https://doi.org/10.1124/jpet.117.246132. Epub 2018 Mar 16
142. Norman P. Investigational p38 inhibitors for the treatment of chronic obstructive pulmonary disease. Expert Opin Investig Drugs. 2015;24(3):383–92. https://doi.org/10.1517/13543784.2015.1006358. Epub 2015 Jan 20
143. Campbell RM, Anderson BD, Brooks NA, Brooks HB, Chan EM, De Dios A, Gilmour R, Graff JR, Jambrina E, Mader M, McCann D, Na S, Parsons SH, Pratt SE, Shih C, Stancato LF, Starling JJ, Tate C, Velasco JA, Wang Y, Ye XS. Characterization of LY2228820 dimesylate, a potent and selective inhibitor of p38 MAPK with antitumor activity. Mol Cancer Ther. 2014;13(2):364–74. https://doi.org/10.1158/1535-7163.MCT-13-0513. Epub 2013 Dec 19
144. Ishitsuka K, Hideshima T, Neri P, Vallet S, Shiraishi N, Okawa Y, Shen Z, Raje N, Kiziltepe T, Ocio EM, Chauhan D, Tassone P, Munshi N, Campbell RM, Dios AD, Shih C, Starling JJ, Tamura K, Anderson KC. p38 mitogen-activated protein kinase inhibitor LY2228820 enhances bortezomib-induced cytotoxicity and inhibits osteoclastogenesis in multiple myeloma; therapeutic implications. Br J Haematol. 2008;141(5):598–606. https://doi.org/10.1111/j.1365-2141.2008.07044.x. Epub 2008 Apr 7
145. Tate CM, Blosser W, Wyss L, Evans G, Xue Q, Pan Y, Stancato L. LY2228820 dimesylate, a selective inhibitor of p38 mitogen-activated protein kinase, reduces angiogenic endothelial cord formation in vitro and in vivo. J Biol Chem. 2013;288(9):6743–53. https://doi.org/10.1074/jbc.M112.425553. Epub 2013 Jan 18. PMID: 23335506; PMCID: PMC3585111
146. Patnaik A, Haluska P, Tolcher AW, Erlichman C, Papadopoulos KP, Lensing JL, Beeram M, Molina JR, Rasco DW, Arcos RR, Kelly CS, Wijayawardana SR, Zhang X, Stancato LF, Bell R, Shi P, Kulanthaivel P, Pitou C, Mulle LB, Farrington DL, Chan EM, Goetz MP. A first-in-human phase I study of the Oral p38 MAPK inhibitor, Ralimetinib (LY2228820 Dimesylate),

in patients with advanced cancer. Clin Cancer Res. 2016;22(5):1095–102. https://doi.org/10.1158/1078-0432.CCR-15-1718. Epub 2015 Nov 18. Erratum in: Clin Cancer Res. 2016 May 15;22(10):2596. https://doi.org/10.1158/1078-0432.CCR-16-0645

147. Cheng HS, Chong YK, Lim EKY, Lee XY, Pang QY, Novera W, Marvalim C, Lee JXT, Ang BT, Tang C, Tan NS. Dual p38MAPK and MEK inhibition disrupts adaptive chemoresistance in mesenchymal glioblastoma to temozolomide. Neuro-Oncology. 2024;26(7):1247–61. https://doi.org/10.1093/neuonc/noae028. PMID: 38366847; PMCID: PMC11226874
148. Biau J, Thivat E, Chautard E, Stefan D, Boone M, Chauffert B, Bourgne C, Richard D, Molnar I, Levesque S, Bellini R, Kwiatkowski F, Karayan-Tapon L, Verrelle P, Godfraind C, Durando X. Phase 1 trial of ralimetinib (LY2228820) with radiotherapy plus concomitant temozolomide in the treatment of newly diagnosed glioblastoma. Radiother Oncol. 2021;154:227–34. https://doi.org/10.1016/j.radonc.2020.09.036. Epub 2020 Sep 22
149. Bendell JC, Bischoff HG, Hwang J, Reinhardt HC, Zander T, Wang X, Hynes S, Pitou C, Campbell R, Iversen P, Farrington DL, Bell-McGuinn K, Thomas M. A phase 1 dose-escalation study of checkpoint kinase 1 (CHK1) inhibitor prexasertib in combination with p38 mitogen-activated protein kinase (p38 MAPK) inhibitor ralimetinib in patients with advanced or metastatic cancer. Investig New Drugs. 2020;38(4):1145–55. https://doi.org/10.1007/s10637-019-00873-6. Epub 2019 Nov 9

Chapter 4
Non-ATP Site Inhibitors and Bi-dentate Inhibitors

Abstract The ATP site of the MAP kinases shares a common amino acid sequence. The ATP-competitive inhibitors are less selective and cause undesirable toxicities. The non-ATP site inhibitors are a class of compounds that inhibit kinase activity by binding to sites other than the ATP-binding pocket. The peptides and small molecule inhibitors are described here. Bi-dentate MAP kinase inhibitors are a specific class of compound that binds to both ATP site and allosteric site.

Keywords Non-ATP site binders · BI-7D3 · BI-87G9 · Doramapimod · Bi-dentate inhibitor · Dual site binding inhibitor

4.1 Introduction

Most MAP kinase inhibitors target the ATP binding site. The ATP site is mostly conserved among more than 518 protein kinases. This conservation creates cross-reactivity and the possibility of undesirable toxicities. Therefore, ATP-competitive inhibitors are less selective than non-ATP site inhibitors. There are two main substrate binding sites: D-recruitment site (DRS) and F-recruitment site (FRS). DRS is a peptide-binding groove residing at the back of MAPK and FRS is a shallow hydrophobic region.

In the last few years for MAPK8, 9, 10, researchers focused on finding JIP site binders that are distinct from the ATP binding site [1–33]. The JNK interacting protein (JIP) scaffolds JIP1–4 bind to multiple components of the JNK signaling pathway, including mixed-lineage kinases, MAP kinase kinase, and serve to enhance its activation in vivo. JIP1 is highly expressed in the brain and in the β cells of the islets of Langerhans.

ATP-noncompetitive MAPK inhibitors are a class of compounds that inhibit kinase activity by binding to sites other than the ATP-binding pocket. Unlike ATP-competitive inhibitors, these compounds do not compete with ATP for binding to the kinase domain but instead target allosteric sites or other regions of the kinase.

Here are some key points about ATP-noncompetitive MAPK inhibitors:

S. K. De, *Drug Design and Discovery*,
https://doi.org/10.1007/978-3-031-75682-5_4

Allosteric Inhibition ATP-noncompetitive inhibitors exert their effects by binding to allosteric sites on the MAPK protein rather than competing with ATP for the active site. This mechanism offers the advantage of potentially higher selectivity and reduced risk of cross-reactivity with other ATP-binding proteins.

Binding Specificity These inhibitors may bind to regions of the kinase outside the ATP-binding pocket, which allows for greater selectivity in targeting specific isoforms or conformations of MAPKs.

Therapeutic Potential ATP-noncompetitive MAPK inhibitors hold promise for therapeutic applications in diseases where kinase signaling is dysregulated, such as cancer, metabolic disorders, neurodegenerative disorders, and inflammatory conditions. By modulating MAPK activity through noncompetitive mechanisms, these inhibitors offer alternative strategies for therapeutic intervention.

Challenges Despite their potential benefits, ATP-noncompetitive MAPK inhibitors may face challenges related to specificity, pharmacokinetics, and off-target effects. Optimization of these compounds is necessary to enhance their therapeutic efficacy and safety profile.

4.2 Small Peptides as ATP-Noncompetitive Inhibitors of MAPK 8, 9, 10

The first ATP-noncompetitive inhibitor is a minimal JIP1 peptide, corresponding to a C-terminal 11-amino acid peptide (JIP1 154–163) [1–6], which is able to inhibit JNK activity with a K_d value of 420 nM against JNK1, compound in Fig. 4.1 and Table 4.1. In order to improve peptide cell permeability and stability, researchers designed the D-JNK1 peptide which is a retro-inverso D-peptide with a linked cell permeable TAT sequence. These peptide JNK inhibitors bind with a JIP site which were confirmed by X-ray crystallography (PDB code: 1UKH) and kinetics studies [4, 5]. Kaoud et al. reported several peptide-based JNK inhibitors using JIP1 peptide with or without TAT sequence [6]. They found a potent JNK2 inhibitor with an IC_{50} value of 92 nM (Table 4.1). This peptide consists of a JIP-10 mer (JIP144–153, PKRPTTLNLF), linker 6-aminohexanoyl and inverso HIV-TAT sequence. This peptide inhibits phosphorylation of c-Jun and cell growth in a breast cancer cell line. Similarly, a potent JNK inhibitor was obtained using JIP-10 mer, the same linker, and nine arginines. These two peptides are very selective over JNK inhibition, more than 160 times selective against p38α and 1700 times selective against ERK2. However, peptide's instability and poor cell permeability may interfere with the advance stage development of JNK inhibitors from the JIP1 peptide.

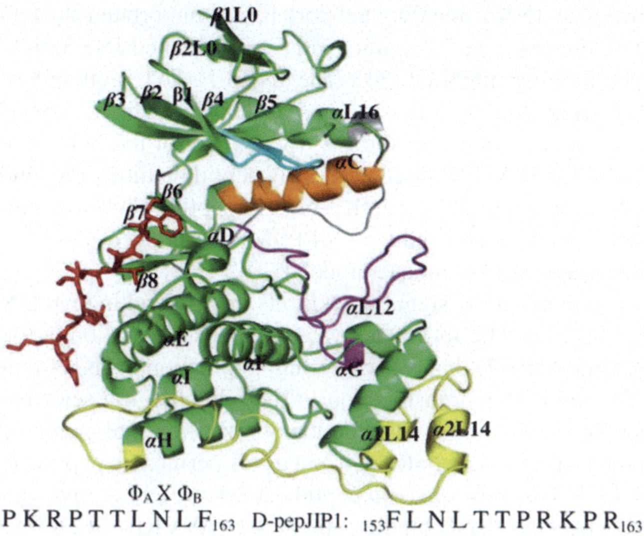

Fig. 4.1 Adapted from Yan et al. with permission from the American Chemical Society publications (ref 18) [Cartoon diagram of the JNK1 structure highlighting the location of the L-pepJIP1 (red and sticks) (PDB code: 1UKI) and the sequences of the scaffolding protein (L-pepJIP1)]. The G loop is indicated in cyan, the MAP kinase insert in yellow, αC helix in brown, the activation loop in magenta, the catalytic loop in blue, the loop connecting β5 and β6, and loop 16 in gray. The secondary structures are labeled based on reference

Table 4.1 JIP-1 site peptide binders

Peptide	Sequence	JNK1, IC_{50} (μM)	JNK2, IC_{50} (μM)	JNK3, IC_{50} (μM)	References
pepJIP 10 mer	Ac-PKRPTTLNLF-amide	1.05	2.15	2.67	Heo et al. [4]
D-JNK-1	dqsrpvqflnlttprkprpprrrqrrkkG	0.05	0.025	0.05	Bonny et al. [33]
L-PYC71N	FFWRLNKIFYFID				Ngoei et al. [2]
L-PYC98	LAYRVIIPCFRKRYFCFL	N.R.	N.R.	N. R.	Ngoei et al. [3]
D-PYC98-TAT	lfcfyrkrfcpiivryalGrrrqrrkkrG	N.R.	N.R.	N.R.	Ngoei et al. [3]
JIP[10]-Δ-TAT (inverso)	Ac-PKRPTTLNLF-aminohexanoyl-RRRQRRKKRG-amide	1.21	0.092	1.27	Kaoud et al. [6]
JIP[10]- Δ-R9	PKRPTTLNLF-aminohexanoyl-RRRRRRRRR	1.13	0.89	1.29	Kaoud et al. [6]

N.R. not reported

The ASK1-JNK3 signaling network has an important role in Parkinson's disease. Pan et al. discovered a 21-amino acid peptide named JNK3-N-TAT as a JNK3 non-ATP site binding inhibitor [30]. The JNK3-N-TAT sequence is Tyr-Gly-Arg-Lys-Lys-Arg-Arg-Gln-Arg-Arg-Arg-Cys-Ser-Glu-Pro-Thr-Leu-Asp-Val-Lys-Ile. It is a 10-amino acid sequence of the C9–18 segment of the N-terminus of JNK3 conjugated with the HIV-TAT sequence. This peptide inhibits the binding of β-arrestin 2 to the N-terminus of JNK3. JNK3-N-TAT peptide shows an improvement of mitochondrial function and reduces MPP^+/MPTP-induced damage in both SH-SY5Y cellular assays and PD mouse models.

Mitochondrial JNK signaling exhibits in vitro and in vivo DNA damage causing various diseases. The JNK is activated by phosphorylation of two upstream kinases (MKK4 and MKK7) that causes a small population of JNK to migrate to mitochondria. To inhibit JNK translocation to the mitochondria selectively, investigators at Scripps Research developed a peptide of mitochondrial membrane protein Sab with small interfering RNAs (siRNAs) and a cell-permeable peptide to the KIM1 domain of Sab [31]. This *retro-inverso* peptide consists of 11 amino acids corresponding to the phosphorylated JNK binding site of Sab (AVVRPGSLDLR), conjugated with an HIV-Tat motif to improve cell permeability. This peptide disrupts JNK translocation to the mitochondria and reduces the phosphorylation of Bcl-2 without any remarkable effect on c-Jun phosphorylation or AP-1 transcription in the nuclear. Regular JIP1 peptide prevents these three functions; there is no selectivity. Tat-SabKIM1 peptide selectively prevents Bcl-2 phosphorylation, cell death, loss of mitochondrial membrane potential, and superoxide generation in anisomycin-stressed HeLa cells. Nevertheless, this peptide protects against 6-OHDA-induced oxidative stress, mitochondrial dysfunction, and neurotoxicity in vitro and in vivo by selectively blocking the JNK mitochondrial translocation.

Laughlin et al. reported three crystal structures [32] for JNK3 bound to JIP1-derived peptide (pepJIP1: RPKRPTTLNLF, PDB code: 4H39, 2.0 Å), Sab-derived peptide (pepSAB: AVVRPGSLDLP, PDB code: 4H3B, 2.1 Å), and a substrate target, ATF2 (pepATF2: KHKHEMTLKFG, PDB code: 4H36, 3.0 Å), respectively. These three are different 11-mer D-motif peptides and bind to the C-terminal lobe of JNK3. The crystal structure pepJIP1 with JNK3 reveals that peptide binds to the carboxy-terminal and moves a rotation of the amino-terminal N-lobe. The Arg160 interacts with the residue Glu367 of JNK3 making a salt bridge. The Arg160 also forms hydrogen bonding interactions with JNK3 Tyr168, the backbone of Trp362, and a network of water. Thr162 and Thr163 form hydrogen bonding networks with Arg165 of JNK3. PepJIP1 forms van der Waals interactions with Tyr171, Asp364, Val156, and Val197 of JNK3. These interactions are not observed with pepATF2. For the binding mode of pepSAB, Arg343 (JIP1 Arg160, ATF2 His49) makes an electrostatic interaction with Glu367 of JNK3 and considerably shifts the L-16 loop and G-loop. The Arg160 of pepJIP1 is important for binding affinity. From ITC, K_d values are for pepJIP1 0.8 μM, pepSAB 17.2 μM, and pepATF2 > 10 μM respectively.

4.3 Small Molecules as ATP-Noncompetitive JNK (MAPK 8, 9, 10) Inhibitors

In 2008, investigators at Sanford-Burnham-Prebys reported a series of small molecules as ATP-noncompetitive JNK inhibitors [7–10]. First small-molecule JIP site inhibitor was discovered by a group of scientists from Sanford-Burnham-Prebys coded **BI-78D3 (compound 91)** [7] which showed promising in vitro and in vivo data. **BI-78D3** inhibited JIP-JNK interaction with the IC_{50} values of 500 nM in DELFIA assay and 280 nM in kinase activity assays respectively. Molecular docking studies indicate that the benzodioxan group of **BI-78D3** binds close to Arg127, an extensive hydrogen bonding network, which occupies the JIP-1-JNK interaction site. Liquid chromatography/mass spectrometry stability analysis suggests that BI-78D3 has favorable microsome and plasma stability ($T_{1/2} = 54$ min). **BI-78D3** is also a selective JNK inhibitor, almost inactive against p38α, PI-3 K, mTOR, and other kinases tested. **BI-78D3** displays the inhibition of TNF-alpha-induced phosphorylation of c-Jun in a dose-dependent manner with an IC_{50} value of 12.5 μM in HeLa cells. **BI-78D3** abolishes ConA-induced liver damage as measured by the release of the liver enzyme alanine-aminotransferase (ALT) to the circulation. Furthermore, it restores insulin sensitivity compared to vehicle control in a mouse mode of type-2 diabetes.

These molecules bind to JIP sites. The structures are shown in Fig. 4.2. These compounds should be more selective than ATP site inhibitors. Further optimization of compound **91** resulted in compound **92** where the central core was replaced from tetrazole to thiadiazole (Fig. 4.2) [8]. Compound **92** was tested in a mouse model of type-2 diabetes, 25 mg/Kg dose, i.p., and showed a significant reduction in blood glucose levels compared to the vehicle control. Along the same line, researchers at Sanford-Burnham reported the discovery of another molecule such as compound **93** that exhibited significant inhibition of JIP-JNK activation in vitro, in cell and favorable microsomal ($T_{1/2} = 70$ min) and plasma stability (80% remaining after 1 h) [9].

Further optimization of primary amine of compound **93** resulted in compound **94** **(BI-90H9)** as shown in Fig. 4.3 as a secondary amine derivative, which had good balance of potency from both JNK activity and DELFIA assays; improved cell permeability (76.3%) compared to **BI-78D3** (30%, PAMPA % of flux), solubility, and plasma stability (68% remaining after 1 h [10]. These interesting results indicate that targeting the protein-protein interaction between JNK and JIP with a small molecule is a new and promising avenue for the development of novel therapeutic drugs that inactivate the JNK pathway biology.

Investigators at Abbott Laboratories identified 68 compounds from an affinity-based high throughput screening of a library of 500,000 small molecules [11]. They extensively characterized allosteric binding site inhibitors using NMR spectroscopy, Protein X-crystallography, surface plasmon resonance, direct enzymatic activity, and activation cascade assays. However, they found many compounds bound to the ATP site, confirmed by high-resolution NMR binding studies using ^{13}C-methyl-labeled inactive JNK1R1 (residues 1–364, Thr183-Glu, Tyr185Glu).

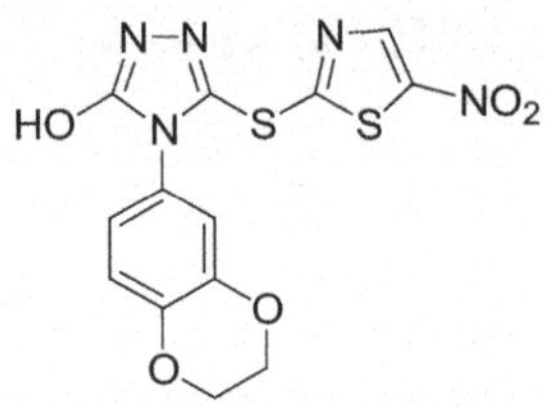

91 (BI-78D3)

IC_{50} (JNK1) = 500 nM (DELFIA assay)

IC_{50} (JNK1) = 280 nM (Kinase assay)

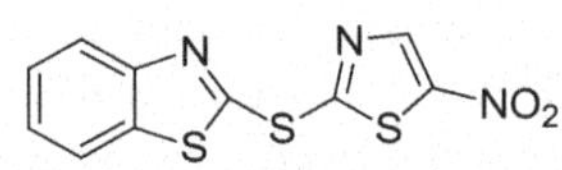

92

IC_{50} (JNK1) = 160 nM (DELFIA assay)

IC_{50} (JNK1) = 1800 nM (Kinase assay

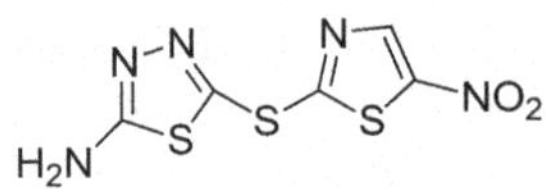

93

IC_{50} (JNK1) = 239 nM (DELFIA assay)

IC_{50} (JNK1) = 700 nM (Kinase assay)

JIP-peptide: Arg-Pro-Lys-Arg-Pro-Thr-Thr-Leu-Asn-Leu-Phe

K_d (JNK1) = 420 nM

Fig. 4.2 Recently reported JIP site inhibitors (non-ATP site or allosteric inhibitors)

Fig. 4.3 Thiadiazole derivative

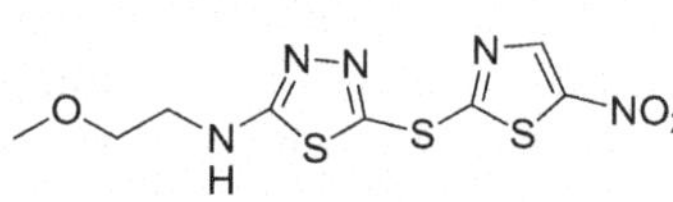

94 (BI-90H9)

JNK1 IC_{50} = 160 nM (DELFIA)

JNK1 IC_{50} = 4.8 μM (Kinase assay)

Only two compounds (**95, 96**) bind to the non-ATP site (Fig. 4.4) as these do not exhibit any activity in biochemical assay using phosphorylated (e.g., active) JNK1 (IC_{50} > 50 μM). These compounds (**95, 96**) bind to a different site on the inactive form of the protein.

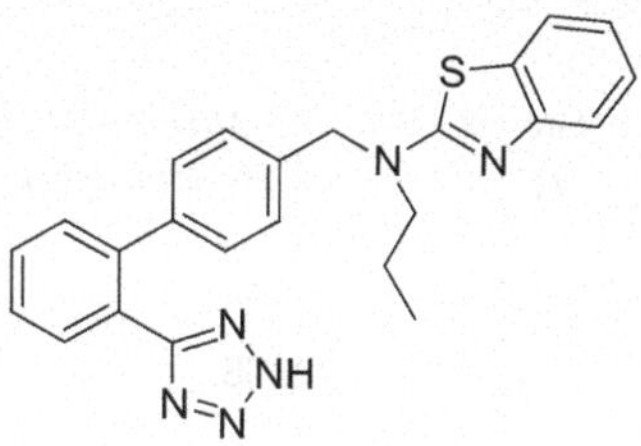

95

IC_{50} (JNK1) > 100 μM

IC_{50} (MKK7-JNK1) = 7.8 μM

K_d (JNK1-u) = 11 μM

K_d (JNK1-a) > 50 μM

K_d (p38-u) = 18 μM

K_d (ERK2-u) = 18 μM

EC_{50} (p-c-Jun) > 30 μM

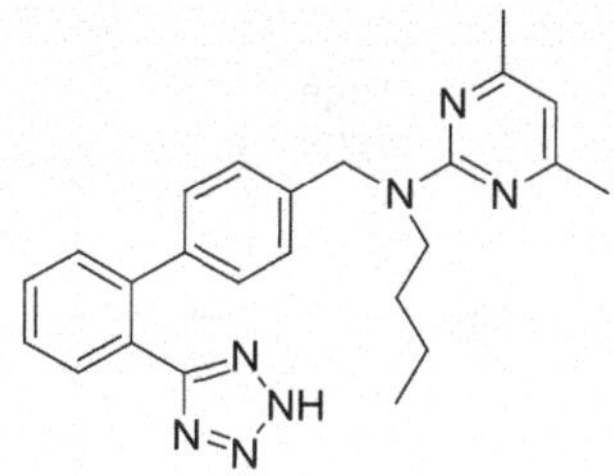

96

IC_{50} (JNK1 = 62 μM

IC_{50} (MKK7-JNK1) = 7.7 μM

K_d (JNK1-u) = 16 μM

K_d (JNK1-a) > 50 μM

K_d (p38-u) = 17 μM

K_d (ERK2-u) = 32 μM

EC_{50} (p-c-Jun) > 30 μM

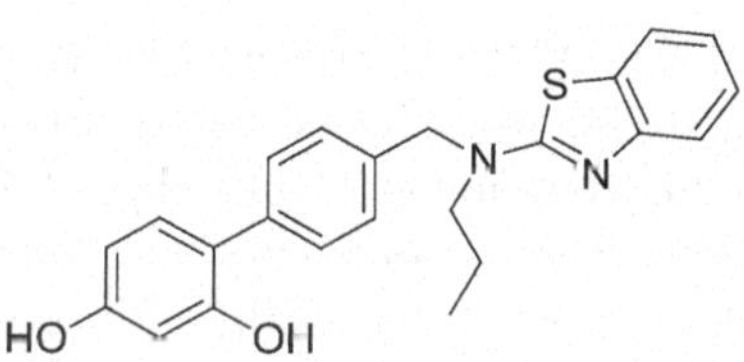

97

IC_{50} (JNK1) > 100 μM

IC_{50} (MKK7-JNK1) = 2.9 μM

EC_{50} (p-c-Jun) > 10 μM

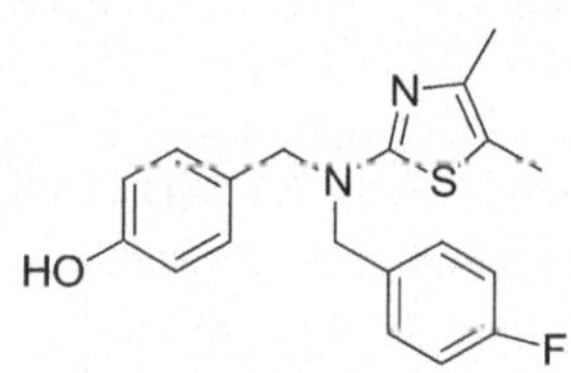

98

IC_{50} (JNK1) > 100 μM

IC_{50} (MKK7-JNK1) = 3.8 μM

EC_{50} (p-c-Jun) = 4 μM

Fig. 4.4 Non-ATP site binder

The X-ray crystallographic studies on compound **96** (Fig. 4.5) in complex with JNK1 (PDB code: 3O2M) reveal that it binds away from the ATP-binding site. It is located in the MAP insertion region on the surface of the protein. The northern binding pocket is occupied in the activation loop containing residues Ser179 to

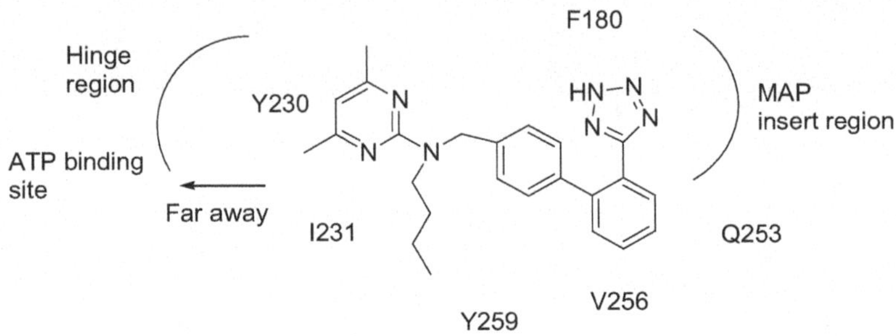

Fig. 4.5 Non-ATP site binder

Glu185 which indicates a different conformation. The southern pocket is placed with residues Gly242 to Ala267 and the western side pocket is occupied with residues Met242 to Trp234. The butyl group is inserted into a hydrophobic pocket close to Trp234. The dimethyl pyrimidine moiety is accommodated between Ile231 and Tyr230 and forms a ring-stacking interaction. The biaryl tetrazole moiety stays in the activation loop. The tetrazole group does not make any specific interaction. Both compounds are inactive in cytokine (TNFα)-stimulated JNK-1-active HepG2 cells. In order to improve cellular potency, they made several analogs of this series such as **97** and **98** by reducing molecular weight and replacing tetrazole moiety, while maintaining lipophilicity and important interactions. Compound **98** inhibits the c-Jun phosphorylation with an IC_{50} value of 4 μM.

The use of kinase inhibitors for treating non-oncology diseases requires great selectivity of the target kinase.

Kauod et al. applied a new approach to find non-ATP site binder [14]. They screened a small molecule library from NCI diversity set using ligand-based virtual screening with **BI-78D3** as a reference compound. They found 11 hits (Fig. 4.6) from which natural product (−) zuonin A is the best. These compounds have some structural features similar to **BI-78D3**, containing either a benzo[*d*][1,3]dioxole moiety or nitro group or an aromatic ring with an electronegative substituent. The key compound lignan (−)-zuonin A is a JNK-JIP site binder and exhibits a selectivity of 100-fold over ERK2 and p38 MAPKα.

4.4 Dual Site Binding MAP Kinase 8, 9, 10 Inhibitors

More recently, Pfizer reported a small molecule (compound **111**) where a compound can bind to the ATP and/or JIP site from their extensive kinetic experiments [12]. This compound had no cellular activity due to poor cell permeability. Similarly, researchers at Sanford-Burnham reported a dual JNK inhibitor [13] denoted as **BI-98F10 (112)** as shown in Fig. 4.7. These compounds are termed as dual site inhibitors or mixed inhibitors. **BI-98F10** was a selective JNK inhibitor as this

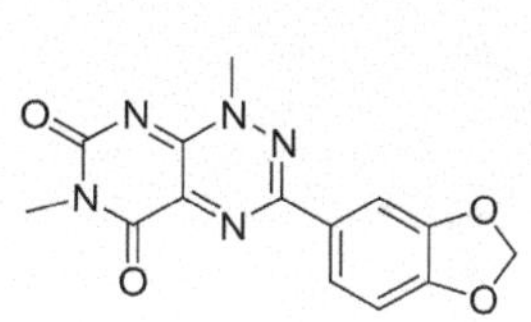

99

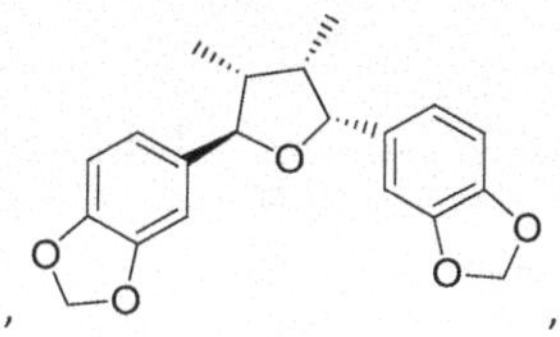

100

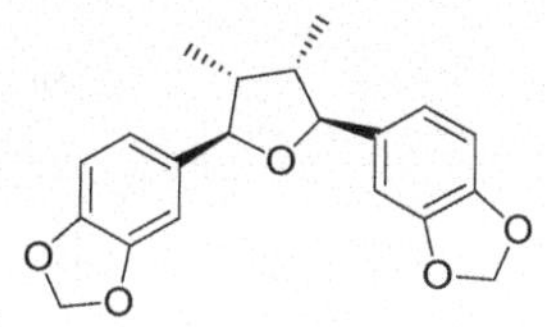

101

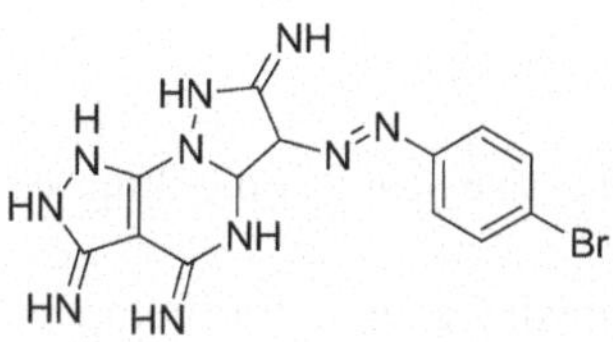

102

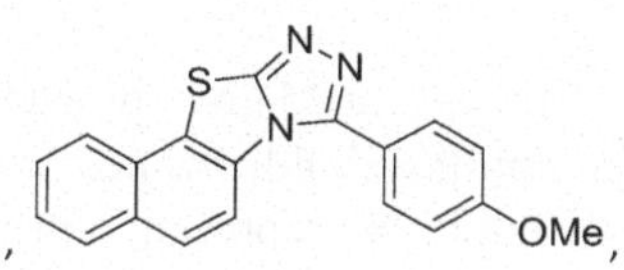

103

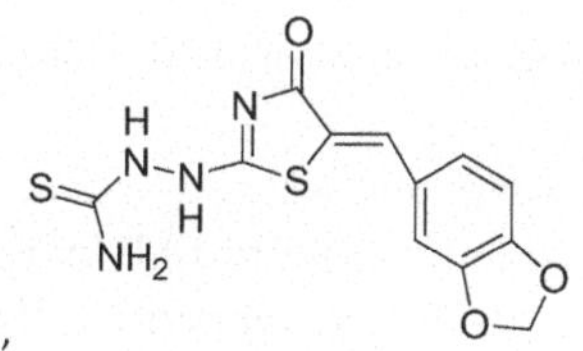

104

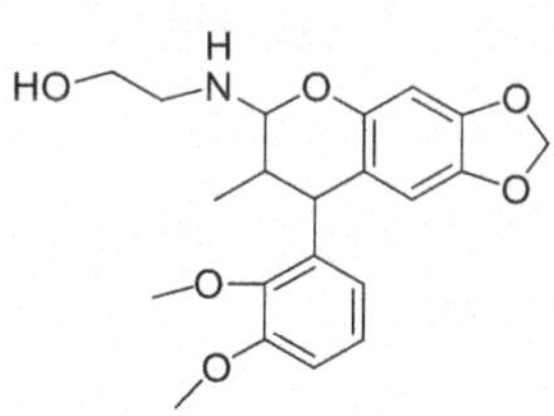

105

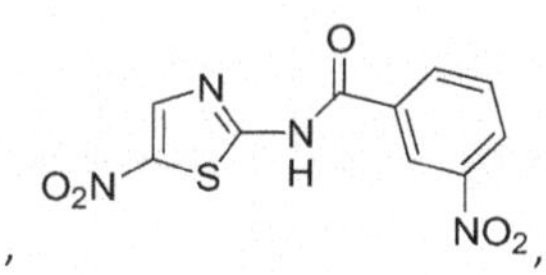

106

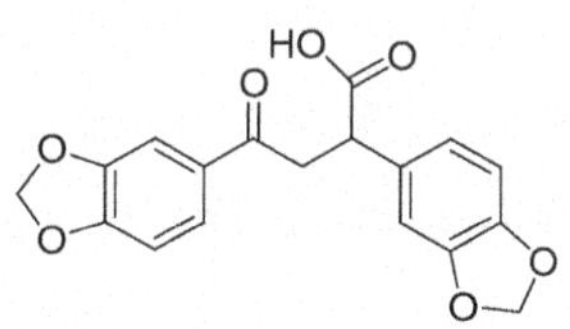

107

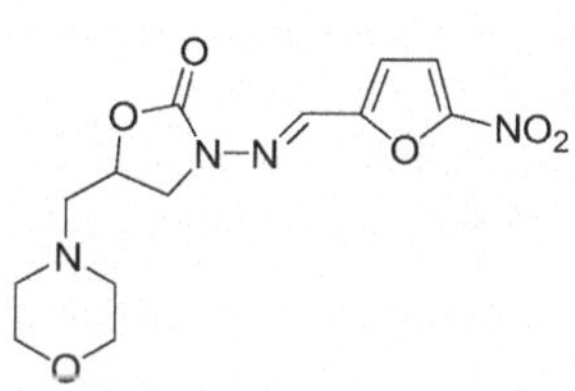

108

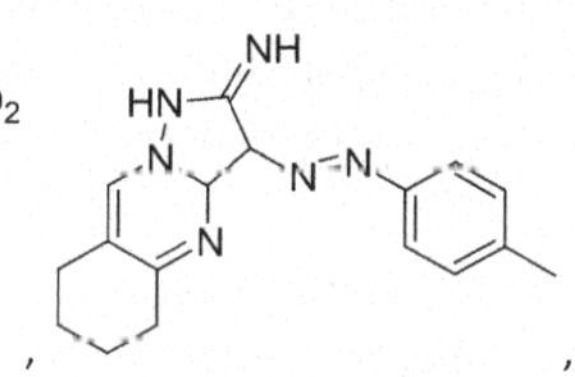

109

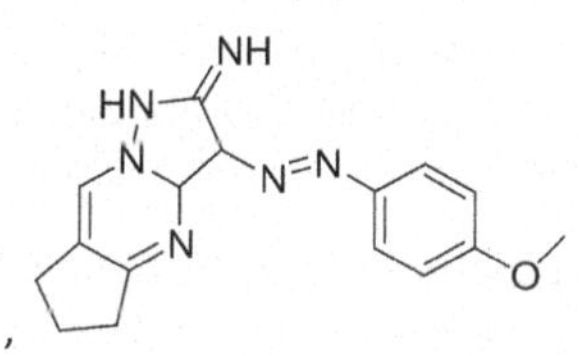

110

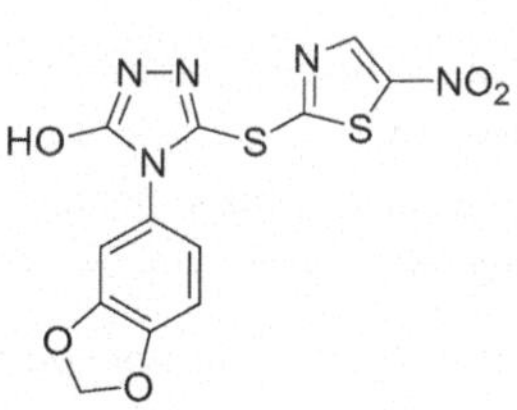

BI-78D3

Fig. 4.6 Non-ATP site JNK inhibitors

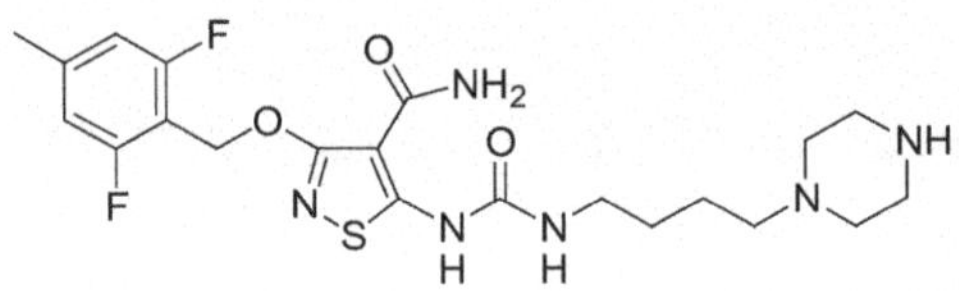

111
IC_{50} (JNK) = 3600 nM

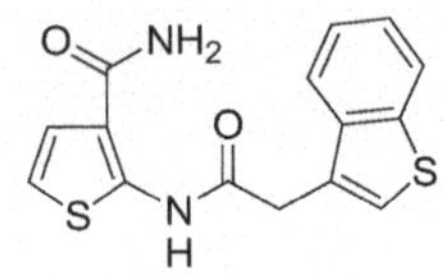

112 (BI-9F10)
IC_{50} (JNK) = 1300 nM

Fig. 4.7 Dual site JNK inhibitors

compound was tested over 56 kinases at 25 μM, shown inactive against other kinases including other MAP kinase families (Table 4.2).

ATP-noncompetitive JNK inhibitors represent a promising avenue for therapeutic intervention in diseases characterized by dysregulated JNK signaling. Their unique mechanism of action and potential for selective targeting offers opportunities for the development of novel treatments with improved efficacy and safety profiles.

4.5 Bi-dentate Inhibitors of MAP Kinase 8, 9, 10

Bi-dentate JNK inhibitors are a specific type of inhibitor that binds to both c-Jun N-terminal kinases (JNKs) ATP site and JIP site (Fig. 4.8). These inhibitors are designed to simultaneously engage multiple binding sites on the JNK protein, enhancing their affinity and potency compared to mono-dentate inhibitors.

Here are some key characteristics and considerations regarding bi-dentate JNK inhibitors:

Dual Binding Sites Bi-dentate JNK inhibitors typically consist of two pharmacophores connected by a linker. Each pharmacophore targets a distinct binding site on the JNK protein, allowing for simultaneous interaction with multiple regions of the kinase.

Enhanced Binding Affinity By engaging multiple binding sites, bi-dentate inhibitors can achieve higher binding affinity and potency compared to mono-dentate inhibitors. This dual-binding approach enhances the stability of the inhibitor-protein complex and improves the inhibitory efficacy against JNK activity.

Selective Targeting Bi-dentate inhibitors can be designed to selectively target specific isoforms or conformations of JNKs, offering opportunities for greater selectivity and specificity in modulating JNK signaling pathways.

Therapeutic Potential Bi-dentate JNK inhibitors hold promise for therapeutic applications in diseases where dysregulated JNK signaling is implicated, such as cancer, inflammatory disorders, and neurodegenerative diseases. Their enhanced potency and selectivity make them attractive candidates for drug development.

Table 4.2 Small molecules JIP site binders or dual inhibitors

Inhibitor	JNK IC_{50} (μM) Binding assay	JNK IC_{50} (μM) Activity assay	References
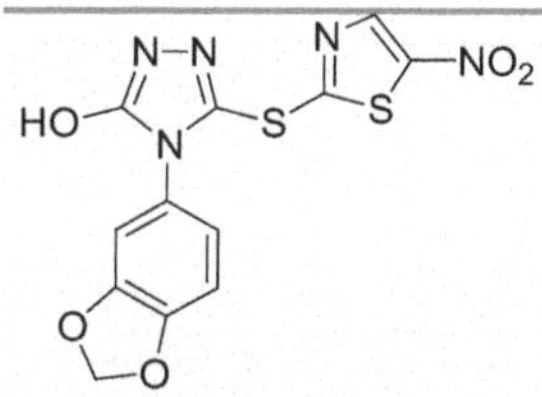	0.5	0.28	Stebbins et al. [7]
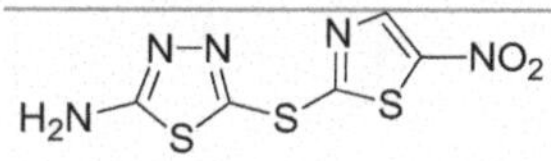	0.23	0.70	De et al. [8]
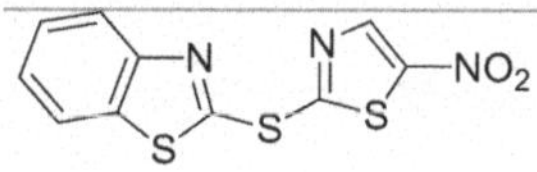	0.16	1.8	De et al. [10]
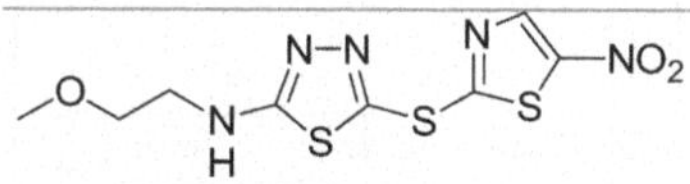	0.16	4.8	De et al. [9]
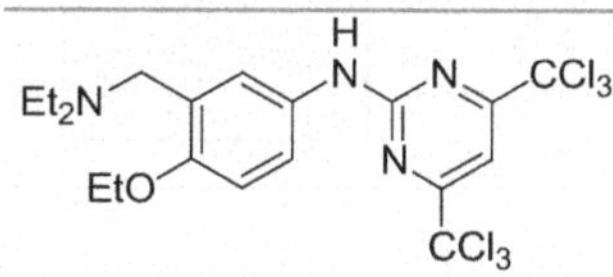	5.7	N.R.	Chen et al. [12]
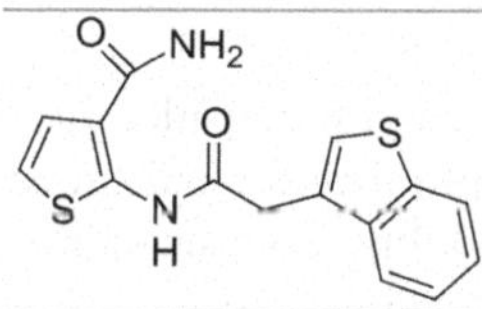	4.6	1.3	De et al. [13]
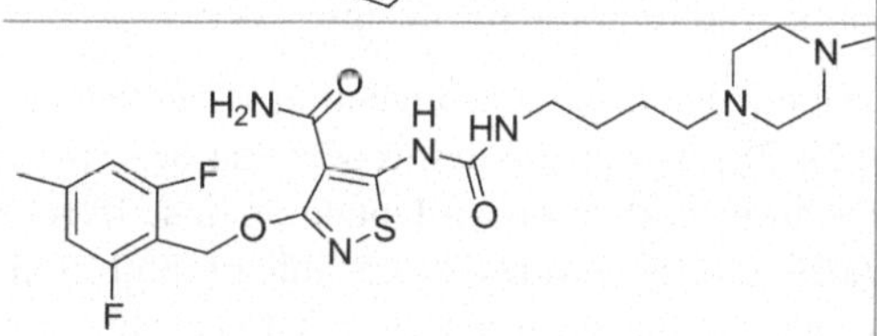	3.60	NR	Chen et al. [12]
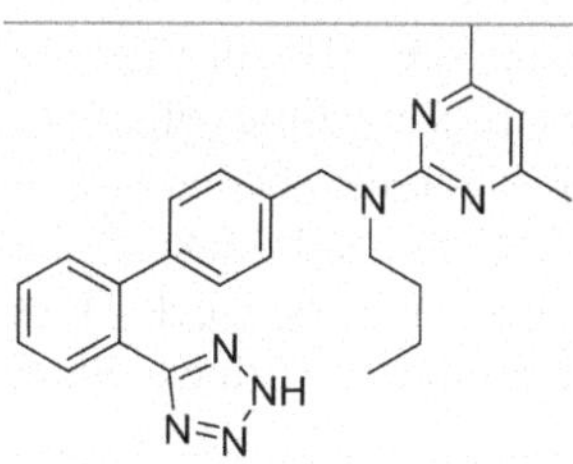	>100	7.8 μM (MKK7-JNK)	Comess et al. [11]

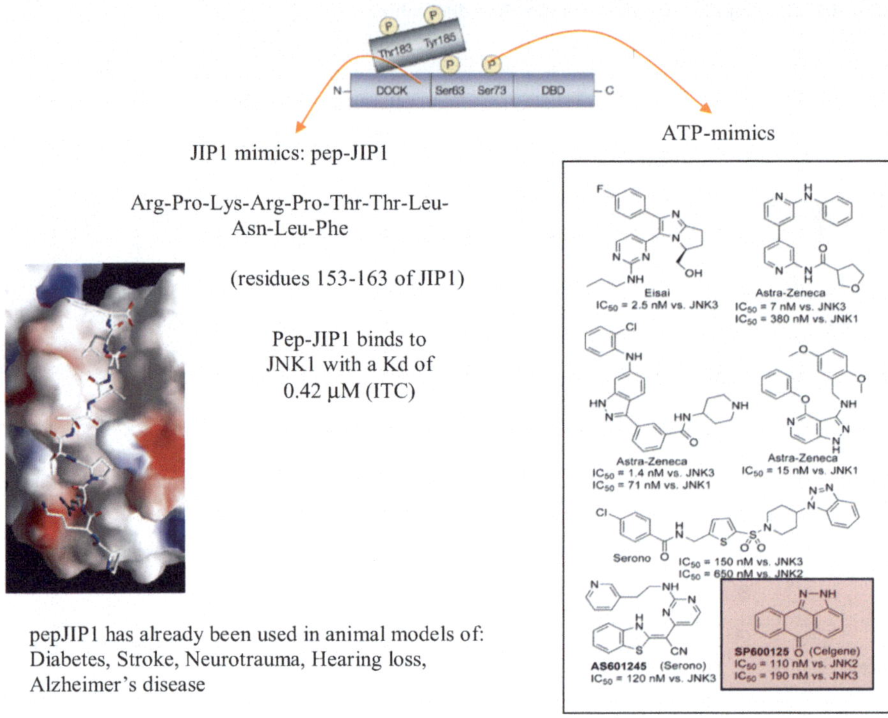

Fig. 4.8 JNK ATP mimics and JIP mimics

Design Challenges Designing bi-dentate inhibitors requires careful consideration of the chemical structure, linker length, and pharmacophore optimization to achieve optimal binding affinity and selectivity. Computational modeling and structure-activity relationship studies are often employed to guide the design process.

The researchers from Sanford-Burnham disclosed the invention of novel bi-dentate compounds as JNK inhibitors [27, 28]. It was the first report that described the synthesis and in vitro and in vivo inhibition data on bi-dentate against JNK1. While some JNK-interacting peptides (JIP) and JIP mimics are capable of inhibition as has been described previously, no compound has been reported that is capable of targeting and inhibiting JNK kinase binding to the docking site (JIP site) for the substrate or scaffolding proteins and the ATP binding site at the same time (Fig. 4.9).

Vazquez et al. developed the NMR-based approach to screen the second site of protein kinase such as JNK. The compounds bind both ATP site and docking site with nanomolar affinity although the individual fragments are very weak. This approach can be applied on other kinases to get bidentate inhibitors with excellent selectivity.

Initial SAR studies indicated the requirement for three residues of the 11-amino acid JIP-derived peptides that were essential for JNK inhibition. Investigators at Sanform-Burnham-Prebys selected a weak ATP mimic and a minimal JIP-derived

Fig. 4.9 NMR screening to find non-ATP site binder

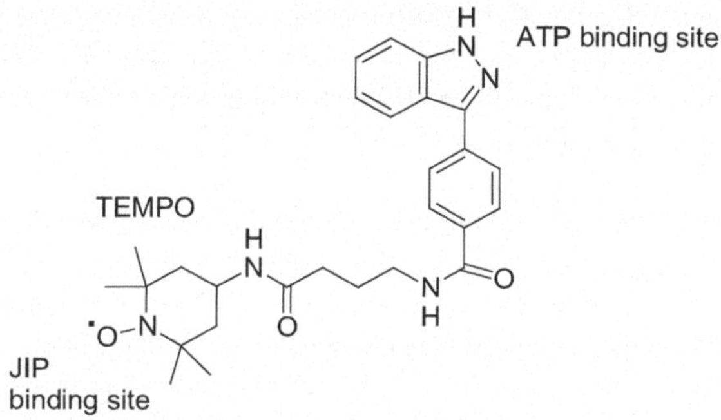

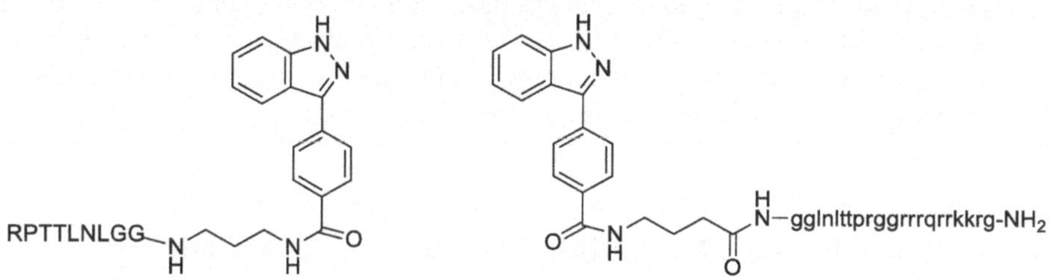

Fig. 4.10 First-generation bi-dentate JNK inhibitor from Sanford-Burnham-Prebys

peptide combined by a linker of a three-carbon chain. The ATP mimic used for this class of compounds was a modification of **SP600125**, in which the keto group was removed to improve solubility and other properties such as reduced heat and light sensitivity that facilitate the subsequent new synthetic efforts. This compound resulted in a relatively weak ATP mimetic and inhibiting JNK1 phosphorylation of the substrate ATF2 in a Time-Resolved Fluorescence Resonance Energy Transfer (TR-FRET)-based LanthaScreen™ kinase activity assay with a 14 μM IC_{50} value. The minimal peptide sequence RPTTLNL was unable to inhibit JNK1 phosphorylation of the substrate ATF2 in a Time-Resolved Fluorescence Resonance Energy Transfer (TR-FRET)-based LanthaScreen™ kinase activity assay at concentrations up to 25 μM. This represents a weakly interacting substrate-binding moiety. Nevertheless, when linked to the indazole moiety via a two-glycine 3-carbon linker the resulting compound (**BI-77B8**) [28] is able to compete with pepJIP1 for JNK1 binding and exhibit the inhibition of JNK1 kinase activity with IC_{50} values 0.9 nM and 0.7 nM respectively. To enhance in vivo stability and favorable pharmacological properties with this compound, similar to the clinical candidate D-JNKi, investigators synthesized an all-D retro-inverso version of **BI-77B8** consisting of the cell-penetrating D-retro-inverso HIV-TAT sequence. It has been reported that a D-peptide is more stable than a peptide with L-amino acids. The lead compound **BI-87G9** (Fig. 4.10) strongly competes with pepJIP1 for binding to JNK1 and inhibits JNK1 kinase activity with IC_{50} values of 18.5 nM and 46 nM respectively. This compound is a selective JNK inhibitor, as they used the same in vitro LanthaScreen™ kinase activity assay and the same ATF2 substrate **BI-87G9** was inactive against p38 at concentrations up to 100 μM which is a member of the MAPK family with highest structural similarity to JNK.

This compound also showed in vivo activity in the diabetes mice model. **BI-87G9** was effective in restoring normoglycemia dosed 4 days at 25 mg /day without causing hypoglycemia in a type 2 diabetes mouse model. The investigators also showed the link between hepatotoxicity and JNK roles using JNK1−/− and JNK2−/− mice. In vivo analysis of **BI-87G9** to block Jo2-induced liver damage which is measured by the release of liver enzymes alanine-aminotransferase (ALT) and aspartate aminotransferase (AST) into the blood has been described.

Nevertheless, by applying the fragment-based drug discovery approach, investigators successfully produced a first class of bi-dentate molecules with superior JNK inhibitory properties. Possibly, this approach can be used in the design and synthesis of other potent and selective bi-dentate kinase inhibitors.

4.6 Small Molecule Bi-dentate Inhibitors of MAK Kinase 8, 9, 10

Feng et al. [29] have reported a small-molecule bi-dentate JNK inhibitor (Fig. 4.11). This compound also inhibited LRRK2 activity. The bi-dentate inhibitor was very potent against JNK3 with an IC_{50} value of 11 nM. The key compound inhibits the phosphorylation of c-Jun with an IC_{50} value of 2.8 μM. The extensive docking studies of the key compound with JNK3 reveal that the compound is placed at both ATP and JIP sites. The indazole moiety forms hydrogen bonding interactions in the hinge region of the ATP site. The chromane (benzodihydropyran) moiety stays in the JIP site. Nevertheless, this compound can be used on Parkinson's disease as both JNK and LRRK2 are potential targets for Parkinson's disease.

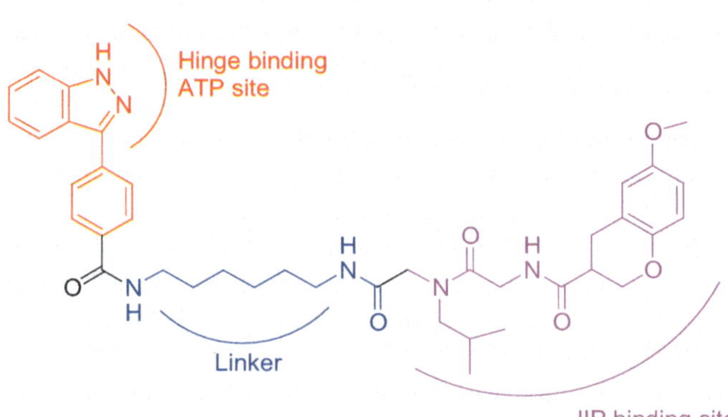

Fig. 4.11 Small-molecule bi-dentate inhibitor

The bi-dentate JNK inhibitors represent a sophisticated approach to modulating c-Jun N-terminal kinases (JNKs), offering enhanced potency and selectivity compared to mono-dentate inhibitors. By simultaneously engaging multiple binding sites on the JNK protein, these inhibitors can achieve higher affinity and stability in inhibiting JNK activity. The bi-dentate JNK inhibitors represent a promising avenue for therapeutic intervention in diseases associated with dysregulated JNK signaling. Further research and development efforts are needed to explore their full therapeutic potential and translate them into clinical applications.

4.7 Allosteric Inhibitors of ERK1/2 (Non-ATP-Competitive ERK Inhibitors)

ERKs have two main substrate-binding sites: D-recruitment site (DRS) and F-recruitment site (FRS). Compounds bound in DRS or FRS sites have significant differences than compounds bound in the ATP site. Since the ATP site is well conserved among the protein kinase families, compounds bound in the allosteric site have more selectivity. Hancock et al. identified 7 non-ATP-competitive ERK2 inhibitors [34]. These compounds inhibit ERK-specific phosphorylation of ribosomal S6 kinase-1 (Rsk-1) or the ternary complex factor Elk-1 (TCF/Elk-1). Several compounds inhibit the proliferation of HeLa cervical carcinoma cells with IC_{50} values in the 25–50 μM range (Fig. 4.12).

Chen et al. identified 4 compounds by the computer-aided drug design (CADD) screen as ERK2 inhibitors [35]. These compounds inhibit ERK-mediated phosphorylation of ribosomal S6 kinase-1 (Rsk-1) and/or the transcription factor Elk-1. Several compounds also inhibit the proliferation of HeLa cervical carcinoma cells with IC_{50} values in the 2–10 μM range (Fig. 4.13).

Li et al. reported the structure–activity relationship (SAR) studies of 3-(2-amino-ethyl)-5-(4-ethoxy-benzylidene)-thiazolidine-2,4-diones as substrate-specific ERK1/2 inhibitors [36]. They found that the shift of the ethoxy group from the 4- to the 2- position on the phenyl ring improves the inhibition of cell proliferation and inducing apoptosis (Fig. 4.14).

The ERK1/2 plays a crucial role in cancer cell proliferation and survival. Boston et al. reported on the characterization of ERK docking domain inhibitors that induce apoptosis by targeting Rsk-1 and Caspase-9 [37]. They synthesized several analogs

Fig. 4.12 Allosteric ERK2 inhibitor

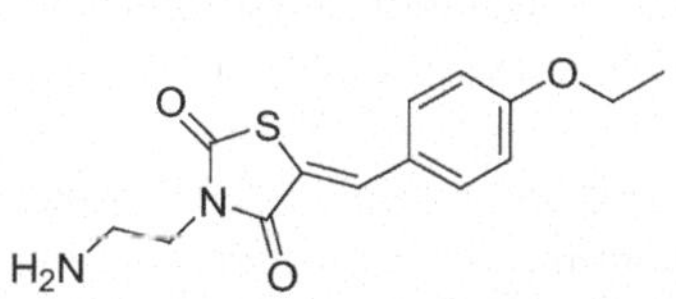

IC_{50} (A549 lung carcinoma cell) = 25 μM

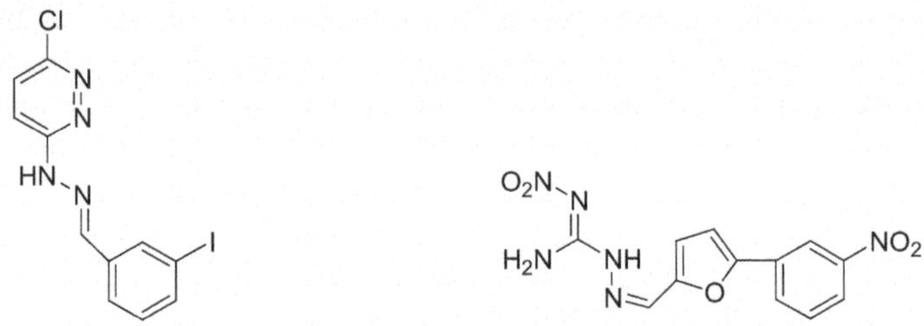

IC_{50} (HeLa cells) = 5 μM IC_{50} (HeLa Cells) = 2 μM

Fig. 4.13 ATP noncompetitive ERK1/2 inhibitors

Fig. 4.14 Thiazolidine 2,4-dione derivatives as ERK1/2 inhibitors

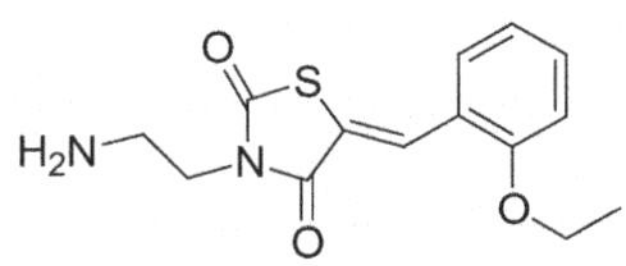

of previously published thiazolidine-2,4-dione derivatives. Several compounds show good anti-proliferative activity in HeLa cells (Fig. 4.15).

Sammons et al. identified 10 compounds with a core structure comprising of a tertiary amine, arginine, and cyclic guanidine [38]. These compounds bind in the D-recruitment site (DRS) and inhibit ERK2 from phosphorylating a DRS-targeting substrate. Nuclear magnetic resonance and X-ray crystallography studies suggest that the neutralizing effect between the acidic Asp319 residue and the basic guanidine group of the compound is a possible mechanism of the inhibition of ERK2. The best compound **2507–1** blocks ERK2 activity with an IC_{50} of 9.9 ± 1.9 μM in vitro and it has a K_d value of 0.38 ± 0.04 μM (Fig. 4.16).

The small-molecule **DEL-22379** inhibits ERK dimerization with an IC_{50} of 0.50 μM (Fig. 4.17). It blocks the sub-localization without inhibiting total kinase activity and phosphorylation [39]. It targets protein-protein interaction, rather than catalytic activities. It binds to ERK2 with surrounding Pro174, Asp175, His176, Asp177, Phe181, and Phe329 residues in the dimerization interface and prevents the interaction between the two ERK molecules. This compound reduces tumor progression in a colorectal cancer patient-derived xenograft model.

A combination of computation and experimental techniques is used to identify small-molecule inhibitors targeting the ERK1/2 docking domain (Fig. 4.18). Thienyl benzenesulfonate derivatives such as **SF-3-029** and **SF-3-030** bind in the F-recruitment site (FRS). These compounds selectively induce apoptosis in melanoma cells containing BRAF mutation or activated ERK1/2 signal. Lead compounds inhibit AP-1 promoter activity with IC_{50} values 5–10 μM [40].

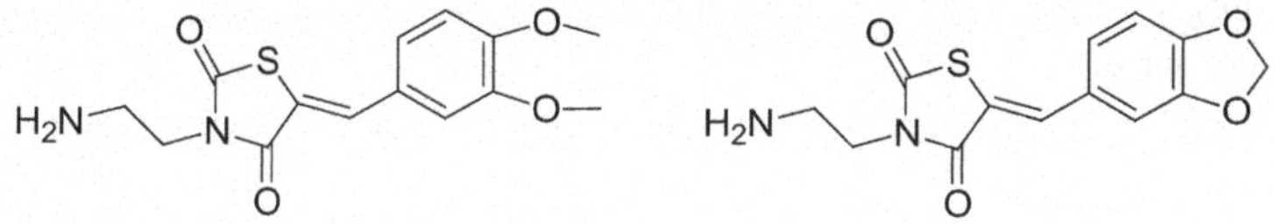

Fig. 4.15 ERK docking domain inhibitors

Fig. 4.16 Cyclic guanidine derivative as ERK2 inhibitor

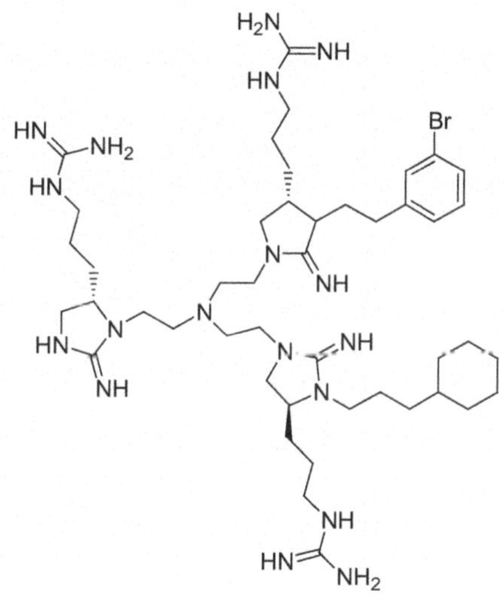

2507-1

IC_{50} (ERK2) = 9.9 μM

4.8 Allosteric Inhibitors of p38 MAPK

Interleukin-1β (IL-1β) is an important cytokine protein that in humans is encoded by the *IL1B* gene. It is a central mediator of inflammatory responses and inflammation. It regulates a variety of cellular activities, including cell proliferation, differentiation, and apoptosis. Excessive production of IL-1β causes a number of different pathologies such as autoinflammatory diseases, metabolic disorders, and malignant tumors. The p38α MAPK is directly involved in the production of IL-1β. Therefore, the inhibition of this MAPK results in inhibition of IL-1β production. Carrasco et al. reported the hit-to-lead optimization technique to get the compound **AIK3–305**, an allosteric inhibitor of p38α MAPK (Fig. 4.19). It is a potent, orally bioavailable, and selective inhibitor of IL-1β with an IC_{50} of 28 nM. It has a long distribution half-life time of 2.1 h and an elimination half-life time of 9 h. It binds in the A-loop of the p38α MAPK with a K_d value of 12 nM. This compound exhibits in vivo efficacy in a LPS endotoxic shock model in male BALB/c mice [41].

Wilson et al. screened 32,000 compounds to find an allosteric p38 MAPK inhibitor [42]. They found a few compounds from their screening campaign. The best

Fig. 4.17 ERK2 dimerization inhibitor

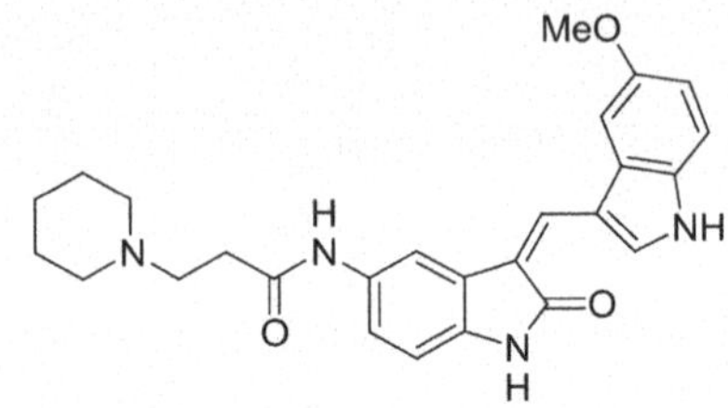

DEL-22379

IC_{50} (ERK2 dimerization) = 0.50 μM

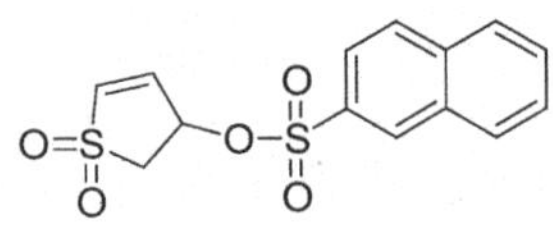

SF-3-030

IC_{50} (AP-1) = 5-10 μM

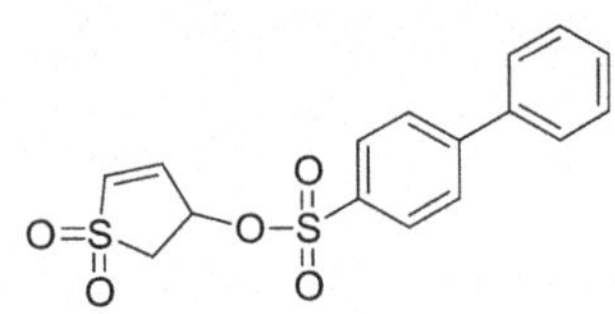

SF-3-029

IC_{50} (AP-1) = 5-10 μM

Fig. 4.18 Thienyl benzenesulfonate derivatives as ERK1/2 docking binders

Fig. 4.19 Trisubstituted pyridine derivatives as p38α

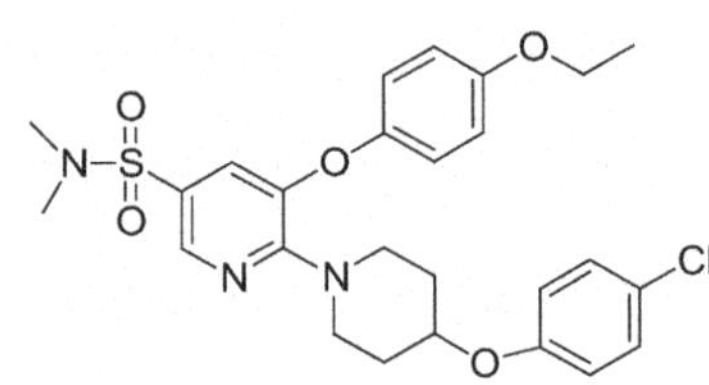

AIK3-305

K_d (p38) = 12 nM

IC_{50} (IL-1β) = 28 nM

compound as shown in Fig. 4.20 inhibits p38 activity with an IC_{50} value of 2.62 μM. This compound binds in an allosteric site of p38 with a Ki value of 46 nM. It is a non-ATP-competitive inhibitor of p38 MAP kinase. This compound reduces p38-dependent proliferation and TNF-α secretion in a dose-dependent manner.

Gomez-Gutierrez et al. identified compound **UPC-K-005** as an inhibitor of p38α [43]. It is a non-ATP-competitive inhibitor with an IC_{50} value of 16 μM (Fig. 4.21). Site-directed mutagenesis studies indicate that it binds to an allosteric site in p38α. A docking study reveals that the tetrazole ring makes a hydrophobic interaction

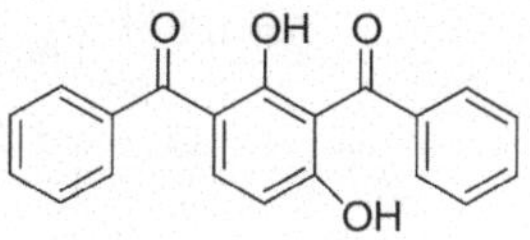

IC_{50} (p38) = 2.62 μM

K_i (p38) = 46 nM

Fig. 4.20 2,4-dibenzoylresorcinol as an allosteric inhibitor of p38

Fig. 4.21 Chemical structure of UPC-K-005

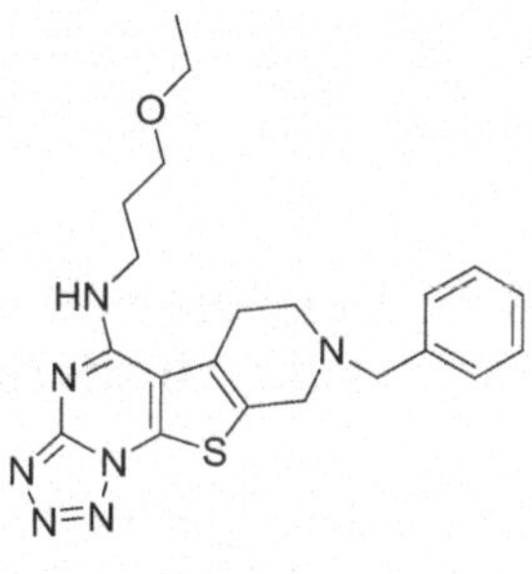

UPC-K-005

IC_{50} (p38α) = 16 μM

Fig. 4.22 Thiazole-urea derivatives as allosteric p38α inhibitor

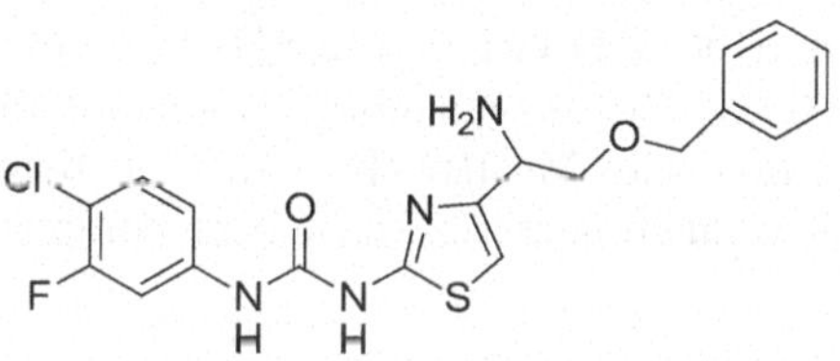

with Leu171 and Val183. The one nitrogen atom of the tetrazole ring forms a hydrogen bonding interaction with Arg189. The oxygen atom of the ether group forms a hydrogen bonding interaction with the guanidine group of Arg186. The nitrogen atom on the tetrahydropyridine ring can be protonated in aqueous solution so it acts as a hydrogen bond donor to the oxygen atom of Asp177.

Getlik et al. reported the structured-based design, synthesis, and biological activity of thiazole-urea derivatives as p38α inhibitors [44]. The lead compound inhibits p38α with an IC_{50} value of 135 nM. One analog of this series (Fig. 4.22) inhibits the phosphorylation of the mitogen-activated protein kinase-activated protein kinase 2 (MK2) in HeLa cells. These compounds bind in the allosteric site of p38α.

Fig. 4.23 Urea derivative as p38 MAP kinase allosteric modulator

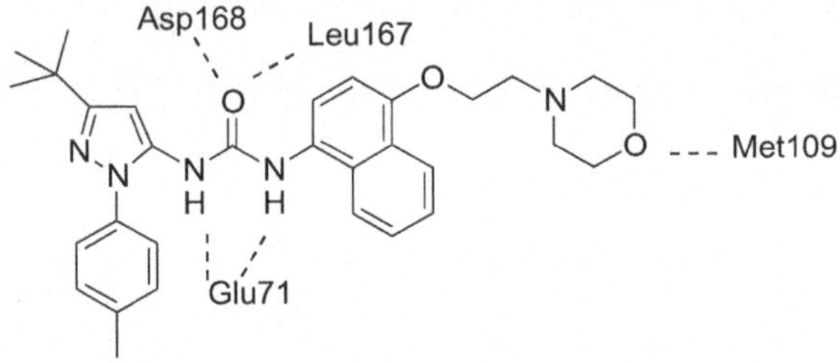

Doramapimod (BIRB-796)
IC_{50} (p38α) = 38 nM
IC_{50} (p38β) = 65 nM
IC_{50} (TNF- α) = 21 nM

4.9 Development of Clinical Candidate BIRB-796 as an Allosteric p38 MAP Kinase Inhibitor

Investigators from Boehringer-Ingelheim developed doramapimod (BIRB-796) as a selective inhibitor of p38α and p38β with IC_{50} values of 35 nM and 65 nM [45–49]. There are no significant inhibitions with other MAP kinases. It inhibits the production of LPS-stimulated cytokine TNF-α with an IC_{50} value of 21 nM. The X-ray crystal structure of BRRB-796 in complex with p38α (PDB code: 1KV2) reveals that the carbonyl group of the urea forms a hydrogen bonding interaction with Asp168 and Leu167. The NH of the urea forms a hydrogen bonding interaction with Glu71. The oxygen atom of the morpholine group forms a hydrogen bonding interaction with Met109 (Fig. 4.23). It binds to the allosteric site of the kinase [47]. Several clinical trials have been completed using this compound.

References

1. Baar RK, Kendrick TS, Bogoyevitch MA. Identification of the critical features of a small peptide inhibitor of JNK activity. J Biol Chem. 2002;277:10987–77.
2. Ngoei KRW, Catimel B, Church N, Lio DS, Dogovski C, Perugini MA, Watt PM, Cheng HC, Ng DCH, Bogoyevitch MA. Characterization of a novel JNK (c-Jun N-terminal kinase) inhibitory peptide. Biochem J. 2011;434:399–413.
3. Ngoei KRW, Catimel B, Milech N, Watt PM, Bogoyevitch MA. A novel retro-inverso peptide is a preferential JNK substrate-competitive inhibitor. Int J Biochem Cell Biol. 2013;45:1939–50.
4. Heo YS, Kim SK, Seo CI, Kim YK, Sung BJ, Lee HS, Lee JI, Park SY, Kim JH, Hwang KY, Hyun YL, Jeon YH, Ro S, Cho JM, Lee TG, Yang CH. Structural basis for the selective inhibition of JNK1 by the scaffolding protein JIP1 and SP600125. EMBO J. 2004;23:2185–95.
5. Niu L, Chang KC, Wilson S, Tran P, Zuo F, Swinney DC. Kinetic characterization of human JNK2alpha2 reaction mechanism using substrate competitive inhibitors. Biochemistry. 2007;46:47754784.

6. Kaoud TS, Mitra S, Lee S, Taliaferro J, Cantrell M, Linse KD, Van Den Berg CL, Dalby KN. Development of JNK2-selective peptide inhibitors that inhibit breast cancer cell migration. ACS Chem Biol. 2011;6:658–66.
7. Stebbins JL, De SK, Machleidt T, Becattini B, Vazquez J, Kuntzen C, Chen LH, Cellitti JF, Riel-Mehan M, Emdadi A, Solinas G, Karin M, Pellecchia M. Identification of a novel JNK inhibitor targeting JNK-JIP interaction. Proc Natl Acad Sci USA. 2008;105:16809–13.
8. De SK, Stebbins JL, Chen L-H, Riel-Mehan M, Machleidt T, Dahl R, Yuan H, Emdadi A, Barile E, Chen V, Murphy R, Pellecchia M. Design, synthesis, and structure-activity relationship of substrate competitive, selective, and *in vivo* active triazole and thiadiazole inhibitors of the c-Jun-N-terminal kinase. J Med Chem. 2009;52:1943–52.
9. De SK, Chen V, Stebbins JL, Chen L-H, Cellitti JF, Machleidt T, Barile E, Riel-Mehan M, Dahl R, Yang L, Emdaddi A, Murphy M, Pellecchia M. Synthesis and optimization of thiadiazole derivatives as a novel class of substrate competitive c-Jun N-terminal kinase inhibitors. Bioorg Med Chem. 2010;18:590–6.
10. De SK, Chen L-H, Stebbins JL, Machleidt T, Riel-Mehan M, Dahl R, Chen V, Yuan H, Barile E, Emdadi A, Murphy R, Pellecchia M. Discovery of 2-(5-nitrothiazol-2yl)thio)benzo[*d*]thiazoles as novel JNK inhibitors. Bioorg Med Chem. 2009; 17:2712–7.
11. Comess KM, Sun C, Abad-Zapatero C, Goedken ER, Gum RJ, Borhani DW, Argiriadi M, Groebe DR, Jia Y, Clampit JE, Haasch DL, Smith HT, Wang S, Song D, Coen ML, Cloutier TE, Tang H, Cheng X, Quinn C, Liu B, Xin Z, Liu G, Fry EH, Stoll V, Ng TI, Banach D, Marcotte D, Burns DJ, Calderwood DJ, Hajduk PJ. Discovery and characterization of non-ATP site inhibitors of the mitogen activated protein (MAP) kinases. ACS Chem Biol. 2011;6:234–44.
12. Chen T, Kablauo N, Little J, Timofeevski S, Tschantz WR, Chen P, Peng J, Charlton M, Stanton R, Bauer P. Identification of small-molecule inhibitors of the JIP-JNK interaction. Biochem J. 2009;420:283–94.
13. De SK, Barile E, Chen V, Stebbins JL, Cellitti JF, Machleidt T, Carlson CB, Yang L, Dahl R, Pellecchia M. Design, synthesis, and structure-activity relationship studies of thiophene-3-carboxamide derivatives as dual inhibitors of the c-Jun-N-terminal kinase. Bioorg Med Chem. 2011;19:2582–8.
14. Kaoud TS, Park H, Mitra S, Yan C, Tseng CC, Shi Y, Jose J, Taliaferro JM, Lee K, Ren P, Hong J, Dalby KN. Manipulating JNK signaling with (−)-zuonin A. ACS Chem Biol. 2012;7:1873.
15. Kaoud TS, Yan C, Mitra S, Tseng CC, Jose J, Taliaferro JM, Tuohetahuntila M, Devkota A, Sammons R, Park J, Park H, Shi Y, Hong J, Ren P, Dalby KN. From in Silico discovery to intracellular activity: targeting JNK-protein interactions with small molecules. ACS Med Chem Lett. 2012;3:721.
16. Kook S, Zhan X, Kaoud TS, Dalby KN, Gurevich VV, Gurevich EV. Arrestin-3 binds c-Jun N-terminal kinase 1 (JNK1) and JNK2 and facilitates the activation of these ubiquitous JNK isoforms in cells via scaffolding. J Biol Chem. 2013;288:37332.
17. Zhan X, Kook S, Kaoud TS, Dalby KN, Gurevich EV, Gurevich VV. Arrestin-3-dependent activation of c-Jun N-terminal kinases (JNKs). Curr Protoc Pharmacol. 2015;68:2.12.1–2.
18. Yan C, Kaoud T, Lee S, Dalby KN, Ren P. Understanding the specificity of a docking interaction between JNK1 and the scaffolding protein JIP1. J Phys Chem B. 2011;115:1491.
19. Zhan X, Kaoud TS, Dalby KN, Gurevich VV. Nonvisual arrestins function as simple scaffolds assembling the MKK4-JNK3α2 signaling complex. Biochemistry. 2011;50:10520–9.
20. Schnieders MJ, Kaoud TS, Yan C, Dalby KN, Ren P. Computational insights for the discovery of non-ATP competitive inhibitors of MAP kinases. Curr Pharm Des. 2012;18:1173–85.
21. Zhan X, Kaoud TS, Kook S, Dalby KN, Gurevich VV. JNK3 enzyme binding to arrestin-3 differentially affects the recruitment of upstream mitogen-activated protein (MAP) kinase kinases. J Biol Chem. 2013;288:28535–47.
22. Dykstra DW, Dalby KN, Ren P. Elucidating binding modes of zuonin A enantiomers to JNK1 via in silico methods. J Mol Graph Model. 2013;45:38–44.
23. Zhan X, Stoy H, Kaoud TS, Perry NA, Chen Q, Perez A, Els-Heindl S, Slagis JV, Iverson TM, Beck-Sickinger AG, Gurevich EV, Dalby KN, Gurevich VV. Peptide mini-scaffold facilitates JNK3 activation in cells. Sci Rep. 2016;6:21025.

24. Xie X, Kaoud TS, Edupuganti R, Zhang T, Kogawa T, Zhao Y, Chauhan GB, Giannoukos DN, Qi Y, Tripathy D, Wang J, Gray NS, Dalby KN, Bartholomeusz C, Ueno NT. c-Jun N-terminal kinase promotes stem cell phenotype in triple-negative breast cancer through upregulation of Notch1 via activation of c-Jun. Oncogene. 2017;36:2599–608.
25. Perry NA, Kaoud TS, Ortega OO, Kaya AI, Marcus DJ, Pleinis JM, Berndt S, Chen Q, Zhan X, Dalby KN, Lopez CF, Iverson TM, Gurevich VV. Arrestin-3 scaffolding of the JNK3 cascade suggests a mechanism for signal amplification. Proc Natl Acad Sci USA. 2019;116:810–5.
26. Vin H, Ojeda SS, Ching G, Leung ML, Chitsazzadeh V, Dwyer DW, Adelmann CH, Restrepo M, Richards KN, Stewart LR, Du L, Ferguson SB, Chakravarti D, Ehrenreiter K, Baccarini M, Ruggieri R, Curry JL, Kim KB, Ciurea AM, Duvic M, Prieto VG, Ullrich SE, Dalby KN, Flores ER, Tsai KY. BRAF inhibitors suppress apoptosis through off-target inhibition of JNK signaling. elife. 2013;2:e00969.
27. Vazquez J, De SK, Chen L-H, Riel-Mehan M, Emdadi A, Cellitti J, Stebbins JL, Rega MF, Pellecchia M. Development of paramagnetic probes for molecular recognition in protein kinases. J Med Chem. 2008;51:3460–5.
28. Stebbins JL, De SK, Pavlickova P, Chen V, Machleidt T, Chen L-H, Kuntzen C, Kitada S, Karin M, Pellecchia M. Design and characterization of a potent and selective dual ATP and substrate-competitive subnanomolar bidentate c-Jun N-terminal kinase inhibitor. J Med Chem. 2011;54:6206–14.
29. Feng Y, Chambers JW, Iqbal S, Koenig M, Park HJ, Cherry L, Hernandez P, Losada MF, LoGrasso PV. A small molecule bi-dentate binding dual inhibitor probe of LRRK2 and JNK kinases. ACS Chem Biol. 2013;16:1747.
30. Pan J, Li H, Zhang B, Xiong R, Zhang Y, Kang WY, Chen W, Zhao ZB, Chen SD. Small peptide inhibitor of JNK3 protects dopaminergic neurons from MPTP induced injury via inhibiting the ASK1-JNK3 signaling pathway. PLoS One. 2015;10(4):e0119204. https://doi.org/10.1371/journal.pone.0119204. PMID: 25856433; PMCID: PMC4391862
31. Chambers JW, Cherry L, Laughlin JD, Figuera-Losada M, Lograsso PV. Selective inhibition of mitochondrial JNK signaling achieved using peptide mimicry of the sab kinase interacting motif-1 (KIM1). ACS Chem Biol. 2011;6(8):808–18. https://doi.org/10.1021/cb200062a. Epub 2011 May 24. PMID: 21563797; PMCID: PMC3158843
32. Laughlin JD, Nwachukwu JC, Figuera-Losada M, Cherry L, Nettles KW, LoGrasso PV. Structural mechanisms of allostery and autoinhibition in JNK family kinases. Structure. 2012;20(12):2174–84. https://doi.org/10.1016/j.str.2012.09.021. Epub 2012 Nov 8. PMID: 23142346; PMCID: PMC3589125
33. Bonny C, Oberson A, Negri S, Sauser C, Schorderet DF. Cell-permeable peptide inhibitors of JNK: novel blockers of beta-cell death. Diabetes. 2001;50(1):77–82. https://doi.org/10.2337/diabetes.50.1.77. PMID: 11147798
34. Hancock CN, Macias A, Lee EK, Yu SY, Mackerell AD Jr, Shapiro P. Identification of novel extracellular signal-regulated kinase docking domain inhibitors. J Med Chem. 2005;48(14):4586–95. https://doi.org/10.1021/jm0501174. PMID: 15999996
35. Chen F, Hancock CN, Macias AT, Joh J, Still K, Zhong S, MacKerell AD Jr, Shapiro P. Characterization of ATP-independent ERK inhibitors identified through in silico analysis of the active ERK2 structure. Bioorg Med Chem Lett. 2006;16(24):6281–7. https://doi.org/10.1016/j.bmcl.2006.09.038. Epub 2006 Sep 26. PMID: 17000106; PMCID: PMC1857279
36. Li Q, Al-Ayoubi A, Guo T, Zheng H, Sarkar A, Nguyen T, Eblen ST, Grant S, Kellogg GE, Zhang S. Structure-activity relationship (SAR) studies of 3-(2-amino-ethyl)-5-(4-ethoxy-benzylidene)-thiazolidine-2,4-dione: development of potential substrate-specific ERK1/2 inhibitors. Bioorg Med Chem Lett. 2009;19(21):6042–6. https://doi.org/10.1016/j.bmcl.2009.09.057. Epub 2009 Sep 18. PMID: 19796943; PMCID: PMC6088847
37. Boston SR, Deshmukh R, Strome S, Priyakumar UD, MacKerell AD Jr, Shapiro P. Characterization of ERK docking domain inhibitors that induce apoptosis by targeting Rsk-1 and caspase-9. BMC Cancer. 2011;11:7. https://doi.org/10.1186/1471-2407-11-7. PMID: 21219631; PMCID: PMC3025971

38. Sammons RM, Perry NA, Li Y, Cho EJ, Piserchio A, Zamora-Olivares DP, Ghose R, Kaoud TS, Debevec G, Bartholomeusz C, Gurevich VV, Iverson TM, Giulianotti M, Houghten RA, Dalby KN. A novel class of common docking domain inhibitors that prevent ERK2 activation and substrate phosphorylation. ACS Chem Biol. 2019;14(6):1183–94. https://doi.org/10.1021/acschembio.9b00093. Epub 2019 May 13. PMID: 31058487; PMCID: PMC7105935
39. Herrero A, Pinto A, Colón-Bolea P, Casar B, Jones M, Agudo-Ibáñez L, Vidal R, Tenbaum SP, Nuciforo P, Valdizán EM, Horvath Z, Orfi L, Pineda-Lucena A, Bony E, Keri G, Rivas G, Pazos A, Gozalbes R, Palmer HG, Hurlstone A, Crespo P. Small molecule inhibition of ERK dimerization prevents tumorigenesis by RAS-ERK pathway oncogenes. Cancer Cell. 2015;28(2):170–82. https://doi.org/10.1016/j.ccell.2015.07.001. PMID: 26267534
40. Samadani R, Zhang J, Brophy A, Oashi T, Priyakumar UD, Raman EP, St John FJ, Jung KY, Fletcher S, Pozharski E, MacKerell AD Jr, Shapiro P. Small-molecule inhibitors of ERK-mediated immediate early gene expression and proliferation of melanoma cells expressing mutated BRaf. Biochem J. 2015;467(3):425–38. https://doi.org/10.1042/BJ20131571. PMID: 25695333; PMCID: PMC4643458
41. Carrasco E, Gomez-Gutierrez P, Campos PM, Vega M, Messeguer A, Perez JJ. Discovery of novel 2,3,5-trisubstituted pyridine analogs as potent inhibitors of IL-1β via modulation of the p38 MAPK signaling pathway. Eur J Med Chem. 2021;223:113620. https://doi.org/10.1016/j.ejmech.2021.113620. Epub 2021 Jun 12. PMID: 34186234
42. Wilson BA, Alam MS, Guszczynski T, Jakob M, Shenoy SR, Mitchell CA, Goncharova EI, Evans JR, Wipf P, Liu G, Ashwell JD, O'Keefe BR. Discovery and characterization of a biologically active non-ATP-competitive p38 MAP kinase inhibitor. J Biomol Screen. 2016;21(3):277–89. https://doi.org/10.1177/1087057115615518. Epub 2015 Nov 4. PMID: 26538432; PMCID: PMC7480010
43. Gomez-Gutierrez P, Campos PM, Vega M, Perez JJ. Identification of a novel inhibitory allosteric site in p38α. PLoS One. 2016;11(11):e0167379. https://doi.org/10.1371/journal.pone.0167379. PMID: 27898710; PMCID: PMC5127581
44. Getlik M, Grütter C, Simard JR, Nguyen HD, Robubi A, Aust B, van Otterlo WA, Rauh D. Structure-based design, synthesis and biological evaluation of N-pyrazole, N'-thiazole urea inhibitors of MAP kinase p38α. Eur J Med Chem. 2012;48:1–15. https://doi.org/10.1016/j.ejmech.2011.11.019. Epub 2011 Nov 18. PMID: 22154891
45. Regan J, Capolino A, Cirillo PF, Gilmore T, Graham AG, Hickey E, Kroe RR, Madwed J, Moriak M, Nelson R, Pargellis CA, Swinamer A, Torcellini C, Tsang M, Moss N. Structure-activity relationships of the p38alpha MAP kinase inhibitor 1-(5-tert-butyl-2-p-tolyl-2H-pyrazol-3-yl)-3-[4-(2-morpholin-4-yl-ethoxy)naph- thalen-1-yl]urea (BIRB 796). J Med Chem. 2003;46(22):4676–86. https://doi.org/10.1021/jm030121k. PMID: 14561087
46. Regan J, Breitfelder S, Cirillo P, Gilmore T, Graham AG, Hickey E, Klaus B, Madwed J, Moriak M, Moss N, Pargellis C, Pav S, Proto A, Swinamer A, Tong L, Torcellini C. Pyrazole urea-based inhibitors of p38 MAP kinase: from lead compound to clinical candidate. J Med Chem. 2002;45(14):2994–3008. https://doi.org/10.1021/jm020057r. PMID: 12086485
47. Pargellis C, Tong L, Churchill L, Cirillo PF, Gilmore T, Graham AG, Grob PM, Hickey ER, Moss N, Pav S, Regan J. Inhibition of p38 MAP kinase by utilizing a novel allosteric binding site. Nat Struct Biol. 2002;9(4):268–72. https://doi.org/10.1038/nsb770. PMID: 11896401
48. Kuma Y, Sabio G, Bain J, Shpiro N, Márquez R, Cuenda A. BIRB796 inhibits all p38 MAPK isoforms in vitro and in vivo. J Biol Chem. 2005;280(20):19472–9. https://doi.org/10.1074/jbc.M414221200. Epub 2005 Mar 8. PMID: 15755732
49. Suplatov D, Kopylov K, Sharapova Y, Švedas V. Human p38α mitogen-activated protein kinase in the Asp168-Phe169-Gly170-in (DFG-in) state can bind allosteric inhibitor Doramapimod. J Biomol Struct Dyn. 2019;37(8):2049–60. https://doi.org/10.1080/07391102.2018.1475260. Epub 2018 Dec 5. PMID: 29749295

Chapter 5
Covalent Inhibitors of MAP Kinases

Abstract Several covalent inhibitors such as aspirin, Penicillin G, and Omeperazole have been used for decades. Covalent inhibitors form a covalent bond with a specific target protein and effectively disrupt the activity of the target protein, which involve disease processes such as cancer progression, inflammatory responses, or infectious diseases. The reaction mechanisms of some FDA-approved covalent inhibitors are depicted. The binding mode of covalent inhibitors of MAP kinases is summarized.

Keywords Covalent kinase inhibitors · Ritlecitinib · Dimethyl fumarate · Afatinib · Ibrutinib · Warheads · Acrylamide · BI-78D3 · CC-90003 · JNK-IN-7 · YL-5084

5.1 Introduction

Covalent inhibitors are rationally designed that bind to their target proteins, in which the protein has lost its function [1–37]. Covalent inhibitors are indeed designed to bind irreversibly to specific target proteins, forming a covalent bond. This bond modifies the protein's structure, often at key functional sites such as the active site or allosteric sites, ultimately rendering the protein non-functional or less effective in its role within cellular processes.

The rational design of covalent inhibitors involves understanding the structure and function of the target protein, as well as identifying reactive sites or residues within the protein that can form stable covalent bonds with the inhibitor molecule. By selectively modifying these sites, covalent inhibitors can effectively disrupt the activity of the target protein, which may be involved in disease processes such as cancer progression, inflammatory responses, or infectious diseases.

The irreversible nature of covalent inhibition can provide advantages in terms of potency and duration of action, but it also requires careful consideration to minimize off-target effects and potential toxicity. Thus, rational design approaches aim to balance specificity and efficacy while minimizing adverse effects, making covalent inhibitors valuable tools in drug discovery and development. Here some

S. K. De, *Drug Design and Discovery*,
https://doi.org/10.1007/978-3-031-75682-5_5

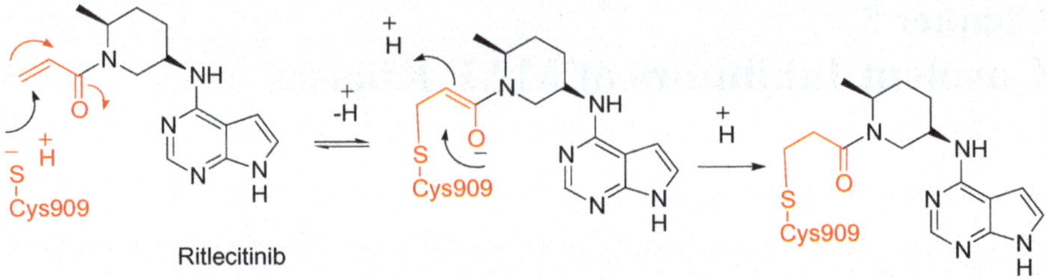

Fig. 5.1 Ritlecitinib forms a covalent bond with Cys909 of JNK3

historical covalent drugs such as ritlecitinib, dimethyl fumarate, wortmannin, afatinib, ibrutinib, bortezomib, and other reaction mechanisms are described.

5.2 Ritlecitinib

Ritlecitinib binds covalently to Cys909 of JAK3. The same position has a serine residue for other JAK isoforms that makes it a highly selective JAK3 inhibitor [31]. The reaction mechanism is shown in Fig. 5.1.

5.3 Dimethyl Fumarate

Dimethyl fumarate is used for the treatment of relapse-remitting multiple sclerosis and psoriasis [32]. It binds covalently to Cys599 of p90 ribosomal S6 kinase 2 (RSK2). The reaction mechanism is shown in Fig. 5.2.

5.4 Wortmannin

Wortmannin is a class of natural products of furano steroids. It is a potent and selective lysine targeted covalent inhibitor of phosphoinositide 3-kinases (PI3Ks). The furan ring in wortmannin is an extended and activated Michael acceptor. It covalently reacts with a conserved active site lysine nucleophile as shown in Fig. 5.3. Furan ring of the wortmannin undergoes ring opening to form a conjugated enamine [35].

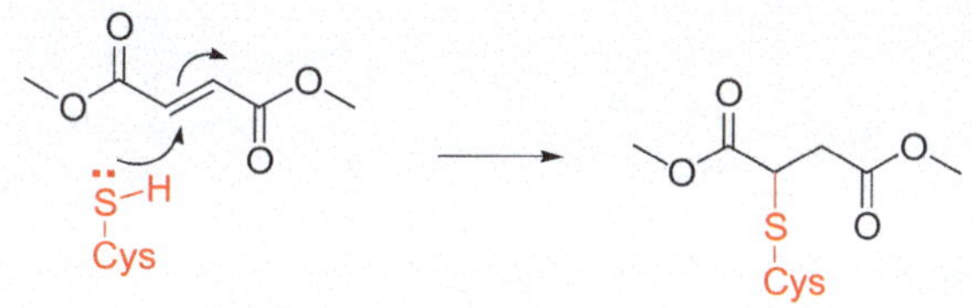

Fig. 5.2 Dimethyl fumarate forms a covalent bond with Cys599 of RSK2

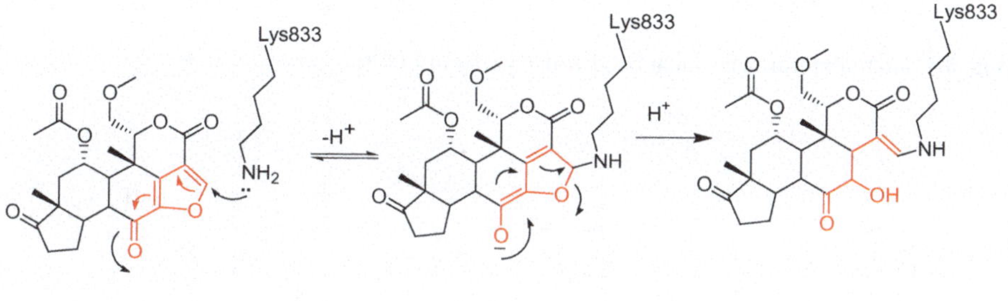

Fig. 5.3 Reaction mechanism of wortmannin that forms covalent bond with Lys833 of PI3Kγ

5.5 Afatinib

Afatinib is an epidermal growth factor receptor (EGFR) covalent inhibitor. It undergoes a hetero-Michael addition reaction with cys797 of EGFR at the active site as shown in Fig. 5.4.

5.6 Ibrutinib

Ibrutinib is a BTK covalent inhibitor. It forms a covalent bond with Cys481 of BTK at the ATP binding site [36]. It undergoes proton transfer, nucleophilic attack, and solvent-assisted tautomerization as shown in Fig. 5.5.

5.7 Bortezomib

Bortezomib is a covalent inhibitor of proteasome. Threonine protonates on amino group and deprotonates from hydroxyl group [37]. Threonine nucleophile attacks bortezomib to form a tetrahedral adduct which is stabilized through a hydrogen bonding network between *N*-terminal amino group of threonine and hydroxy group of boronic acid as shown in Fig. 5.6.

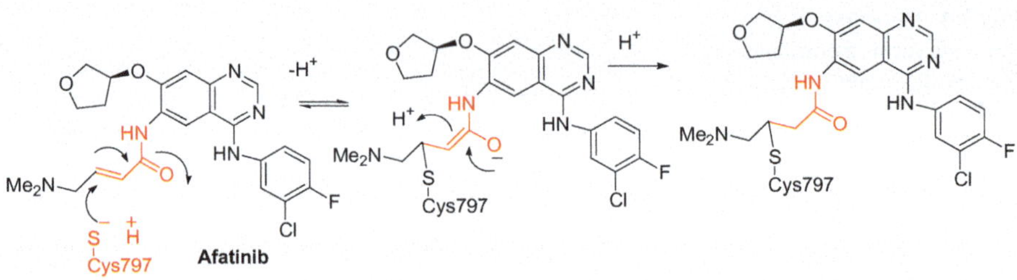

Fig. 5.4 Afatinib forms a covalent bond with Cys797 of EGFR

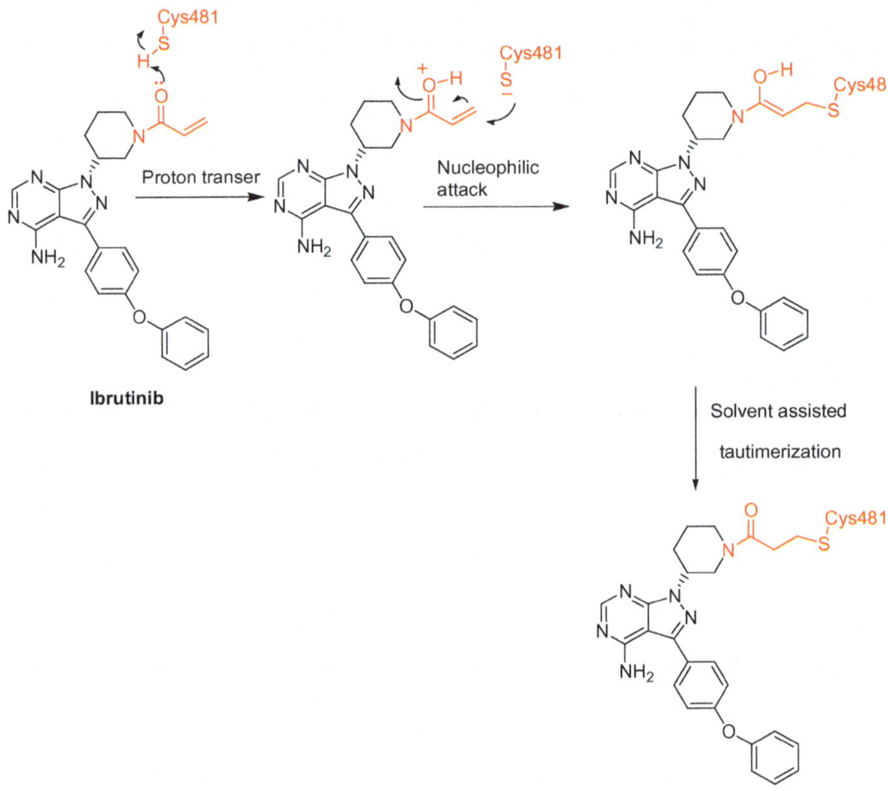

Fig. 5.5 Ibrutinib forms a covalent bond with Cys481 of BTK

5.8 Irreversible Covalent Inhibitors

Irreversible covalent inhibitors represent a class of compounds designed to bind irreversibly to specific target proteins through the formation of covalent bonds [38–43]. These inhibitors play a crucial role in drug discovery and development, particularly in the treatment of diseases such as cancer, inflammatory disorders, and

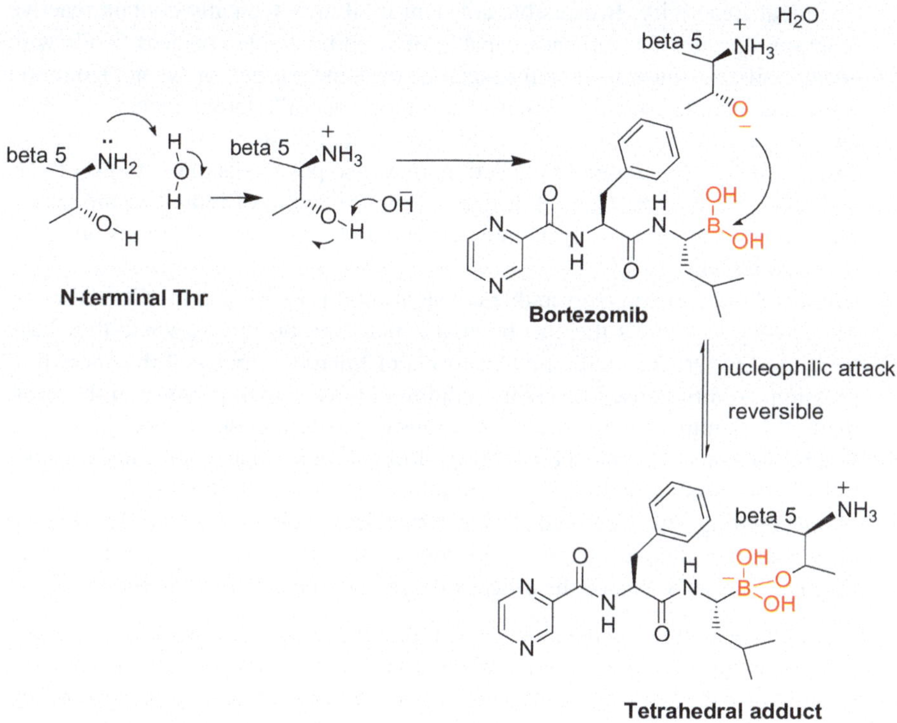

Fig. 5.6 Bortezomib forms a covalent bond with threonine of 20 S proteasome

infectious diseases. Here are some key characteristics and considerations regarding irreversible covalent inhibitors:

A. Mechanism of Action: Irreversible covalent inhibitors bind to their target proteins by forming covalent bonds with specific amino acid residues within the active site or allosteric sites. Once formed, these covalent bonds are typically stable and irreversible, leading to permanent inactivation or modification of the target protein.
B. Potency and Selectivity: Irreversible covalent inhibitors can exhibit high potency and selectivity by targeting specific amino acid residues or structural features within the target protein. This targeted approach minimizes off-target effects and reduces the risk of toxicity compared to non-specific inhibitors.
C. Duration of Action: The irreversible nature of covalent binding results in sustained inhibition of target protein function, leading to prolonged pharmacological effects. This prolonged duration of action can be advantageous in the treatment of chronic diseases or conditions requiring sustained target engagement.

D. Chemical Reactivity: Irreversible covalent inhibitors typically contain reactive functional groups or warheads capable of forming stable covalent bonds with nucleophilic residues in proteins, such as cysteine, serine, or lysine. Common warheads include epoxides, Michael acceptors, and alkylating agents.
E. Off-Target Effects: While irreversible covalent inhibitors offer advantages in terms of potency and duration of action, they also pose risks of off-target effects and potential toxicity. Careful design and optimization of inhibitors are necessary to minimize off-target interactions and enhance selectivity for the intended target.
F. Clinical Applications: Irreversible covalent inhibitors have demonstrated clinical efficacy in various therapeutic areas, including oncology, where they have been used to target specific mutant forms of kinases implicated in cancer progression. Additionally, irreversible inhibitors have shown promise in the treatment of inflammatory disorders, infectious diseases, and other conditions.
G. Drug Resistance: Despite their potency, drug resistance can emerge due to mutations in the target protein that prevent inhibitor binding or alter the accessibility of the binding site. Rational design strategies, such as structure-based drug design and medicinal chemistry optimization, can help overcome resistance mechanisms and enhance the efficacy of irreversible covalent inhibitors.

Overall, irreversible covalent inhibitors represent a valuable class of therapeutics with the potential to address unmet medical needs and improve patient outcomes across a range of diseases. However, careful consideration of factors such as potency, selectivity, safety, and resistance mechanisms is essential in the design and development of effective inhibitors for clinical use.

5.9 Reversible Covalent Inhibitors

Reversible covalent inhibitors represent a class of compounds designed to bind reversibly to a target protein through the formation of covalent bonds, which can be broken under certain conditions. These inhibitors offer a balance between the specificity and potency of irreversible covalent inhibitors and the reversibility of non-covalent inhibitors [13, 15, 44–49].

Here's how reversible covalent inhibitors work:

A. *Binding Mechanism*: Reversible covalent inhibitors bind to their target proteins by forming covalent bonds with specific amino acid residues within the active site or allosteric sites of the protein. The formation of these bonds alters the protein's conformation, inhibiting its activity.
B. *Reversibility*: Unlike irreversible covalent inhibitors, which form bonds that are difficult to break, reversible covalent inhibitors form bonds that can be broken under certain conditions, such as changes in pH, temperature, or the presence of specific enzymes or molecules.

C. *Specificity and Potency*: Reversible covalent inhibitors can be designed to target specific amino acid residues within the target protein, allowing for increased selectivity and potency compared to non-covalent inhibitors. This targeted approach minimizes off-target effects and reduces the risk of toxicity.
D. *Drug Development*: Reversible covalent inhibitors have gained attention in drug discovery and development due to their ability to modulate the activity of specific proteins involved in disease pathways. They are being explored as potential therapeutics for various diseases, including cancer, inflammatory disorders, and infectious diseases.
E. *Challenges*: Despite their advantages, reversible covalent inhibitors present challenges in terms of optimization of binding kinetics, selectivity, and off-target effects. Designing inhibitors with the right balance of potency, selectivity, and reversibility requires careful consideration of the chemical structure and properties of both the inhibitor and the target protein.

Overall, reversible covalent inhibitors represent a promising approach in drug discovery and development, offering the potential for improved efficacy and reduced toxicity compared to traditional non-covalent inhibitors. Their reversible nature allows for fine-tuning of pharmacokinetic and pharmacodynamic properties, making them valuable tools in the development of novel therapeutics.

5.10 Irreversible Covalent Binding Warheads

Irreversible covalent binding warheads are chemical groups or moieties designed to form permanent, irreversible covalent bonds with specific target proteins. These warheads lead to the permanent inactivation or modification of the target protein, thereby exerting their pharmacological effects [50–54]. Here are some examples of irreversible covalent binding warheads commonly used in drug discovery and development:

Epoxides Epoxides are highly reactive three-membered cyclic ethers that can form irreversible covalent bonds with nucleophilic residues, such as cysteine, histidine, or lysine, in proteins. Epoxides react via ring-opening reactions, leading to the formation of stable covalent adducts with the target protein.

Haloacetamides Haloacetamides contain a halogen atom (e.g., chlorine, bromine) adjacent to a carbonyl group. These compounds react irreversibly with nucleophilic residues, such as cysteine thiols, through nucleophilic substitution reactions, forming stable covalent bonds.

Alkylating Agents Alkylating agents contain electrophilic groups that can react with nucleophilic residues in proteins, such as cysteine, lysine, or histidine. Once formed, these alkyl adducts are generally stable and irreversible, leading to permanent inactivation of the target protein.

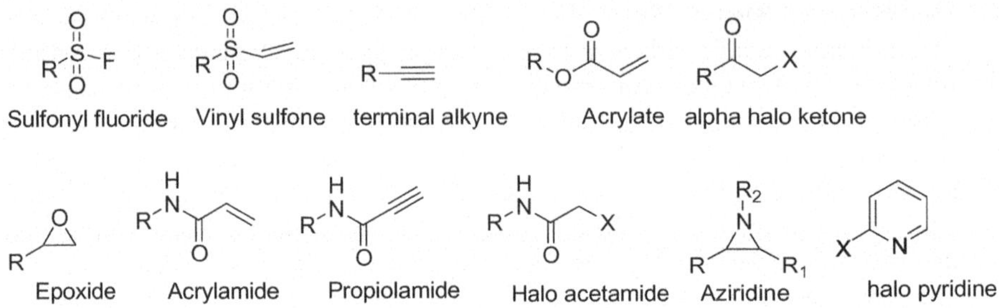

Fig. 5.7 Irreversible covalent binding warheads

Acrylamide Acrylamide is a mild electrophile. Thiol of cysteine in the protein reacts with acrylamide to form an irreversible permanent bond.

Nitrogen Mustards Nitrogen mustards are alkylating agents that contain nitrogen atoms capable of forming highly reactive alkyl groups. These compounds irreversibly alkylate nucleophilic residues in proteins, leading to irreversible inhibition or modification of the target.

Some examples are shown in Fig. 5.7.

The covalent bond formation of the target protein with acrylamide is depicted in Fig. 5.8.

5.11 Reversible Covalent Binding Warheads

Reversible covalent binding warheads are chemical groups or moieties designed to form temporary, reversible covalent bonds with specific target proteins. These warheads allow for modulation of protein function without permanently disabling the target, providing a degree of control over the duration and reversibility of inhibition [55–72]. Here are some examples of reversible covalent binding warheads commonly used in drug discovery and development:

Ketone The ketone moiety activated by an adjacent electron-withdrawing group also works as a reversible electrophilic warhead and shows broad-range reactivity toward nucleophilic amino acid residues, such as cysteine and serine. Ketone-based peptidomimetics serve as reversible covalent inhibitors of serine and cysteine proteases mimicking a tetrahedral transition state through the formation of hemiketal and thiohemiketal complex respectively.

Aldehyde Aldehydes are reversible covalent warheads. The aldehyde reacts with cysteine or lysine reversibly to make the thiohemiacetal or Schiff base.

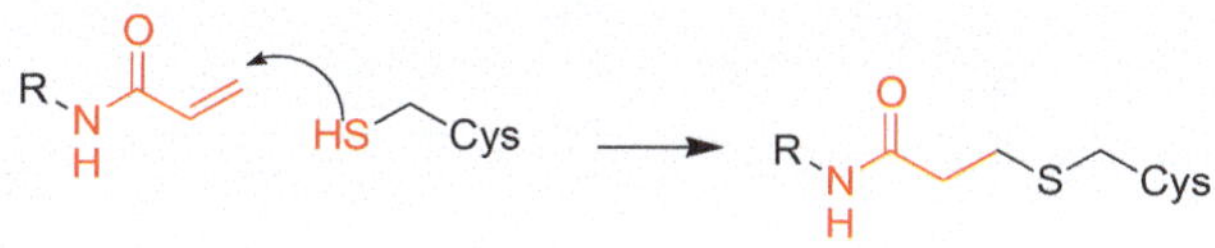

Fig. 5.8 Acrylamide; irreversible thiol-Michael addition

Hydroxamic Acid Hydroxamic acids contain the hydroxamate functional group, which forms reversible covalent bonds with metal ions, particularly zinc ions found in metalloenzymes. This mechanism of action is commonly utilized in the design of metalloenzyme inhibitors.

Disulfide Compounds containing disulfide bonds (-S-S-) can undergo reversible oxidation and reduction reactions with cysteine residues in proteins. These reversible interactions enable modulation of protein function in response to changes in cellular redox conditions.

Sulfonamide Sulfonamide-containing compounds can form reversible covalent bonds with amino groups of lysine residues in proteins. This interaction has been exploited in the design of reversible covalent inhibitors targeting a variety of protein classes.

Imine/Enamine Formation Compounds containing imine or enamine functional groups can reversibly react with nucleophilic residues, such as lysine or cysteine, through Schiff base formation. This reversible covalent interaction has been utilized in the development of reversible covalent inhibitors.

Boronic Acid Boronic acids can form covalent adducts with nucleophiles such as serine, lysine, threonine, tyrosine, and cysteine residues in target proteins. Bortezomib contains boronic acid as an electrophilic warhead which targets the *N*-terminal threonine of 26S proteasome, and is the first-in-class reversible covalent inhibitor for the treatment of multiple melanomas and mantle cell lymphoma [38].

Nitrile The nitrile is used as a reversible covalent binding warhead due to its lower electrophilicity compared with other reactive groups. Nirmatrelvir reacts with Cys145 in the main protease of SARS-CoV-2.

α-Cyanoacrylamide Addition of an electron-withdrawing group such as nitrile group at the α-position of acrylamide makes it reversible. After thia-Michael addition, the α-position becomes sufficiently acidic which facilitates a retro-Michael reaction as shown in Fig. 5.9. This type of reversible covalent inhibitor will decrease the on-target residence time in comparison to irreversible covalent inhibitors.

These reversible covalent binding warheads offer advantages in terms of selectivity, potency, and reversibility compared to irreversible covalent inhibitors. By forming temporary covalent bonds with target proteins, reversible covalent inhibitors provide a dynamic approach to modulating protein function and offer potential benefits in terms of therapeutic efficacy and safety profiles. However, careful consideration of factors such as target protein structure, binding kinetics, and off-target effects is

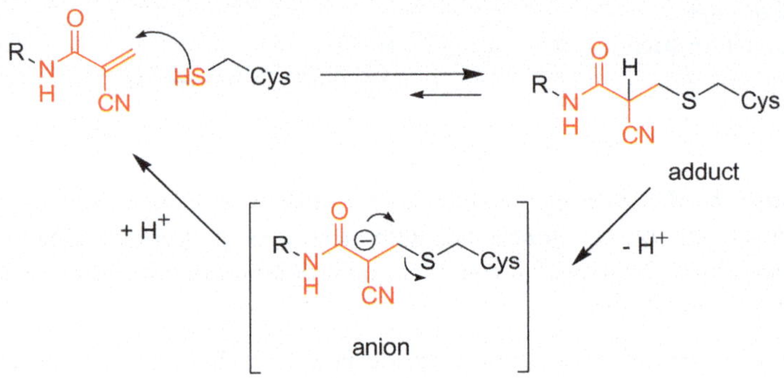

Fig. 5.9 α-Cyanoacrylamide; reversible thiol-Michael addition

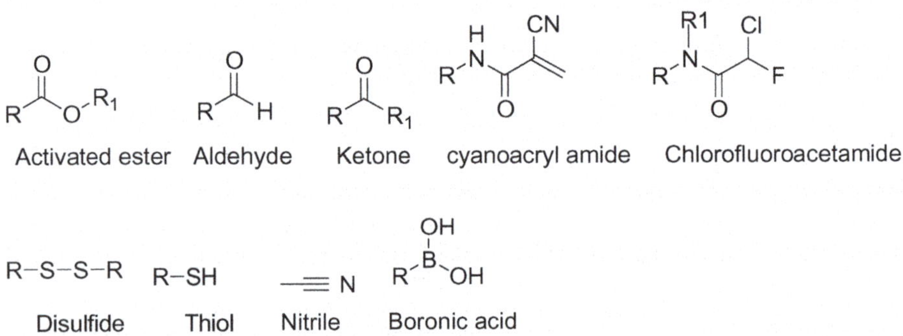

Fig. 5.10 Reversible covalent binding warheads

essential in the design and optimization of reversible covalent inhibitors for specific therapeutic applications.

Some reversible covalent binding warheads are shown in Fig. 5.10.

These reversible covalent binding warheads offer advantages in terms of selectivity, potency, and reversibility compared to irreversible covalent inhibitors. By forming temporary covalent bonds with target proteins, reversible covalent inhibitors provide a dynamic approach to modulating protein function and offer potential benefits in terms of therapeutic efficacy and safety profiles. However, careful consideration of factors such as target protein structure, binding kinetics, and off-target effects is essential in the design and optimization of reversible covalent inhibitors for specific therapeutic applications.

Most covalent inhibitors are targeting the specific cysteine of chosen proteins. But different warheads can make a covalent bond with other amino acid residues such as serine, threonine, lysine, and tyrosine. Warhead electrophile could be a low reactive functional group such as acrylamide which can react with targeted specific non-catalytic cysteine residue, also described as a targeted covalent inhibitor (TCI). Unsaturated aldehyde or ketone or ester should be avoided as these are highly reactive functional groups that can react with other residues of amino acids including serine, threonine, lysine, tyrosine, or others instead of selective cysteine. Therefore, side effects or toxicity can be a major concern using unsaturated carbonyl compounds.

5.12 Representative Targeted Covalent Inhibitors and Their Target and Indication

Representative targeted covalent inhibitors are depicted in Fig. 5.11 and their applications are summarized in Table 5.1.

5.13 Covalent Inhibitors of MAPK 8, 9, 10 (JNKs)

Having success with other kinase covalent inhibitors [14, 25, 42, 44, 52, 73–88], the Gray laboratory made the first JNK covalent inhibitors [47]. They used the core phenylaminopyrimidine of imatinib (FDA-approved drug) which has favorable drug properties. Through structure-activity relationship studies they found several lead compounds (Fig. 5.12) such as JNK-IN-7 and JNK-IN-8 with low nanomolar IC_{50} values in both kinase in vitro assay and cell-based assay. JNK-IN-11 is the most potent with IC_{50} values of 0.5–1.3 nM against JNK1/2/3 in the Z'-LYTE assay. These inhibitors are selective against 200 kinases tested. They co-crystallized JNK-IN-2 and JNK-IN-7 in complex with JNK3 (PDB code 3V6R) which provided the binding modes and targeting Cys154 as shown in Fig. 5.13. Indeed, a covalent bond formation between acrylamide and Cys154 was confirmed. The inhibitor forms two hydrogen bonds with Met149 in the hinge region of the ATP binding site. Without an acrylamide motif, the potency was reduced 100 times. These inhibitors contain an additional dimethylamino group which enhances water solubility. These inhibitors have balanced potency in both in vitro and cell-based, water solubility, and drug-like properties.

Muth et al. [89] reported another class of pyridinylimidazole as cysteine-targeted covalent inhibitor of JNK3 from previously reported dual inhibitor of JNK and p-38α scaffold, as shown in Fig. 5.14. Methylation of imidazole-*N*-1 position increased the activity and selectivity. Introducing an acrylamide group in the right position, targeting Cys154 of JNK3 produced a very potent covalent inhibitor with the IC_{50} values of 0.3 nM and 36 nM against JNK3 and p-38 respectively. Replacing the 4-fluorophenyl ring with a methyl group resulted in almost 1000 times JNK3 selectivity over p38 with the IC_{50} values of 2 nM and 1952 nM respectively.

Key compounds were tested against 410 kinases at 1 μM and 0.5 μM respectively. Only 7 to 15 kinases showed moderate inhibitions, hence compounds are very selective covalent JNK inhibitors. Compounds are relatively stable in human liver microsomes.

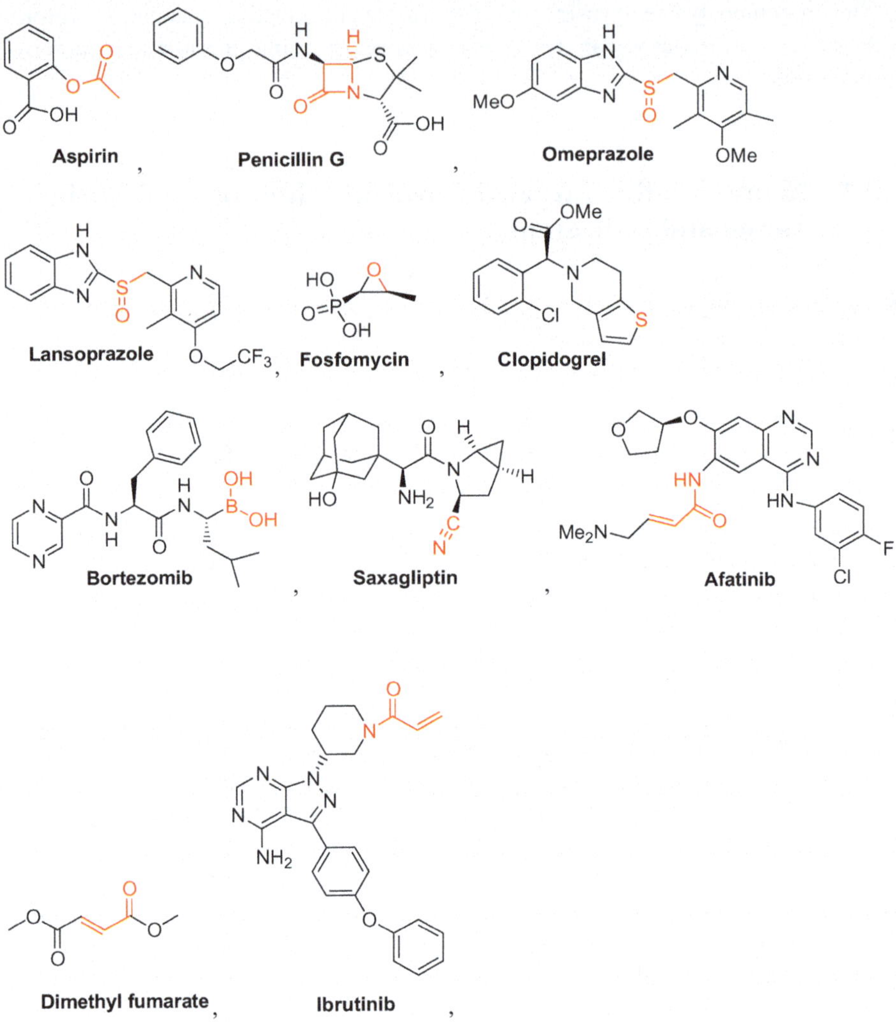

Fig. 5.11 Chemical structures of some covalent inhibitors

The covalent bond formation with Cys154 of JNK3 was confirmed by mass analysis (Fig. 5.15). The inhibitor forms two hydrogen bonding interactions with Met149 at the hinge region.

Recently, Zimmermann et al. [90] reported the screening chemical library results from the construction of a DNA-encoded chemical containing scaffold potentially capable of forming covalent bond through cysteine residue of the target proteins. They found two fragments. Linking two fragments with a short PEG linker resulted in a good covalent inhibitor (compound **115**) of JNK as shown in Fig. 5.16, confirmed by LC-MS. There was no potency or selectivity against other kinases mentioned.

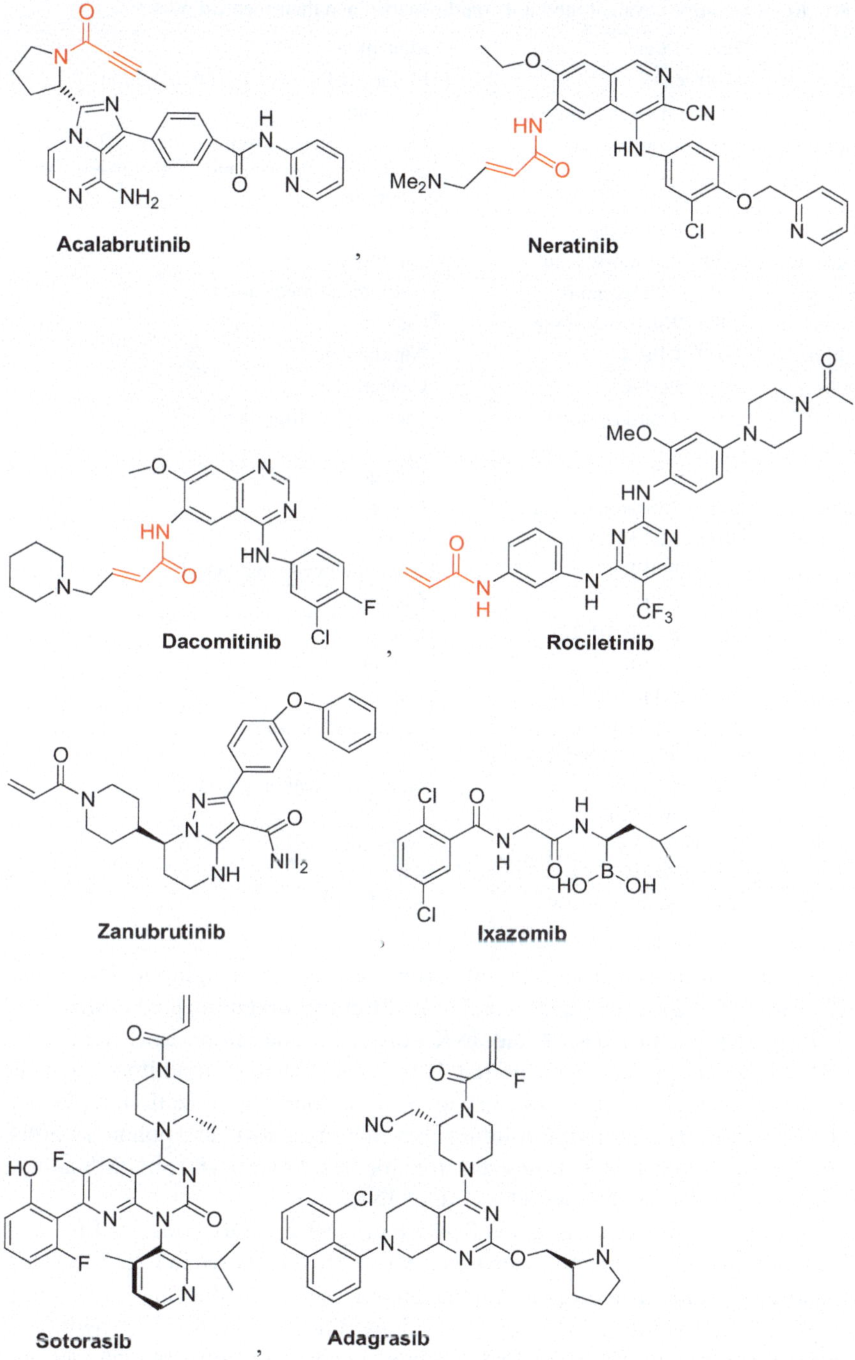

Fig. 5.11 (continued)

Table 5.1 Representative covalent inhibitors on the market and their medical uses

Drug	Year	Target	Indication
Aspirin	1890	COX	Reduce pain, fever, or inflammation
Penicillin G	1928	DD-transpeptidase	Antibiotic
Omeperazole	1979	Proton pump	Gastroesophageal reflux disease (GERD), peptic ulcer disease, and Zollinger–Ellison syndrome
Lansoprazole	1992	Proton pump	Reduces stomach acid
Fosfomycin	1996	Peptido-glycan	Antibiotic
Clopidorgrel	1997	ADP receptor	Antiplatelet medication
Botezomib	2003	26S proteosome	Cancer
Saxagliptin	2009	DPP-4	Antidiabetic
Afatinib	2013	EGFR	Cancer
Dimethyl fumarate	2013	Transcription factor	Psoriasis, multiple sclerosis
Ibrutinib	2014	BTK	Cancer
Ixazomib	2015	26S proteosome	Cancer
Osimertinib	2015	EGFR	Cancer
Rociletinib	2016	EGFR	Non-small cell lung cancer
Acalabrutinib	2017	BTK	Cancer
Neratinib	2017	EGFR/HER2	Cancer
Dacomitinib	2018	EGFR	Cancer
Zanubrutinib	2019	BTK	Cancer
Voxelator	2019	Hemoglobin oxygen-affinity modulator	Sickle cell disease
Sotorasib	2021	KRAS (G12C)	Non-small-cell lung cancer
Adagrasib	2023	KRAS (G12C)	Non-small-cell lung cancer
Nirmatrelvir	2021	Main protease	COVID-19

Making an isoform selective JNK inhibitor is always challenging as JNK1 and JANK2 share 98% sequence identity and only differ by two amino acids (Met77 vs Leu77, Ile106 vs leu106). JNK2 and JNK3 differ by one amino acid (Leu77 vs Met115). Gray lab previously reported [91] some covalent inhibitors such as JNK-IN-8 and JNK-IN-11 as shown in Fig. 5.17. Through optimization, replacing phenyl group with (*R*)-3-aminopyrrolidine, they identify a novel and potent **YL5084** JNK covalent inhibitor. It is 31 times selective for JNK2 over JNK1. LC/MS analysis confirms a covalent adduct with Cys116 of JNK2.

One analog of this series is crystallized with JNK2 (PDB code: 8ELC) and reveals a covalent bond formation with Cys116 of JNK 2 as shown in Fig. 5.18. This compound also forms a hydrogen bonding interaction with Met111 at the hinge region of the ATP-binding site. The pyrazolopyrdine moiety forms a hydrogen bonding interaction with Lys55 of JNK2. There are some hydrophobic interactions with Ala53, Val54, Lys55, Met108, Tyr185, Gln117, and Leu168.

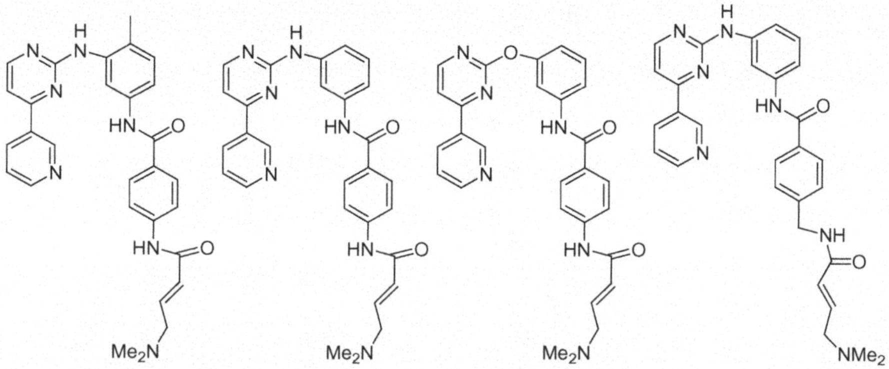

JNK-IN-1 **JNK-IN-2** **JNK-IN-3** **JNK-IN-4**

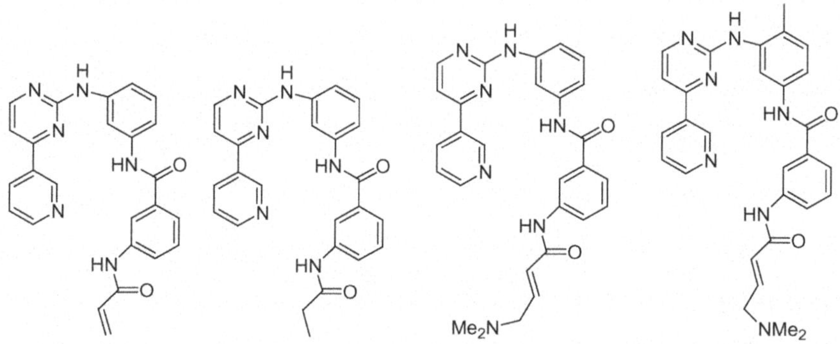

JNK-IN-5 **JNK-IN-6** **JNK-IN-7** **JNK-IN-8**

JNK-IN-7	JNK-IN-8
IC_{50} (JNK1) = 1.54 nM	IC_{50} (JNK1) = 4.67 nM
IC_{50} (JNK2) = 1.99 nM	IC_{50} (JNK2) = 1.7 nM
IC_{50} (JNK3) = 0.75 nM	IC_{50} (JNK3) = 0.98 NM
EC_{50} (c-Jun, Hela) = 130 nM	EC_{50} (c-Jun) = 486 nM

Fig. 5.12 Covalent inhibitors of JNK from Gray lab

5.13.1 Covalent Inhibitors of ERK1, 2

Investigators at AstraZeneca identified several small molecules as ERK1/2 reversible inhibitors from their in-house compounds' library screening [92]. One analog of this series is compound **116** and it has moderate potency in ERK2 biochemical assays. To understand the binding mode of this series, they solved a crystal structure

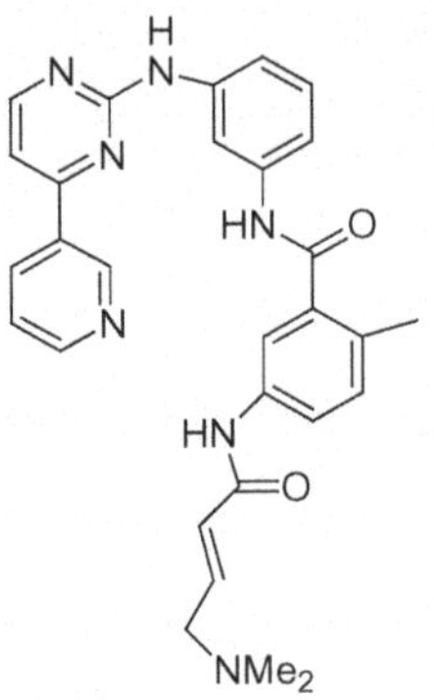

JNK-IN-9

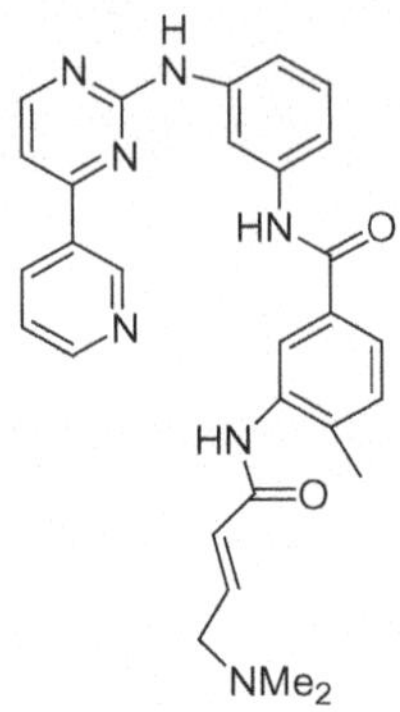

JNK-IN-10

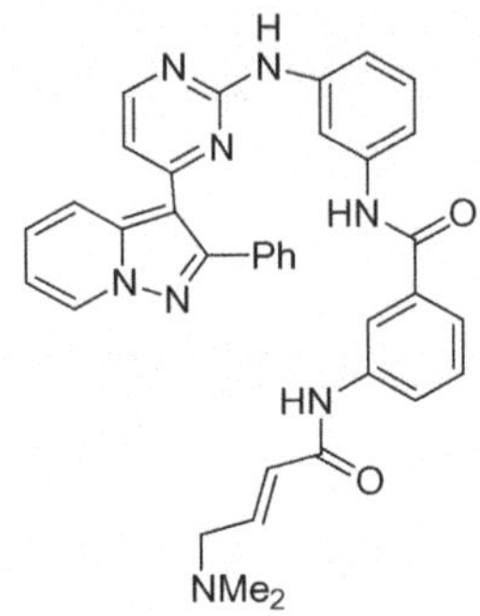

JNK-IN-11

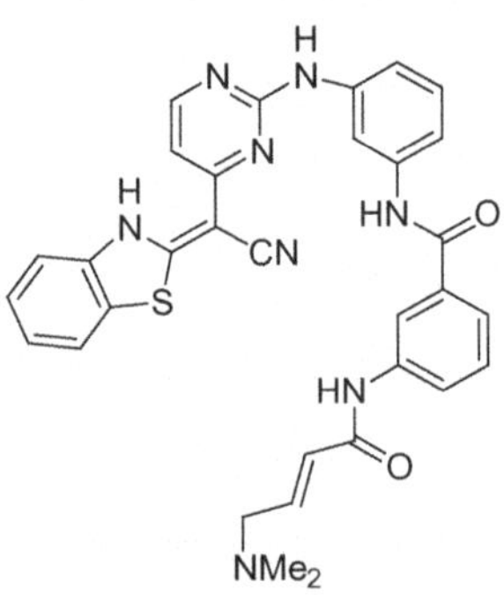

JNK-IN-12

Fig. 5.12 (continued)

of **116** in complex with ERK2 (PDB code: 4ZZN). The crystal structure reveals that 2-amino pyridyl moiety forms two hydrogen bonding interactions with Met108 at the hinge region. The amide group of **116** forms a network of hydrogen bonds with Glu-71, Gln-105, and Asp-167. The Cys166 is very close to the amide moiety which can be utilized for covalent bonding. To introduce an acrylamide, the amide bond is reversed from **116**, resulting in compound **117** (Fig. 5.19). Compound **117** showed a good potency both in biochemical and cellular assays.

For the SAR studies on scaffold **118** (Fig. 5.20), they synthesized several analogs such as **75–77**. Replacing pyridine core with pyrimidine hinge binding group results in compound **119**, similar pyrimidine group is present on GDC-0994 another ERK2 inhibitor. Compound **119** shows 10 times more potent than compound **117**. Compound **120** without the Michael acceptor shows less potent than **119**. The covalent bond formation of compound **119** with Cys166 of ERK2 was confirmed by the mass spectrometry, which was not observed with compound **120**. A crystal structure was solved with compound **119** in complex with ERK2 and it revealed that binding

Fig. 5.13 Covalent bond formation of JNK-IN-7 with Cys154 of JNK3 and hydrogen bonding network

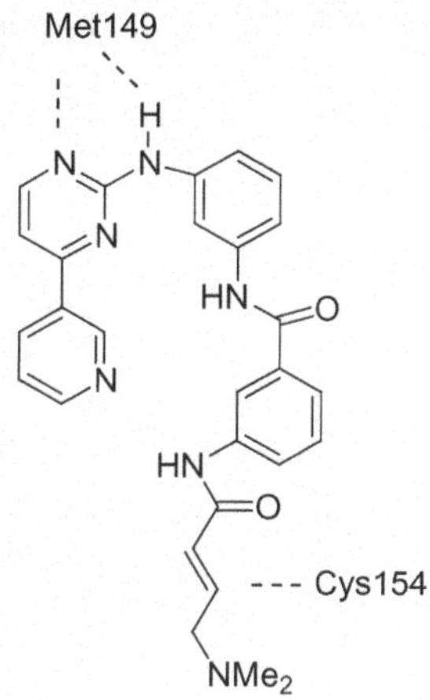

mode was similar to the reversible inhibitor **116**. Replacing chloro on thc 5-position of pyrimidine with methyl and trifluoromethyl is well tolerated. Indeed, compound **121** showed more potent in biochemical assays and broader kinase selectivity. Compound **121** demonstrates potent anti-proliferative activity in A375 cell lines with an IC_{50} value of 53 nM and favorable physicochemical properties for further in vivo studies.

5.13.2 Fragment-Based Discovery of Covalent ERK1/2 Inhibitor

Scientists from Astex Pharmaceuticals developed a high-throughput protein crystallography method as the primary hit-finding technology for covalent ERK2 inhibitors [93]. Several acrylamide electrophilic fragments are identified to target Cys166 in the ATP binding site. From their campaign, compound **122** is the best ATP-competitive covalent inhibitor with an IC50 value of 7.8 μM. Considering this a very small fragment, there is a scope to modify it to obtain a more potent ERK2 covalent inhibitor targeting Cys166 (Fig. 5.21).

5.13.3 Covalent ERK1/2 Inhibitor Targeting Outside of ATP Binding Site

Kaoud et al. developed a new ERK2 inhibitor which binds to the D-recruitment site (DRS) of ERK to block the binding of ERK to its substrate p90RSK [94]. The DRS is a conserved site away from the ATP site. BI-78D3 tightly binds to ERK2 with a

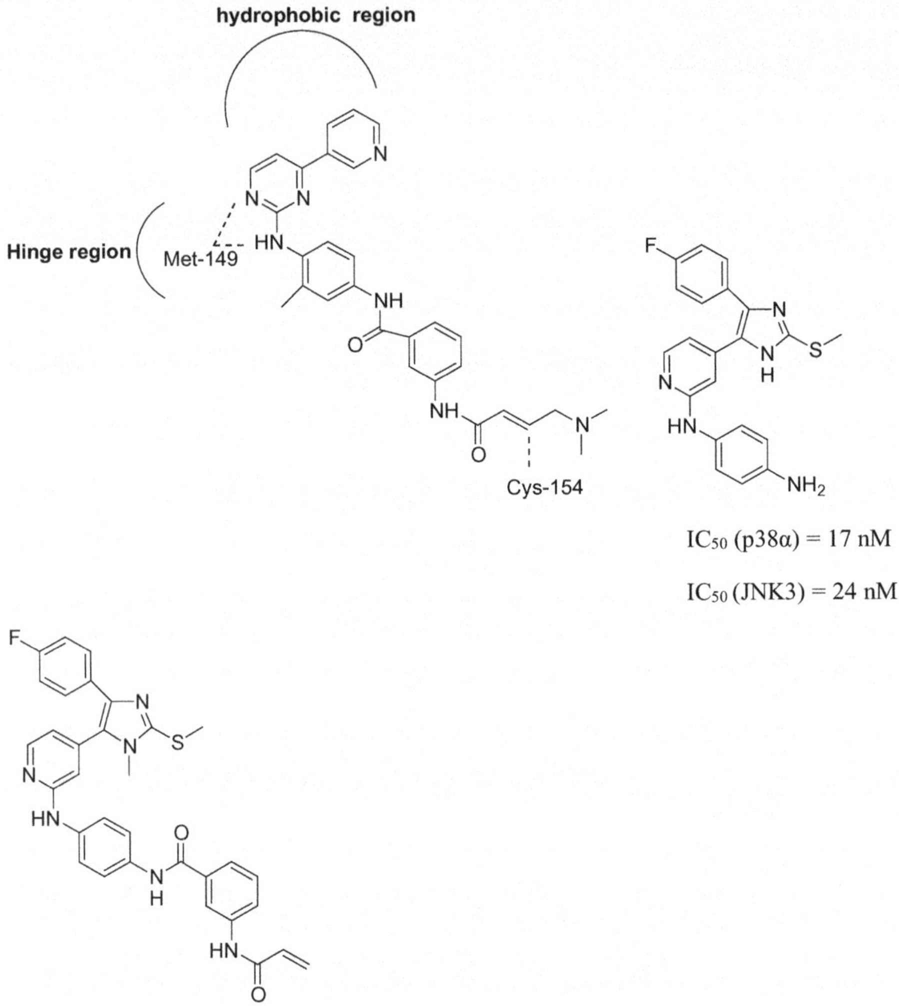

Fig. 5.14 Covalent inhibitor from non-covalent scaffold

Ki value of 0.80 μM. The conserved Cys159 of ERK2 reacts with BI-78D3 to form a tetrahedral adduct which slowly undergoes hydrolysis. They proposed a reaction mechanism which is shown in Fig. 5.22. The nucleophilic attack of the Cys159 thiolate anion to C5 of the 1,2,4-triazol-3-one of **BI-78D3** which is confirmed by NMR and mass spectrometric. BI-78D3 does not covalently modify p38 MAPK, JNK, or ERK5 due to steric hindrance by the nitrothiazole and benzodioxane rings in BI-78D3. Notably, this compound inhibits melanoma cell growth in vitro and in vivo. It overcomes acquired resistance to b-RAF and MEK inhibitors.

Fig. 5.15 Covalent bond formation with Cys154 of JNK3

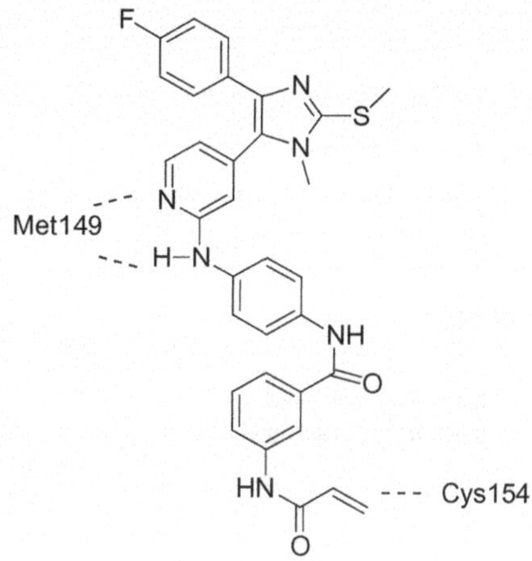

Fig. 5.16 Covalent inhibitor of JNK

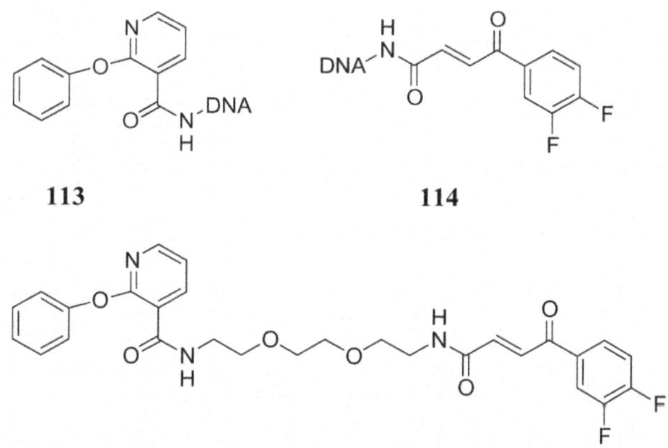

113 **114**

115

5.13.3.1 Development of Covalent Clinical Candidate CC-90003 as ERK1/2 Inhibitor

Investigators from Celgene Corporation developed a covalent inhibitor of ERK1/2. Compound (**CC-90003**) forms a covalent bond with Cys164 of ERK2 and Cys183 of ERK1 (Fig. 5.23) in the A-loop of the ATP binding site [95–97]. Covalent bond formation is confirmed by mass spectrometry. It is similar to compound **75** from AstraZeneca. The equivalent cysteine is present in 48 of 538 protein kinases. With substitution of the ligand, it shows selectivity in cell-based assays. It is an irreversible covalent inhibitor of ERK1 and ERK2 with IC_{50} values of 10–20 nM. It strongly inhibits the proliferation of PDX models such as the pancreatic adenocarcinoma model PAXF-2059 (IC_{50} = 0.62 μM), the lung cancer model LXFA-983

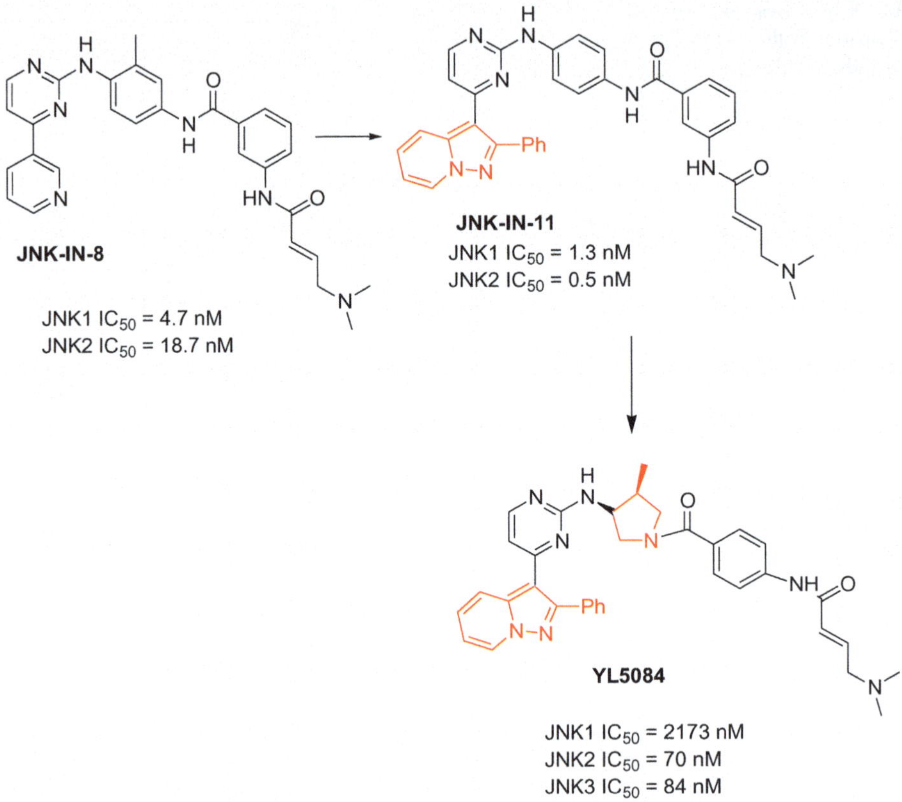

Fig. 5.17 New covalent inhibitor starting from previously reported JNK-IN-8

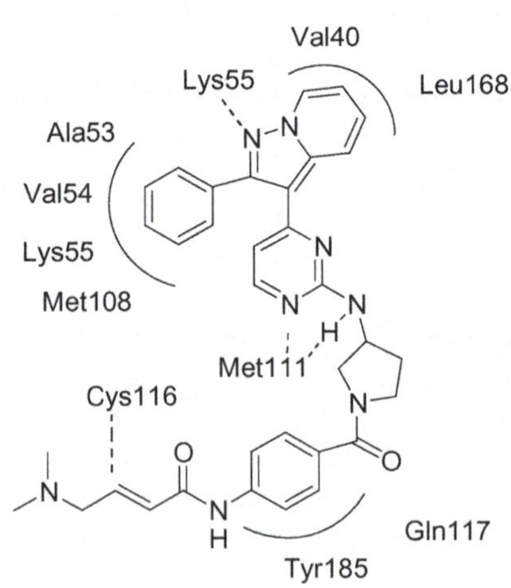

Fig. 5.18 Schematic of ligand-protein interactions

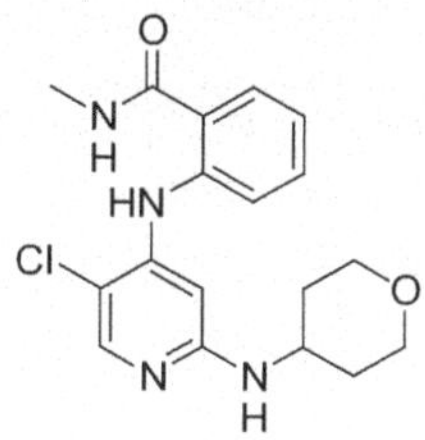

116

Km (ERK2) = 0.26 μM

IC_{50} (A375 pERK) = 7.5 μM

IC_{50} (A375 pRSK) = 10 μM

LogD (pH 7.4) = 2.8

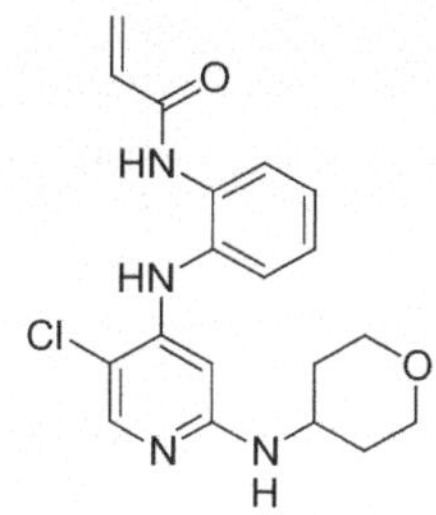

117

Km (ERK2) = 0.042 μM

IC_{50} (A375 pERK) = 0.12 μM

IC_{50} (A375 pRSK) = 0.4

LogD (pH 7.4) = 2.5

Fig. 5.19 Reversible ERK2 inhibitor to irreversible covalent ERK2 inhibitor

(IC_{50} = 0.80 μM), and the colorectal cancer model CXF-243 (IC_{50} = 0.586 μM). **CC-90003** in combination with docetaxel shows complete tumor regression in lung cancer and prevents its regeneration in the lung cancer PDX model LXFA-983. It is currently in a phase 1 clinical trial (NCT02313012) for the treatment of patients with relapsed or refractory BRAF or RAS mutation solid tumors.

5.14 Covalent Inhibitors of p38 MAP Kinases

Li et al. designed a series of 1,2-dialkynylimidazole derivatives as potential irreversible p38 MAP kinase α-isoform (p38α) inhibitors [98]. Based on known selective p38 α/β inhibitors such as **RWJ-67657** [99] and **SB-203580** [100], they modified the structure resulting in a more potent compound **123** compared to **RWJ-67657** with an IC_{50} value of 200 nM (p38α) (Fig. 5.24). Terminal alkyne is a good Michael acceptor and forms a covalent bond with cysteine. The covalent bond formation of compound **123** with a cysteine of p38α was confirmed by mass spectrometry. The pyridinylimidazole cores are prone to inhibit CYP450 enzymes that make hepatotoxicity. However, compound **123** shows no significant inhibition of CYP450 2D6 (4% inhibition at 10 μM) in comparison with **SB-203580** (78% inhibition at 10 μM). There is no cellular activity on this compound reported.

Along the same line, the same group identified compound **124** (rooperol) as a covalent inhibitor of p38α targeting its D-recruitment site (DRS). Rooperol is a natural product found in *H. rooperi* corms [101]. It forms a covalent bond with the conserved Cys119 of p38α in its D-recruitment site [102]. It inhibits p38α with an

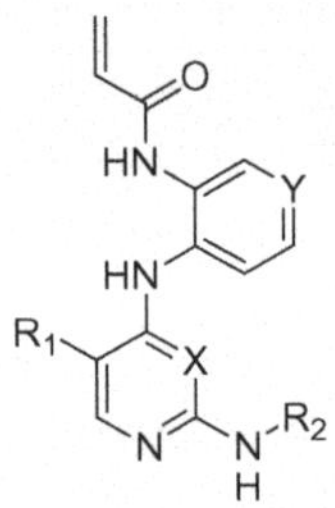

118 (X = N or CH;

Y = N or CH)

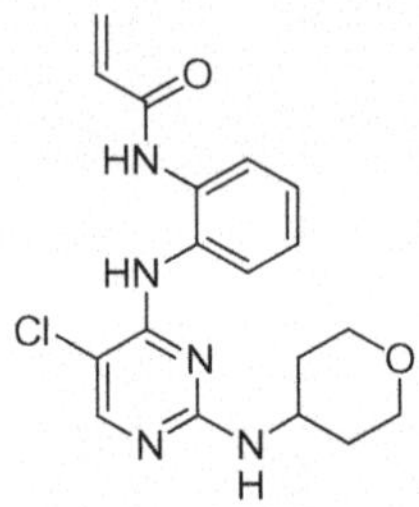

119

Km (ERK2) = 0.0085 μM

IC_{50} (A375 pERK) = 0.019 μM

IC_{50} (A375 pRSK) = 0.069 μM

LogD (pH 7.4) = 2.7

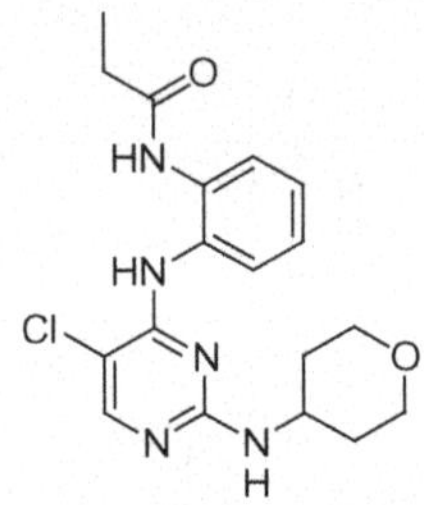

120

Km (ERK2) = 0.48 μM

IC_{50} (A375 pERK) > 30 μM

IC_{50} (A375 pRSK) > 30 μM

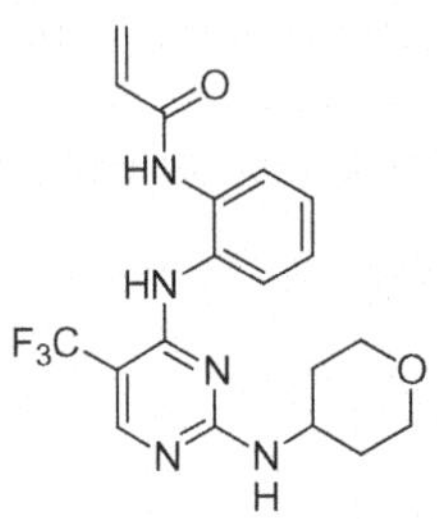

121

Km (ERK2) = 0.0069 μM

IC_{50} (A375 pERK) = 0.022 μM

IC_{50} (A375 pRSK) = 0.065 μM

LogD (pH 7.4) = 2.9

Fig. 5.20 Pyrimidine derivatives for covalent inhibitors of ERK2

IC_{50} value of 18.60 μM (Fig. 5.25). The covalent bond formation with Cys119 is confirmed by mass spectrometry.

MAP kinase covalent inhibitors refer to a class of compounds designed to irreversibly bind to specific residues within the MAP kinase (mitogen-activated protein kinase) signaling pathway. MAP kinases play crucial roles in cellular processes like

Fig. 5.21 Pyrazole-based ERK2 covalent inhibitor

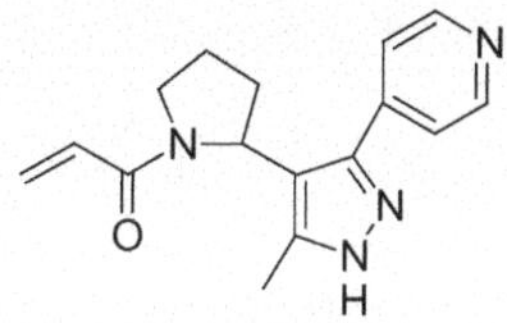

122

IC_{50} (ERK2) = 7.8 μM

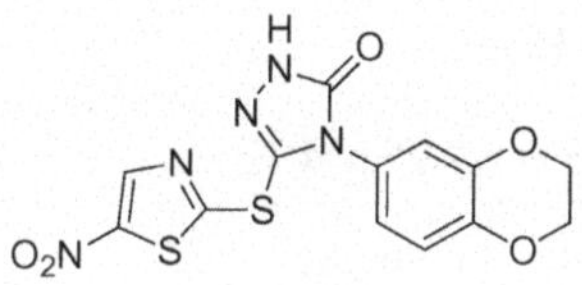

BI-78D3

Ki (ERK2) = 0.8 μM

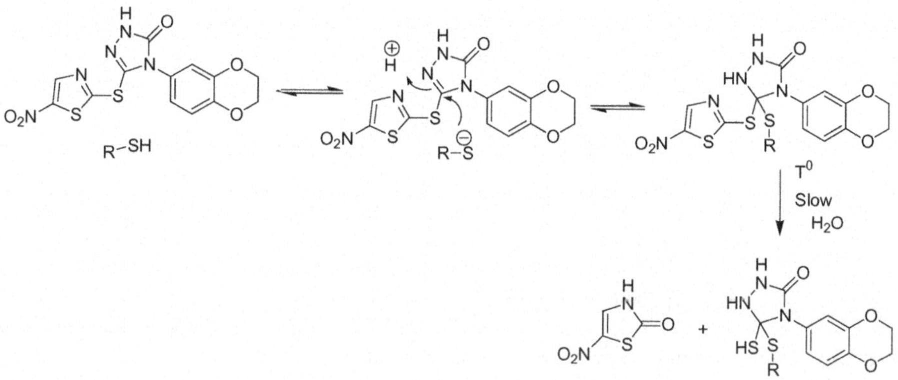

Fig. 5.22 Possible covalent adduct of Cys159 of ERK2

Fig. 5.23 CC-90003 forms a covalent bond with Cys166 of ERK2 and Cys183 of ERK1

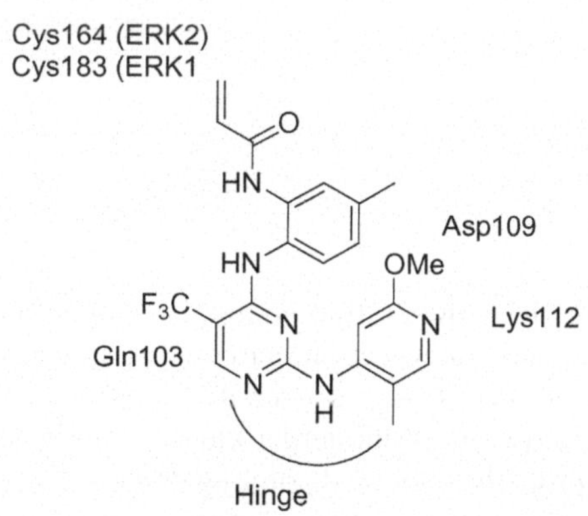

CC-90003

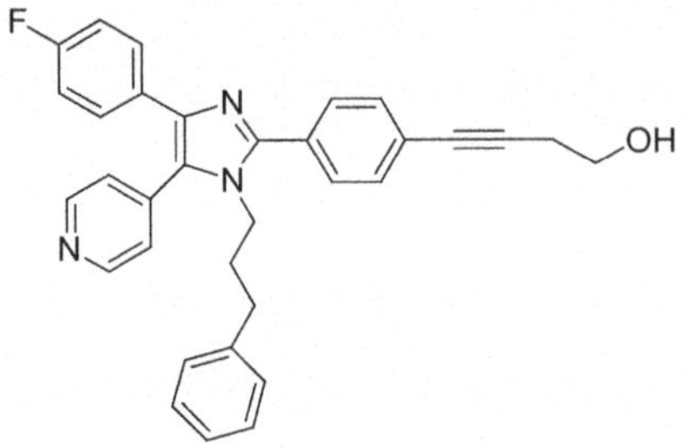

RWJ- 67657

IC_{50} (p38α) = 1 μM

IC_{50} (p38β) = 11 μM

IC_{50} (TNF-α) = 0.003 μM

IC_{50} (IL-1β) = 0.011 μM

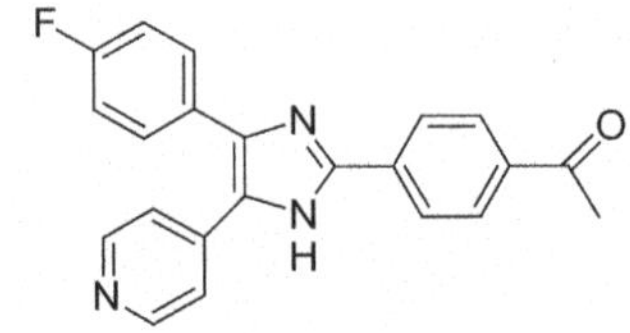

SB-203580

IC_{50} (p38α) = 0.05 μM

IC50 (p38β) = 0.50 μM

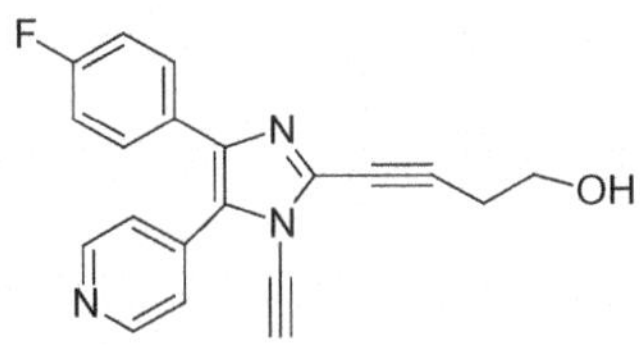

123

IC_{50} (p38α) = 0.20 μM

Fig. 5.24 4,5-Diaryl-imidazole derivatives as p38α inhibitors

Fig. 5.25 Rooperol as a covalent inhibitor of p38α

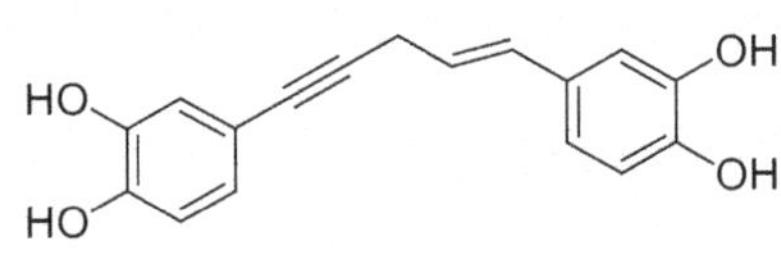

124

IC_{50} (p38α) = 18.60 μM

proliferation, differentiation, and apoptosis, making them attractive targets for therapeutic intervention, particularly in cancer treatment. Covalent inhibitors work by forming strong, irreversible bonds with specific amino acid residues within the active site of the target kinase. This binding prevents the kinase from functioning properly, thereby disrupting downstream signaling pathways involved in cell growth and survival. These inhibitors offer potential advantages in terms of potency, selectivity, and duration of action compared to reversible inhibitors. However, the irreversible nature of covalent inhibition also raises concerns about off-target effects and potential toxicity.

Despite challenges, researchers continue to explore the development of MAP kinase covalent inhibitors as potential therapeutics, aiming for improved efficacy and safety profiles in the treatment of various diseases, especially cancers driven by dysregulated MAP kinase signaling.

References

1. Jackson PA, Widen JC, Hark DA, Brummond KM. Covalent modifiers: a chemical perspective on the reactivity of α, β-unsaturated carbonyls with thiols via hetero-Michael addition reactions. J Med Chem. 2017;60:839.
2. Sutanto F, Konstantinidou M, Domling A. Covalent inhibitors: a rational approach to drug discovery. RSC Med Chem. 2020;11:876.
3. Gehringer M, Laufer SA. Emerging and re-emerging warheads for targeted covalent inhibitors: applications in medicinal chemistry and chemical biology. J Med Chem. 2019;62:5673.
4. Singh J, Petter RC, Baillie TA, Whitty A. The resurgence of covalent drugs. Nat Rev Drug Discov. 2011;10:307.
5. Cohen MS, Zhang C, Shokat KM, Taunton J. Structural bioinformatics-based design of selective, irreversible kinase inhibitors. Science. 2005;308(5726):1318.
6. Liu Q, Sabnis Y, Zhao Z, Zhang T, Buhrlage SJ, Jones LH, Gray NS. Developing irreversible inhibitors of the protein kinase cysteinome. Chem Biol. 2013;2013(20):146.
7. Bauer RA. Covalent inhibitors in drug discovery: from accidental discoveries to avoided liabilities and designed therapies. Drug Discov Today. 2015;20:1061.
8. Boike L, Henning NJ, Nomura DK. Advances in covalent drug discovery. Nat Rev Drug Discov. 2022;21(12):881–98. https://doi.org/10.1038/s41573-022-00542-z. Epub 2022 Aug 25. PMID: 36008483; PMCID: PMC9403961
9. Schaefer D, Cheng X. Recent advances in covalent drug discovery. Pharmaceuticals (Basel). 2023;16(5):663. https://doi.org/10.3390/ph16050663. PMID: 37242447; PMCID: PMC10220821
10. Xiao YC, Chen FE. The vinyl sulfone motif as a structural unit for novel drug design and discovery. Expert Opin Drug Discov. 2023:1–13. https://doi.org/10.1080/17460441.2023.2284201. Epub ahead of print
11. Wu X, Song W, Cheng C, Liu Z, Li X, Cui Y, Gao Y, Li D. Small molecular inhibitors for KRAS-mutant cancers. Front Immunol. 2023;14:1223433. https://doi.org/10.3389/fimmu.2023.1223433. PMID: 37662925; PMCID: PMC10470052
12. Javle M, King G, Spencer K, Borad MJ. Futibatinib, an irreversible FGFR1-4 inhibitor for the treatment of FGFR-aberrant tumors. Oncologist. 2023;28(11):928–43. https://doi.org/10.1093/oncolo/oyad149. PMID: 37390492; PMCID: PMC10628593
13. Feral A, Martin AR, Desfoux A, Amblard M, Vezenkov LL. Covalent-reversible peptide-based protease inhibitors. Design, synthesis, and clinical success stories. Amino Acids. 2023; https://doi.org/10.1007/s00726-023-03286-1. Epub ahead of print
14. Huang H, Jones LH. Covalent drug discovery using sulfur(VI) fluoride exchange warheads. Expert Opin Drug Discov. 2023;18(7):725–35. https://doi.org/10.1080/17460441.2023.2218642. Epub 2023 May 27
15. Mons E, Kim RQ, Mulder MPC. Technologies for direct detection of covalent protein-drug adducts. Pharmaceuticals (Basel). 2023;16(4):547.

16. De SK. Futibatinib: a potent and irreversible inhibitor of fibroblast growth factor receptors for treatment of the bile duct cancer. Curr Med Chem. 2024;31(6):666–70. https://doi.org/10.2174/0929867330666230416152913.
17. De SK. Sotorasib: first approved KRAS mutation inhibitor for the treatment of non-small cell lung cancer. Curr Med Chem. 2023;30(9):1000–2. https://doi.org/10.2174/0929867329666220907161505.
18. De SK. First approval of Adagrasib for the treatment of non-small cell lung cancer harboring a KRASG12C mutation. Curr Med Chem. 2024;31(3):266–72. https://doi.org/10.2174/0929867330666230330122000.
19. De SK. Pirtobrutinib: first non-covalent tyrosine kinase inhibitor for treating relapsed or refractory mantle cell lymphoma in adults. Curr Med Chem. 2023; https://doi.org/10.2174/0109298673251030231004052822. Epub ahead of print
20. De SK. Discovery of novel cysteine protease inhibitors for the treatment of coronavirus (COVID-19). Curr Med Chem. 2024;31(9):1165–9. https://doi.org/10.2174/0929867330666230519163305.
21. De SK. Recent discovery of Peptidomimetics for the treatment of coronavirus (COVID-19), human coronavirus, and enteroviruses. Curr Med Chem. 2023; https://doi.org/10.2174/0929867330666230504150758. Epub ahead of print
22. De SK. Novel 1,4-Dihydropyrido[2,3-B]Pyrazine-2,3-Dione derivatives for treating cancer and other disorders associated with KRAS activity. Anti Cancer Agents Med Chem. 2023;23(8):979–80. https://doi.org/10.2174/1871520623666221128110638.
23. Rathod LS, Dabhade PS, Mokale SN. Recent progress in targeting KRAS mutant cancers with covalent G12C-specific inhibitors. Drug Discov Today. 2023;28(5):103557. https://doi.org/10.1016/j.drudis.2023.103557. Epub 2023 Mar 18
24. Sattler M, Mambetsariev I, Fricke J, Tan T, Liu S, Vaidehi N, Pisick E, Mirzapoiazova T, Rock AG, Merla A, Sharma S, Salgia R. A closer look at EGFR inhibitor resistance in non-small cell lung cancer through the lens of precision medicine. J Clin Med. 2023;12(5):1936. https://doi.org/10.3390/jcm12051936. PMID: 36902723; PMCID: PMC10003860
25. Bonatto V, Lameiro RF, Rocho FR, Lameira J, Leitão A, Montanari CA. Nitriles: an attractive approach to the development of covalent inhibitors. RSC Med Chem. 2022;14(2):201–17. https://doi.org/10.1039/d2md00204c. PMID: 36846367; PMCID: PMC9945868
26. Lam KK, Wong SH, Cheah PY. Targeting the 'Undruggable' driver protein, KRAS, in epithelial cancers: current perspective. Cells. 2023;12(4):631. https://doi.org/10.3390/cells12040631. PMID: 36831298; PMCID: PMC9954350
27. Wang H, Chi L, Yu F, Dai H, Gao C, Si X, Wang Z, Liu L, Zheng J, Shan L, Liu H, Zhang Q. Annual review of KRAS inhibitors in 2022. Eur J Med Chem. 2023;249:115124. https://doi.org/10.1016/j.ejmech.2023.115124. Epub 2023 Jan 16
28. Gai C, Harnor SJ, Zhang S, Cano C, Zhuang C, Zhao Q. Advanced approaches of developing targeted covalent drugs. RSC Med Chem. 2022;13(12):1460–75. https://doi.org/10.1039/d2md00216g. PMID: 36561076; PMCID: PMC9749957
29. Lee J, Park SB. Extended applications of small-molecule covalent inhibitors toward novel therapeutic targets. Pharmaceuticals (Basel). 2022;15(12):1478. https://doi.org/10.3390/ph15121478. PMID: 36558928; PMCID: PMC9786830
30. Huang F, Han X, Xiao X, Zhou J. Covalent warheads targeting cysteine residue: the promising approach in drug development. Molecules. 2022;27(22):7728. https://doi.org/10.3390/molecules27227728. PMID: 36431829; PMCID: PMC9694382
31. Telliez JB, Dowty ME, Wang L, Jussif J, Lin T, Li L, Moy E, Balbo P, Li W, Zhao Y, Crouse K, Dickinson C, Symanowicz P, Hegen M, Banker ME, Vincent F, Unwalla R, Liang S, Gilbert AM, Brown MF, Hayward M, Montgomery J, Yang X, Bauman J, Trujillo JI, Casimiro-Garcia A, Vajdos FF, Leung L, Geoghegan KF, Quazi A, Xuan D, Jones L, Hett E, Wright K, Clark JD, Thorarensen A. Discovery of a JAK3-selective inhibitor: functional differentiation of JAK3-selective inhibition over pan-JAK or JAK1-selective inhibition. ACS Chem Biol. 2016;11(12):3442–51. https://doi.org/10.1021/acschembio.6b00677. Epub 2016 Nov 10

32. Majkutewicz I. Dimethyl fumarate: a review of preclinical efficacy in models of neurodegenerative diseases. Eur J Pharmacol. 2022;926:175025. https://doi.org/10.1016/j.ejphar.2022.175025. Epub 2022 May 13
33. Barile E, Marconi GD, De SK, Baggio C, Gambini L, Salem AF, Kashyap MK, Castro JE, Kipps TJ, Pellecchia M. hBfl-1/hNOXA interaction studies provide new insights on the role of Bfl-1 in cancer cell resistance and for the design of novel anticancer agents. ACS Chem Biol. 2017;12(2):444–55. https://doi.org/10.1021/acschembio.6b00962. Epub 2016 Dec 27. PMID: 28026162; PMCID: PMC5320539
34. Stebbins JL, Santelli E, Feng Y, De SK, Purves A, Motamedchaboki K, Wu B, Ronai ZA, Liddington RC, Pellecchia M. Structure-based design of covalent Siah inhibitors. Chem Biol. 2013;20(8):973–82. https://doi.org/10.1016/j.chembiol.2013.06.008. Epub 2013 Jul 25. PMID: 23891150; PMCID: PMC3763817
35. Walker EH, Pacold ME, Perisic O, Stephens L, Hawkins PT, Wymann MP, Williams RL. Structural determinants of phosphoinositide 3-kinase inhibition by wortmannin, LY294002, quercetin, myricetin, and staurosporine. Mol Cell. 2000;6(4):909–19. https://doi.org/10.1016/s1097-2765(05)00089-4.
36. Voice AT, Tresadern G, Twidale RM, van Vlijmen H, Mulholland AJ. Mechanism of covalent binding of ibrutinib to Bruton's tyrosine kinase revealed by QM/MM calculations. Chem Sci. 2021;12(15):5511–6. https://doi.org/10.1039/d0sc06122k. PMID: 33995994; PMCID: PMC8097726
37. Groll M, Berkers CR, Ploegh HL, Ovaa H. Crystal structure of the boronic acid-based proteasome inhibitor bortezomib in complex with the yeast 20S proteasome. Structure. 2006;14(3):451–6. https://doi.org/10.1016/j.str.2005.11.019.
38. Perbandt M, Werner N, Prester A, Rohde H, Aepfelbacher M, Hinrichs W, Betzel C. Structural basis to repurpose boron-based proteasome inhibitors Bortezomib and Ixazomib as β-lactamase inhibitors. Sci Rep. 2022;12(1):5510. https://doi.org/10.1038/s41598-022-09392-6. PMID: 35365689; PMCID: PMC8976068
39. Paulussen FM, Grossmann TN. Peptide-based covalent inhibitors of protein-protein interactions. J Pept Sci. 2023;29(1):e3457. https://doi.org/10.1002/psc.3457. Epub 2022 Nov 9. PMID: 36239115; PMCID: PMC10077911
40. La Monica G, Bono A, Lauria A, Martorana A. Targeting SARS-CoV-2 main protease for treatment of COVID-19: covalent inhibitors structure-activity relationship insights and evolution perspectives. J Med Chem. 2022;65(19):12500–34. https://doi.org/10.1021/acs.jmedchem.2c01005. Epub 2022 Sep 28. PMID: 36169610; PMCID: PMC9528073
41. Das D, Wang J, Hong J. Next-generation Bruton's tyrosine kinase (BTK) inhibitors potentially targeting BTK C481S mutation- recent developments and perspectives. Curr Top Med Chem. 2022;22(20):1674–91. https://doi.org/10.2174/1568026622666220801101706.
42. Robak E, Robak T. Bruton's kinase inhibitors for the treatment of immunological diseases: current status and perspectives. J Clin Med. 2022;11(10):2807. https://doi.org/10.3390/jcm11102807. PMID: 35628931; PMCID: PMC9145705
43. Lu X, Smaill JB, Patterson AV, Ding K. Discovery of cysteine-targeting covalent protein kinase inhibitors. J Med Chem. 2022;65(1):58–83. https://doi.org/10.1021/acs.jmedchem.1c01719. Epub 2021 Dec 28
44. Tang D, Kroemer G, Kang R. Oncogenic KRAS blockade therapy: renewed enthusiasm and persistent challenges. Mol Cancer. 2021;20(1):128. https://doi.org/10.1186/s12943-021-01422-7. PMID: 34607583; PMCID: PMC8489073
45. Hoyt KW, Urul DA, Ogboo BC, Wittlinger F, Laufer SA, Schaefer EM, May EW, Heppner DE. Pitfalls and considerations in determining the potency and mutant selectivity of covalent epidermal growth factor receptor inhibitors. J Med Chem. 2024;67(1):2–16. https://doi.org/10.1021/acs.jmedchem.3c01502. Epub 2023 Dec 22
46. Hartung IV, Rudolph J, Mader MM, Mulder MPC, Workman P. Expanding chemical probe space: quality criteria for covalent and degrader probes. J Med Chem. 2023;66(14):9297–312.

https://doi.org/10.1021/acs.jmedchem.3c00550. Epub 2023 Jul 5. PMID: 37403870; PMCID: PMC10388296

47. Faridoon NR, Zhang G, Li JJ. An update on the discovery and development of reversible covalent inhibitors. Med Chem Res. 2023;32(6):1039–62. https://doi.org/10.1007/s00044-023-03065-3. Epub 2023 Apr 29. PMID: 37305209; PMCID: PMC10148018
48. Zhang T, Inesta-Vaquera F, Niepel M, Zhang J, Ficarro SB, Machleidt T, Xie T, Marto JA, Kim ND, Sim T, Laughlin JD, Park H, LoGrasso PV, Patricelli M, Nomanbhoy TK, Sorger PK, Alessi DR, Gray NS. Discovery of potent and selective covalent inhibitors of JNK. Chem Biol. 2012;19:140.
49. Yang J, Tabuchi Y, Katsuki R, Taki M. bioTCIs: middle-to-macro biomolecular targeted covalent inhibitors possessing both semi-permanent drug action and stringent target specificity as potential antibody replacements. Int J Mol Sci. 2023;24(4):3525. https://doi.org/10.3390/ijms24043525. PMID: 36834935; PMCID: PMC9968108
50. Brogi S, Ibba R, Rossi S, Butini S, Calderone V, Gemma S, Campiani G. Covalent reversible inhibitors of cysteine proteases containing the nitrile warhead: recent advancement in the field of viral and parasitic diseases. Molecules. 2022;27(8):2561. https://doi.org/10.3390/molecules27082561. PMID: 35458759; PMCID: PMC9029279
51. Péczka N, Orgován Z, Ábrányi-Balogh P, Keserű GM. Electrophilic warheads in covalent drug discovery: an overview. Expert Opin Drug Discov. 2022;17(4):413–22. https://doi.org/10.1080/17460441.2022.2034783. Epub 2022 Feb 6
52. Liu Y, Liang C, Xin L, Ren X, Tian L, Ju X, Li H, Wang Y, Zhao Q, Liu H, Cao W, Xie X, Zhang D, Wang Y, Jian Y. The development of coronavirus 3C-like protease (3CLpro) inhibitors from 2010 to 2020. Eur J Med Chem. 2020;206:112711. https://doi.org/10.1016/j.ejmech.2020.112711. Epub 2020 Aug 6. PMID: 32810751; PMCID: PMC7409838
53. Abdeldayem A, Raouf YS, Constantinescu SN, Moriggl R, Gunning PT. Advances in covalent kinase inhibitors. Chem Soc Rev. 2020;49(9):2617–87. https://doi.org/10.1039/c9cs00720b. Epub 2020 Mar 30
54. Zhang T, Hatcher JM, Teng M, Gray NS, Kostic M. Recent advances in selective and irreversible covalent ligand development and validation. Cell Chem Biol. 2019;26(11):1486–500. https://doi.org/10.1016/j.chembiol.2019.09.012. Epub 2019 Oct 17. PMID: 31631011; PMCID: PMC6886688
55. Kumalo HM, Bhakat S, Soliman ME. Theory and applications of covalent docking in drug discovery: merits and pitfalls. Molecules. 2015;20(2):1984–2000. https://doi.org/10.3390/molecules20021984. PMID: 25633330; PMCID: PMC6272664
56. Fernández-de-la-Pradilla A, Royo S, Schirmeister T, Barthels F, Świderek K, González FV, Moliner V. Impact of the warhead of dipeptidyl Keto Michael acceptors on the inhibition mechanism of cysteine protease Cathepsin L. ACS Catal. 2023;13(20):13354–68. https://doi.org/10.1021/acscatal.3c02748. PMID: 37881790; PMCID: PMC10594577
57. Bajaj T, Wehri E, Suryawanshi RK, King E, Pardeshi KS, Behrouzi K, Khodabakhshi Z, Schulze-Gahmen U, Kumar GR, Mofrad MRK, Nomura DK, Ott M, Schaletzky J, Murthy N. Mercapto-pyrimidines are reversible covalent inhibitors of the papain-like protease (PLpro) and inhibit SARS-CoV-2 (SCoV-2) replication. RSC Adv. 2023;13(26):17667–77. https://doi.org/10.1039/d3ra01915b. PMID: 37312993; PMCID: PMC10259201
58. Chen Z, Du R, Cooper L, Achi JG, Dong M, Ran Y, Zhang J, Zhan P, Rong L, Cui Q. Sulforaphane is a reversible covalent inhibitor of 3-chymotrypsin-like protease of SARS-CoV-2. J Med Virol. 2023;95(3):e28609. https://doi.org/10.1002/jmv.28609.
59. Ai Y, Yang Z, Yang Z, Wan S, Huang C, Huang C, Li M, Li Z, Zhang J, Zhang T. Discovery and computational studies of potent covalent kinase inhibitors with α-substituent electrophiles targeting cysteine. J Chem Inf Model. 2023;63(2):493–506. https://doi.org/10.1021/acs.jcim.2c00458. Epub 2023 Jan 12
60. Che J, Jones LH. Covalent drugs targeting histidine – an unexploited opportunity? RSC Med Chem. 2022;13(10):1121–6. https://doi.org/10.1039/d2md00258b. PMID: 36325394; PMCID: PMC9579939

61. Parenti MD, Naldi M, Manoni E, Fabini E, Cederfelt D, Talibov VO, Gressani V, Guven U, Grossi V, Fasano C, Sanese P, De Marco K, Shtil AA, Kurkin AV, Altieri A, Danielson UH, Caretti G, Simone C, Varchi G, Bartolini M, Del Rio A. Discovery of the 4-aminopiperidine-based compound EM127 for the site-specific covalent inhibition of SMYD3. Eur J Med Chem. 2022;243:114683. https://doi.org/10.1016/j.ejmech.2022.114683. Epub 2022 Sep 8
62. Zheng M, Chen FJ, Li K, Reja RM, Haeffner F, Gao J. Lysine-targeted reversible covalent ligand discovery for proteins via phage display. J Am Chem Soc. 2022;144(34):15885–93. https://doi.org/10.1021/jacs.2c07375. Epub 2022 Aug 17. PMID: 35976695; PMCID: PMC9440474
63. Borsari C, Keles E, McPhail JA, Schaefer A, Sriramaratnam R, Goch W, Schaefer T, De Pascale M, Bal W, Gstaiger M, Burke JE, Wymann MP. Covalent proximity scanning of a distal cysteine to target PI3Kα. J Am Chem Soc. 2022;144(14):6326–42. https://doi.org/10.1021/jacs.1c13568. Epub 2022 Mar 30. PMID: 35353516; PMCID: PMC9011356
64. Zhang Z, Wang Y, Chen X, Song X, Tu Z, Chen Y, Zhang Z, Ding K. Characterization of an aromatic trifluoromethyl ketone as a new warhead for covalently reversible kinase inhibitor design. Bioorg Med Chem. 2021;50:116457. https://doi.org/10.1016/j.bmc.2021.116457. Epub 2021 Oct 5
65. Bonatto V, Shamim A, Rocho FDR, Leitão A, Luque FJ, Lameira J, Montanari CA. Predicting the relative binding affinity for reversible covalent inhibitors by free energy perturbation calculations. J Chem Inf Model. 2021;61(9):4733–44. https://doi.org/10.1021/acs.jcim.1c00515. Epub 2021 Aug 30
66. Deng W, Chen X, Jiang K, Song X, Huang M, Tu ZC, Zhang Z, Lin X, Ortega R, Patterson AV, Smaill JB, Ding K, Chen S, Chen Y, Lu X. Investigation of covalent warheads in the design of 2-Aminopyrimidine-based FGFR4 inhibitors. ACS Med Chem Lett. 2021;12(4):647–52. https://doi.org/10.1021/acsmedchemlett.1c00052. PMID: 33859803; PMCID: PMC8040253
67. Gao P, Song S, Frutos-Beltrán E, Li W, Sun B, Kang D, Zou J, Zhang J, Pannecouque C, De Clercq E, Menéndez-Arias L, Zhan P, Liu X. Novel indolylarylsulfone derivatives as covalent HIV-1 reverse transcriptase inhibitors specifically targeting the drug-resistant mutant Y181C. Bioorg Med Chem. 2021;30:115927. https://doi.org/10.1016/j.bmc.2020.115927. Epub 2020 Dec 10
68. Guo WH, Qi X, Yu X, Liu Y, Chung CI, Bai F, Lin X, Lu D, Wang L, Chen J, Su LH, Nomie KJ, Li F, Wang MC, Shu X, Onuchic JN, Woyach JA, Wang ML, Wang J. Enhancing intracellular accumulation and target engagement of PROTACs with reversible covalent chemistry. Nat Commun. 2020;11(1):4268. https://doi.org/10.1038/s41467-020-17997-6. PMID: 32848159; PMCID: PMC7450057
69. Gehringer M. Covalent inhibitors: back on track? Future Med Chem. 2020;12(15):1363–8. https://doi.org/10.4155/fmc-2020-0118. Epub 2020 Jun 29
70. Dražić T, Kopf S, Corridan J, Leuthold MM, Bertoša B, Klein CD. Peptide-β-lactam inhibitors of dengue and West Nile virus NS2B-NS3 protease display two distinct binding modes. J Med Chem. 2020;63(1):140–56. https://doi.org/10.1021/acs.jmedchem.9b00759. Epub 2019 Dec 24. Erratum in: J Med Chem. 2020 63(18):10530
71. Stubba D, Bensinger D, Steinbacher J, Proskurjakov L, Salcedo Gómez Á, Schmidt U, Roth S, Schmitz K, Schmidt B. Cell-based optimization of covalent reversible Ketoamide inhibitors bridging the unprimed to the primed site of the proteasome β5 subunit. ChemMedChem. 2019;14(23):2005–22. https://doi.org/10.1002/cmdc.201900472. Epub 2019 Nov 12. PMID: 31675179; PMCID: PMC6916368
72. Palazzesi F, Grundl MA, Pautsch A, Weber A, Tautermann CS. A fast ab initio predictor tool for covalent reactivity estimation of acrylamides. J Chem Inf Model. 2019;59(8):3565–71. https://doi.org/10.1021/acs.jcim.9b00316. Epub 2019 Jul 15
73. Shindo N, Fuchida H, Sato M, Watari K, Shibata T, Kuwata K, Miura C, Okamoto K, Hatsuyama Y, Tokunaga K, Sakamoto S, Morimoto S, Abe Y, Shiroishi M, Caaveiro JMM, Ueda T, Tamura T, Matsunaga N, Nakao T, Koyanagi S, Ohdo S, Yamaguchi Y, Hamachi I,

Ono M, Ojida A. Selective and reversible modification of kinase cysteines with chlorofluoroacetamides. Nat Chem Biol. 2019;15(3):250–8. https://doi.org/10.1038/s41589-018-0204-3. Epub 2019 Jan 14
74. Lee H, Lee JY, Jang H, Cho HY, Kang M, Bae SH, Kim S, Kim E, Jang J, Kim JY, Jeon YH. Discovery of proteolysis-targeting chimera targeting undruggable proteins using a covalent ligand screening approach. Eur J Med Chem. 2024;263:115929. https://doi.org/10.1016/j.ejmech.2023.115929. Epub 2023 Nov 2
75. Jelisejevs D, Bula AL, Kinena L. Pyrazolidinone-based peptidomimetic SARS-CoV-2 Mpro inhibitors. Bioorg Med Chem Lett. 2023;96:129530. https://doi.org/10.1016/j.bmcl.2023.129530. Epub 2023 Oct 20
76. Liu M, Li J, Liu W, Yang Y, Zhang M, Ye Y, Zhu W, Zhou C, Zhai H, Xu Z, Zhang G, Huang H. The S1'-S3' pocket of the SARS-CoV-2 main protease is critical for substrate selectivity and can be targeted with covalent inhibitors. Angew Chem Int Ed Engl. 2023;62(41):e202309657. https://doi.org/10.1002/anie.202309657. Epub 2023 Sep 4
77. Martin JS, Mackenzie CJ, Lin D, Homeyer N, Gray DW, Zuccotto F, Gilbert IH. Design and synthesis of covalent inhibitors of FabA. ACS Omega. 2023;8(14):12787–804. https://doi.org/10.1021/acsomega.2c08031. PMID: 37065080; PMCID: PMC10099128
78. Chen X, Li H, Lin Q, Dai S, Yue S, Qu L, Li M, Guo M, Wei H, Li J, Jiang L, Xu G, Chen Y. Structure-based design of a dual-warhead covalent inhibitor of FGFR4. Commun Chem. 2022;5(1):36. https://doi.org/10.1038/s42004-022-00657-9. PMID: 36697897; PMCID: PMC9814781
79. Najmi A, Thangavel N, Mohanan AT, Qadri M, Albratty M, Ashraf SE, Saleh SF, Nayeem M, Mohan S. Structural complementarity of Bruton's tyrosine kinase and its inhibitors for implication in B-cell malignancies and autoimmune diseases. Pharmaceuticals (Basel). 2023;16(3):400. https://doi.org/10.3390/ph16030400. PMID: 36986499; PMCID: PMC10051736
80. Singh J. The Ascension of targeted covalent inhibitors. J Med Chem. 2022;65(8):5886–901. https://doi.org/10.1021/acs.jmedchem.1c02134. Epub 2022 Apr 19
81. Zain R, Vihinen M. Structure-function relationships of covalent and non-covalent BTK inhibitors. Front Immunol. 2021;12:694853. https://doi.org/10.3389/fimmu.2021.694853. PMID: 34349760; PMCID: PMC8328433
82. Gabizon R, London N. The rise of covalent proteolysis targeting chimeras. Curr Opin Chem Biol. 2021;62:24–33. https://doi.org/10.1016/j.cbpa.2020.12.003. Epub 2021 Feb 4
83. Roskoski R Jr. Orally effective FDA-approved protein kinase targeted covalent inhibitors (TCIs). Pharmacol Res. 2021;165:105422. https://doi.org/10.1016/j.phrs.2021.105422. Epub 2021 Jan 9
84. Das D, Hong J. Irreversible kinase inhibitors targeting cysteine residues and their applications in cancer therapy. Mini Rev Med Chem. 2020;20(17):1732–53. https://doi.org/10.2174/1389557520666200513121524.
85. Wang L, Zhao J, Yao Y, Wang C, Zhang J, Shu X, Sun X, Li Y, Liu K, Yuan H, Ma X. Covalent binding design strategy: a prospective method for discovery of potent targeted anticancer agents. Eur J Med Chem. 2017;142:493–505. https://doi.org/10.1016/j.ejmech.2017.09.024. Epub 2017 Sep 20
86. Long MJC, Aye Y. Privileged electrophile sensors: a resource for covalent drug development. Cell. Chem Biol. 2017;24(7):787–800. https://doi.org/10.1016/j.chembiol.2017.05.023. Epub 2017 Jun 22. PMID: 28648380; PMCID: PMC5528179
87. Han W, Du Y. Recent development of the second and third generation irreversible epidermal growth factor receptor inhibitors. Chem Biodivers. 2017;14(7) https://doi.org/10.1002/cbdv.201600372. Epub 2017 Jun 9
88. Hossam M, Lasheen DS, Abouzid KA. Covalent EGFR inhibitors: binding mechanisms, synthetic approaches, and clinical profiles. Arch Pharm (Weinheim). 2016;349(8):573–93. https://doi.org/10.1002/ardp.201600063. Epub 2016 Jun 3
89. Ebelt ND, Kaoud TS, Edupuganti R, Van Ravenstein S, Dalby KN, Van Den Berg CL. A c-Jun N-terminal kinase inhibitor, JNK-IN-8, sensitizes triple negative breast cancer cells to lapatinib. Oncotarget. 2017;8:104894.

90. Ansideri F, Lange A, El-Gokha A, Boeckler FM, Koch P. Fluorescence polarization-based assays for detecting compounds binding to inactive c-Jun N-terminal kinase 3 and p38α mitogen-activated protein kinases. Anal Biochem. 2016;503:28.
91. Zimmermann G, Rieder U, Bajik D, Vanetti S, Chaikuad A, Knapp S, Scheuermann J, Mattarella M, Neri D. A specific and covalent JNK-1 ligand selected from an encoded self-assembling chemical library. Chem. 2017;23:8152.
92. Lu W, Liu Y, Gao Y, Geng Q, Gurbani D, Li L, Ficarro SB, Meyer CJ, Sinha D, You I, Tse J, He Z, Ji W, Che J, Kim AY, Yu T, Wen K, Anderson KC, Marto JA, Westover KD, Zhang T, Gray NS. Development of a covalent inhibitor of c-Jun N-terminal protein kinase (JNK) 2/3 with selectivity over JNK1. J Med Chem. 2023;66(5):3356–71. https://doi.org/10.1021/acs.jmedchem.2c01834. Epub 2023 Feb 24
93. Ward RA, Colclough N, Challinor M, Debreczeni JE, Eckersley K, Fairley G, Feron L, Flemington V, Graham MA, Greenwood R, Hopcroft P, Howard TD, James M, Jones CD, Jones CR, Renshaw J, Roberts K, Snow L, Tonge M, Yeung K. Structure-guided design of highly selective and potent covalent inhibitors of ERK1/2. J Med Chem. 2015;58(11):4790–801. https://doi.org/10.1021/acs.jmedchem.5b00466. Epub 2015 May 28. PMID: 25977981
94. St Denis JD, Chessari G, Cleasby A, Cons BD, Cowan S, Dalton SE, East C, Murray CW, O'Reilly M, Peakman T, Rapti M, Stow JL. X-ray screening of an electrophilic fragment library and application toward the development of a novel ERK 1/2 covalent inhibitor. J Med Chem. 2022;65(18):12319–33. https://doi.org/10.1021/acs.jmedchem.2c01044. Epub 2022 Sep 13
95. Kaoud TS, Johnson WH, Ebelt ND, Piserchio A, Zamora-Olivares D, Van Ravenstein SX, Pridgen JR, Edupuganti R, Sammons R, Cano M, Warthaka M, Harger M, Tavares CDJ, Park J, Radwan MF, Ren P, Anslyn EV, Tsai KY, Ghose R, Dalby KN. Modulating multi-functional ERK complexes by covalent targeting of a recruitment site in vivo. Nat Commun. 2019;10(1):5232. https://doi.org/10.1038/s41467-019-12996-8. PMID: 31745079; PMCID: PMC6863825
96. Aronchik I, Dai Y, Labenski M, Barnes C, Jones T, Qiao L, Beebe L, Malek M, Elis W, Shi T, Mavrommatis K, Bray GL, Filvaroff EH. Efficacy of a covalent ERK1/2 inhibitor, CC-90003, in KRAS-mutant cancer models reveals novel mechanisms of response and resistance. Mol Cancer Res. 2019;17(2):642–54.
97. Pan X, Pei J, Wang A, Shuai W, Feng L, Bu F, Zhu Y, Zhang L, Wang G, Ouyang L. Development of small molecule extracellular signal-regulated kinases (ERKs) inhibitors for cancer therapy. Acta Pharm Sin B. 2022;12(5):2171–92. https://doi.org/10.1016/j.apsb.2021.12.022. Epub 2022 Jan 4. PMID: 35646518; PMCID: PMC9136582
98. Fu L, Chen S, He G, Chen Y, Liu B. Targeting extracellular signal-regulated protein kinase 1/2 (ERK1/2) in cancer: an update on pharmacological small-molecule inhibitors. J Med Chem. 2022;65(20):13561–73. https://doi.org/10.1021/acs.jmedchem.2c01244. Epub 2022 Oct 7
99. Li J, Kaoud TS, Laroche C, Dalby KN, Kerwin SM. Synthesis and biological evaluation of p38alpha kinase-targeting dialkynylimidazoles. Bioorg Med Chem Lett. 2009;19(22):6293–7. https://doi.org/10.1016/j.bmcl.2009.09.094. Epub 2009 Sep 27. PMID: 19822424; PMCID: PMC2789032
100. Wadsworth SA, Cavender DE, Beers SA, Lalan P, Schafer PH, Malloy EA, Wu W, Fahmy B, Olini GC, Davis JE, Pellegrino-Gensey JL, Wachter MP, Siekierka JJ. RWJ 67657, a potent, orally active inhibitor of p38 mitogen-activated protein kinase. J Pharmacol Exp Ther. 1999;291(2):680–7.
101. Birkenkamp KU, Tuyt LM, Lummen C, Wierenga AT, Kruijer W, Vellenga E. The p38 MAP kinase inhibitor SB203580 enhances nuclear factor-kappa B transcriptional activity by a non-specific effect upon the ERK pathway. Br J Pharmacol. 2000;131(1):99–107. https://doi.org/10.1038/sj.bjp.0703534. PMID: 10960075; PMCID: PMC1572293
102. Owira PM, Ojewole JA. 'African potato' (Hypoxis hemerocallidea corm): a plant-medicine for modern and 21st century diseases of mankind? – a review. Phytother Res. 2009;23(2):147–52. https://doi.org/10.1002/ptr.2595.

Index

S. K. De, *Drug Design and Discovery*,
https://doi.org/10.1007/978-3-031-75682-5

The manufacturer's authorised representative in the EU is Springer Nature Customer Service Centre GmbH, Europaplatz 3, 69115 Heidelberg, Germany. If you have any concerns regarding our products, please contact ProductSafety@springernature.com

Printed and bound by CPI Group (UK) Ltd, Croydon, CR0 4YY
30/09/2026
02225562-0002